Early Life History of Marine Fishes

Early Life History of Marine Fishes

Bruce S. Miller
Arthur W. Kendall, Jr.

UNIVERSITY OF CALIFORNIA PRESS
BERKELEY   LOS ANGELES   LONDON

University of California Press, one of the most distinguished university presses in the United States, enriches lives around the world by advancing scholarship in the humanities, social sciences, and natural sciences. Its activities are supported by the UC Press Foundation and by philanthropic contributions from individuals and institutions. For more information, visit www.ucpress.edu.

University of California Press
Berkeley and Los Angeles, California

University of California Press, Ltd.
London, England

© 2009 by the Regents of the University of California

Library of Congress Cataloging-in-Publication Data

Miller, Bruce S.
    Early life history of marine fishes / Bruce S. Miller,
      Arthur W. Kendall, Jr.
      p. cm.
    Includes bibliographical references and index.
    ISBN 978-0-520-24972-1 (cloth : alk. paper)
    1. Marine fishes—Eggs. 2. Marine fishes—Larvae. I. Kendall, Arthur W., Jr. II. Title.
QL620.M52 2009
597.156—dc22
                                                        2008043143

Manufactured in the United States of America

27 26 25 24 23 22 21 20
10 9 8 7 6 5 4 3

The paper used in this publication meets the minimum requirements of ANSI/NISO Z39.48-1992 (R 1997) (*Permanence of Paper*).

Cover illustration (reversed) by Mike Fahay of a 13.4 mm SL pelagic juvenile of the Spanish flag (*Gonioplectrus hispanicus*) is used with permission of the artist.

CONTENTS

PREFACE / *vii*

ACKNOWLEDGMENTS / *ix*

THE COVER ART WORK / *xi*

INTRODUCTION / *1*

1. Fish Reproduction / *9*
   What Are Fishes? / *10*
   Sexuality / *11*
   Reproductive Patterns / *12*
   Gonadal Development / *20*
   Gross Maturation Stages / *22*
   Fecundity / *25*
   Spawning / *29*

2. Development of Eggs and Larvae / *39*
   Embryonic Development / *40*
   Larval Development / *45*
   Juvenile Development / *52*

3. Fish Egg and Larval Identification and Systematics / *55*
   Establishing the Identity of Fish Eggs and Larvae / *56*
   Methods and Equipment for Morphological Identification of Fish Eggs and Larvae / *59*
   Identification and Staging of Fish Eggs / *62*

vi   Contents

    Identification of Fish Larvae / 66
    Egg and Larval Stages in Systematic Studies / 80

4. Ecology of Fish Eggs and Larvae / 125
    Ecology Defined / 125
    The Position of Fish Eggs and Larvae in the Ecosystem / 126
    Egg Ecology / 126
    Functional Morphology of Larvae / 131
    Spring Bloom / 134
    Feeding and Condition / 135
    Growth / 142
    Predation / 143

5. Sampling Fish Eggs and Larvae / 147
    Purposes / 148
    Methods / 152
    Choosing Sampling Methods / 165
    Shipboard Procedures / 166
    Shoreside Sample Processing / 166
    Data Processing / 167
    Reporting Results of Sampling / 171

6. Population Dynamics and Recruitment / 181
    Population Dynamics / 182
    Population Fluctuations / 190
    Spawner-recruit Relationship / 194
    Factors Affecting Survival of Eggs and Larvae / 209
    Importance of Juvenile Stage / 213
    Recruitment Studies / 217

7. Habitat, Water Quality, and Conservation Biology / 229
    Habitats / 230
    Human Impacts and Water Quality / 232
    Conservation Biology and Marine Protected Areas (MPAs) / 236

8. Rearing and Culture of Marine Fishes / 245
    Rearing Marine Fishes to Enhance or Replenish Wild Stocks / 245
    Experimental Culture Methods / 249
    Types of Experimental Studies / 252

LABORATORY EXERCISES / 259

GLOSSARY / 293

LITERATURE CITED / 303

TAXANOMICAL INDEX / 341

GENERAL INDEX / 349

PREFACE

WHY STUDY EARLY LIFE HISTORY (ELH) OF FISHES?

The diversity of reproductive patterns seen in fishes far surpasses that seen in any other group of vertebrates. At one extreme, some fishes broadcast large numbers of small, free-floating eggs into the sea to be fertilized externally and develop as part of the plankton with no further involvement by the parents. At the other extreme, some fishes develop within the female and receive nourishment from her to be born as small adults. Between these extremes are fishes that build nests and protect the eggs and young fish as they develop there. Some fishes brood the young in their mouths, or in pouches in their bodies. Males as well as females care for the young in different species. Thus the ELH of fishes is a fascinating subject for study on its own.

In addition, there are many practical reasons for studying fish ELH. Many fishes are economically important, either as objects of recreational pursuits, or as objects of commercial fisheries. Understanding reproduction and recruitment of these fishes is vital if they are to be managed properly. Large variations in success of reproduction of fishes are the norm, particularly in marine fishes that spawn pelagic eggs. Determining safe levels of catches depends on measuring the success of reproduction. Understanding variations in success of reproduction has proven to be an elusive goal of considerable research. During their early stages, fishes are most vulnerable to environmental problems. Egg and larval development often occurs in nearshore, or estuarine areas that are subject to being degraded by pollution and anthropogenic changes in geography (e.g., dredging and filling, dumping, bulkheading, damming). Understanding the reproductive pattern of fishes in affected areas is essential to assessing the impact of such environmental changes.

CURRENT STATE OF ELH STUDIES

Early life history of fishes has been studied since the late 1800s, and fish culture has been practiced even longer. Current methods of data collection and analysis are permitting investigations that were not possible until the last few decades. Methods for determining the age of fish larvae, and their physiological condition, and identifying organisms that prey on them have been developed recently. Environmental conditions can now be measured at scales and in ways that are relevant to the lives of larvae, through the use of such devices as satellites and instrumented buoys. Computers are available to analyze these data, and produce simulations that allow testing of hypotheses. There is renewed interest in rearing young marine fishes to replenish depleted stocks, and methods for doing so are available for a number of species. Eggs and larvae are being used for bioassay of pollutants.

PURPOSE OF THIS BOOK

The literature on ELH studies of fishes is scattered in numerous scientific journals and various special publications such as symposium volumes. There are a few reviews of various aspects of ELH studies, but no single reference for the entire area of study. No journal is devoted exclusively to ELH studies. This book is intended to provide a significant source reference to the ELH of fishes. It can be used as a textbook for an upper division or graduate lecture/laboratory course on ELH of fishes, or as a reference in a professional library. It serves as a basic introduction to the discipline, and more information can be found in the primary sources that are cited. The subjects in the book are ordered so that they can be read in the sequence written. However, each chapter can stand alone, and if the book is used as a text, material can be assigned in other orders to suit special circumstances. The Laboratory Exercises provide detailed methods for several studies associated with ELH investigations. These can be used to direct actual investigations, or to design laboratory projects as part of a college course.

ACKNOWLEDGMENTS

This book is a culmination of careers by both authors studying the ELH of fishes. Although our careers took different paths, our interests in ELH studies started at the University of Washington (UW) where we were both graduate students in the early 1960s; Bruce in the College of Fisheries and Art in the Department of Oceanography. Dr. Alan DeLacy, Bruce's major professor in Fisheries, was instrumental in developing this interest in both of us. Art studied with Dr. T. S. English in Oceanography. Toward the end of our time as students at UW, we both studied and worked at the UW Friday Harbor Laboratories (FHL), and developed a lasting friendship that eventually included collaborating in teaching courses at UW on ELH, and authoring this book.

Along the way numerous people were influential and helpful in our studies and teaching. Both of us took the month-long larval fish identification course offered by E. H. Ahlstrom in La Jolla, and Art went on to get his PhD under him.

Dr. DeLacy had long taught a course on ELH of fishes at UW, and Bruce continued this course when he joined the faculty in 1975, and taught versions of it until he retired in 2002. Besides this course, students at UW were exposed to ELH studies for several years through a lecture series sponsored by Washington Sea Grant and given by leading scientists in various aspects of ELH studies. The lectures were edited and published as Sea Grant publications. This series started in 1975 with Gotthilf Hempel talking about the egg stage (Hempel 1979), a compilation which included some of Bruce's lecture material, and some of which is also included in this book. Hempel's initial effort was to be followed in 1979 by Reuben Lasker presenting lectures on the larval stage of fishes, but illness required that several other members of the staff (Paul Smith, John Hunter, and Geoff Moser) of the Southwest Fisheries Center in La Jolla step in to give most of the lectures. However, Reuben gave the final lecture and edited the papers that resulted from this endeavor (Lasker [ed.] 1981). Other lectures in this series dealing with EHL studies include those given by Sinclair (1988), MacCall (1990), and Mullin (1993). These Washington Sea Grant publications are often cited and have become valuable ELH references.

The idea for this book originated with an ELH course that Bruce taught at FHL in 1992. Bruce always invited guest speakers to present lectures in their specialties to his classes. For this class he invited several members of the newly assembled research team with the NOAA Alaska Fisheries Science Center (AFSC) Fisheries Oceanography Coordinated Investigations (FOCI). The team, made up of leading researchers in several subdisciplines of ELH studies, presented a series of cutting-edge lectures and laboratory exercises. At the conclusion of this course we realized that if this information were put into book form it would be a much needed major source reference on ELH of fishes. The researchers graciously shared their lecture notes and references with us, and these formed the backbone of this book. Included in that group of researchers were Kevin Bailey, Ann Matarese, Gail Theilacker, Morgan Busby, Annette Brown, Mike Canino, and Ric Brodeur. Gary Stauffer and Bill Aron of the AFSC gave both encouragement and administrative support to this effort and subsequent efforts to develop this book. We wish to thank Marcus Duke UW School of Aquatic and Fishery Sciences (SAFS) for editing a draft version of this book, and Dr. David Armstrong, Director of the SAFS for administrative support of our effort to publish this book. Chuck Crumly, editor for University of California Press, was good at keeping after us to get this book done and we thank Scott Norton and other personnel at UC Press for their excellent technical help.

As this book was being developed, Dr. Suam Kim, then of the Korean Oceanographic Research and Development Institute, published an ELH book in Korean (Kim and Zhang 1994). He translated portions of his book for us and commented on early drafts of our manuscript, as did the graduate students Erin MacDonald, Jake Gregg, and Kelly Van Wormer who took the 2001 ELH course at SAFS. Several people helped by giving lectures, advising students, and sharing their notes with us. Among these are Bob Lauth, Carla Stehr, Janet Duffy-Anderson, Bill MacFarland, Lyle Britt, Sarah Hinckley, Dan Cooper, and Susanne McDermott. We thank Dan Cooper and Sarah Hinckley for their help with Lab Exercise 2. SAFS undergraduate student Andrew Cheung was helpful in helping us prepare the manuscript.

We appreciate the helpful review of an early manuscript of this book by Steve Cadrin.

Finally, we would like to thank our wives, Aase Marie Miller and Anne Kendall, for indulging our "hobby" of writing this book.

THE COVER ART WORK

Off the coast of North Carolina on 31 July 1967, we hauled a Cobb midwater trawl onto the deck of the RV Dolphin, as we had done so many times before. What could have become routine was always exciting, as we eagerly dumped the cod end to get our first glimpse of the fish larvae and juveniles that we caught.

This time we were rewarded by spotting a strikingly beautiful and unusual pelagic juvenile fish (13.4 mm SL) that we could not come close to identifying until years later. It was so unusual that Wally Smith, the chief scientist on the cruise immediately awakened Mike Fahay, who was off watch, to show it to him. Once back at the Sandy Hook Laboratory, Mike pulled the larva out of the collection bottle and drew it on a day off while listening to a World Series baseball game. He used his unique wash style that closely reproduces the actual appearance of fish larvae. The red pigment, which usually disappears in formalin-preserved fish larvae, was present because an antioxidant had been added to the formalin.

Mike framed the drawing and hung it in a hallway outside his office. For years I (A.W.K.) passed this drawing on my way to my office. As I was finishing my dissertation work on larvae of seabasses (Serranidae), I realized that this fish looked somewhat like those larvae, but I still couldn't identify it. After considerable research Mike and I identified it as the Spanish flag (*Gonioplectrus hispanicus*) and found it to contribute to our analysis of the relationships among seabasses based on larval and adult characters (Kendall and Fahay 1979). It has larval characters of two seabass subfamilies (Anthinae and Epinephelinae), but in combination with its adult characters clearly belongs with the groupers (tribe Epinephelini).

This drawing was available to use on this cover because Mike later displayed it in his house, saving it from the arson fire that destroyed the Sandy Hook lab in 1985. We feel that this drawing and its story illustrate the excitement and uncertainty so often encountered in studying the early life history of fishes. Many equally intriguing stories underlie the studies summarized in this book.

INTRODUCTION

LIFE-HISTORY STAGES OF FISHES
EARLY LIFE-HISTORY STUDIES
  Types of Early Life-history Studies
  Recruitment Fluctuations
  Population Assessment
  Aquaculture
  Conservation Studies
  Integration of Basic and Applied Studies
  The Study of Early Life History of Fishes

LIFE-HISTORY STAGES OF FISHES

The life history of fishes can be broken down into five developmental stages. The **egg** or **embryonic stage** begins at fertilization and ends at hatching. During the egg stage development from a single cell to a complex organism occurs. Although the pattern of development is similar in all fishes, the site of development varies considerably. The egg stage is spent within the female in live-bearing fishes, and externally in oviparous fishes. In oviparous fishes development may take place in nests, or attached to some substrate, or the eggs may be broadcast spawned and develop as part of the plankton.

At hatching from the egg most fishes enter the **larval stage,** which lasts from hatching until the fish transform into juveniles. Larvae are usually quite distinct morphologically from the adults. At hatching, larvae of most species are quite poorly developed and still possess a yolk sac for nourishment. During the larval stage fishes develop morphologically so that they can feed and avoid predators and then generally reside as part of the plankton. They often increase in length by 10-fold during the larval period. They often migrate vertically on a diel basis. Development that typically occurs during the larval stage may take place in the female in viviparous fishes, or in the spawned eggs of some fishes that lay large demersal eggs (e.g., Pacific salmon [*Onchorynchus* spp.], toadfishes [*Opsanus* spp.]).

The end of the larval stage occurs when all the fin rays have formed and scales start to form on the body (squamation). At this point the larvae are said to transform or metamorphose into **juveniles.** Juveniles look roughly like the adults, although in the young stages of some species pigment and body proportions are still quite distinct. In some species the young juveniles are so different that they go through a separate transitional stage. The morphological transition from the larval to the juvenile stage is often accompanied by an ecological change from a planktonic to a schooling pelagic or a demersal existence. Juveniles often live in habitats that are distinct from those occupied by the adults. Fishes remain juveniles until they reach sexual maturity.

The **adult stage** follows the juvenile stage, and begins when the gonads first mature. The adult stage is a period of active reproduction, when the gonads go through maturation cycles (annual or more frequent). Most fishes reproduce at least annually as adults (semelparous), but in some (e.g., Pacific salmon, freshwater eels [Anguillidae]) maturation occurs only once near the end of life (iteroparous). A period of **senescence,** in which there is little growth and the gonads degenerate, may follow the adult stage. Most fishes die before reaching senescence.

This book covers the egg and larval stages, otherwise known as the **early life-history** portion of the life history of fishes. Early life history and adult fisheries biology are quite separate fields of study since the ecological role and habitat requirements of fish eggs and larvae differ drastically and are quite distinct from those of adults. The study of early life history deals more with plankton research and oceanography than it does with traditional approaches to fisheries biology. A considerable amount of fisheries research has focused on early life history because of its importance in recruitment, marine aquaculture, and pollution studies.

EARLY LIFE-HISTORY STUDIES

Types of Early Life-history Studies

The objectives of early life-history studies vary widely, but can be grouped into several broad, somewhat overlapping, categories. Most studies have something to do with fisheries science. Developing an understanding of recruitment variation has been a major theme for early life-history studies since the early 1900s. This has required diverse studies on the ecology of fish eggs and larvae. A census of eggs and larvae to estimate the abundance of the population that produced them has been the focus of considerable research. The basic biology (e.g., developmental biology, physiology, taxonomy) of early life-history stages of fishes has been a fruitful subject of considerable research. Much effort has also been expended on aquaculture.

Conservation studies, whether dealing with pollution or other human impacts such as habitat change, introduction of exotic species, and effects of harvesting, often involve working with early stages of fishes.

Recruitment Fluctuations. Johan Hjort (1914, 1926) was the first to realize that fluctuations in fish stocks were primarily due to interannual variations in reproductive success. He then realized that events during the early life history of fishes provided the key to understanding these fluctuations. Putting together the facts that such species as Atlantic cod (*Gadus morhua*) produce planktonic eggs, that the larvae probably feed on planktonic invertebrates, that these food items are not distributed evenly in the ocean, that currents could move the eggs and larvae, and that reared larvae died after their yolk was exhausted, he concluded that year-class variations could be caused by inadequate food for the larvae when they switched from yolk to exogenous food, and that the larvae could be carried away from suitable nursery areas by currents. This reasoning and these hypotheses guided early life-history research through most of the 20th century. Indeed, the "Recruitment Problem" was realized as the dominant impediment to understanding fish population dynamics throughout this period.

Efforts to understand recruitment fluctuations are often based on studies of the ecology of the young stages of fishes. Field studies sample for eggs and larvae using plankton nets. Drift, growth, mortality, and condition are examined as indicators of variations in survival through the sensitive egg, larval, and juvenile stages. These studies are often accompanied by physical oceanographic studies: water properties and currents help explain distribution of fish eggs and larvae. Biological oceanographic studies investigate food production for the larvae, and the distribution and production of competitors and predators of fish larvae.

Additionally, experimental laboratory studies investigate development, growth, energetics, and behavior of larvae. Effects of variations in kind and amount of food of larvae have been the subject of many laboratory experiments. Methods to assess larval condition in field samples are developed in the laboratory (see discussion of otolith studies).

Besides the considerable field and laboratory research conducted for a variety of species investigating the recruitment problem, the hypotheses of Hjort were expanded upon or challenged by a number of scientists. Among them Cushing (1990) developed the match/mismatch hypothesis, Miller et al. (1988) the bigger-is-better hypothesis, Lasker (1975) the stable ocean hypothesis, and Sinclair (1988) the member/vagrant theory.

Population Assessment. As early as 1909 it was realized that a census of planktonic fish egg abundance along with estimates of reproductive parameters of adult fishes could be used to estimate adult fish population abundance.

Early estimates were made for plaice (*Pleuronectes platessa*) in the North Sea (Buchanan-Wollaston 1926), for Atlantic mackerel (*Scomber scombrus*) off the East Coast of the United States (Sette 1943a), and for Pacific sardine (*Sardinops sagax*) off the California coast (Sette and Ahlstrom 1948). Such assessments have improved through advances in sampling and analytical techniques. Available sampling techniques now include an underway pumping system to continuously collect fish eggs as the research ship steams at full speed along transects in the study area (Checkley et al. 1997). Analytical methods include calculating daily as well as seasonal egg production to estimate biomass of the spawning population (Hunter and Lo 1993).

Aquaculture. Although fish culture can probably be traced back to prehistoric times, scientific studies involving marine fish eggs and larvae originated in the late 1800s in northern Europe. In 1865, the Norwegian G. O. Sars discovered that eggs of Atlantic cod are planktonic. This was followed by artificial fertilization and rearing experiments with a wide variety of marine fishes, and by 1900 marine fish hatcheries were established in several countries on both sides of the North Atlantic. In spite of a lack of evidence of success, hatchery programs released ever-increasing numbers of early larvae into the ocean on both sides of the North Atlantic. The inability to rear the larvae beyond the yolk-sac stage was a major problem with these programs. Proper food and feeding conditions could not be found, and the larvae died of starvation after the yolk was expended. These efforts were finally discontinued in the United States in 1954, when they could no longer be justified in the face of the lack of evidence of success and the lack of theoretical basis for them.

The breakthrough in rearing marine fish larvae was the publication by Shelbourne (1964) on his work with plaice. This paved the way for successful rearing of other species elsewhere, although efforts were still hampered by lack of a suitable food source. With the development of culture techniques for the rotifer *Brachionus plicatilis* this problem was largely solved, and reared marine fish larvae became available for a host of experimental studies. Now a large variety of marine species are reared in laboratories throughout the world for experimental purposes, to supply fish for stocking, and for aquaculture (Tucker 1998).

Although rearing of most marine fishes beyond the yolk-sac stage proved to be intractable for many years, rearing of freshwater fishes is relatively easy. Freshwater hatcheries successfully rear a variety of species and introduce them to waters where they were not found previously. Also, reared fishes are used widely to supplement natural reproduction. Rearing programs for various sunfishes (centrarchids) and salmon and trout (salmonids) are widespread and successful in providing young fishes for these purposes.

During the late 1800s in the United States, large-scale federal programs sought to introduce "desirable" fishes to areas where they did not occur naturally. Eggs, larvae, and early juveniles from hatcheries were transported in railroad cars across the country for distribution to anyone who wanted them. The common carp (*Cyprinus carpio*) was among the fishes so introduced, and within a few years after the program started, this Euroasian native was widespread throughout the United States. These days, programs that release young, cultured fishes into the wild are much more careful to ensure that they will not have adverse impacts on natural populations.

The many ongoing commercial aquaculture operations around the world can trace their beginnings to research leading to successful rearing of eggs and larvae of the species they cultivate.

Conservation Studies. Conservation studies involving early life-history stages of fishes have become increasingly common and important. Perhaps the first research involved oil spill pollution and the possible effects on fish eggs and larvae. For example, Peterson (2001) reported the effects of the *Exxon Valdez* oil spill in Prince William Sound, Alaska, and found detrimental effects to both eggs and larvae. Another common problem is the entrainment of eggs and larvae in the incoming water used as a coolant for power plants, such as the case at San Onofre, California, on the Pacific Ocean (Watson and Davis 1989) and in the Hudson River (Wallace 1978). Early life-history stages must be considered when human-induced habitat changes are proposed, so that spawning and nursery grounds of fishes are maintained, including those used by the wide variety of ecologically and economically important saltwater and freshwater fishes that have demersal eggs requiring very specific substrate. However, it is also recognized that even species with pelagic eggs probably always have very specific spawning sites that must be utilized with requisite biological (e.g., food, predators) and physical (e.g., currents, temperature) attributes. Pelagic egg and larval stages also pose the very real threat of unwanted exotic introductions of fishes by way of ballast water carried and discharged by ships around the world. Although the threat of fish introductions may not seem as ominous as invertebrate and plant introductions, nevertheless, the threat is real since fish eggs and larvae have been observed in the ballast water of ships traveling across the Pacific and entering Puget Sound, and from San Francisco Bay to Valdez, Alaska (Jeff Cordell, University of Washington, pers. comm. 2003).

Marine Protected Areas are presently considered to be potentially one of the most important tools for maintaining biodiversity and abundance of many marine fishes. As a means to deal with overharvest of fisheries resources, the strategy of closing areas to all fishing is being considered in many places around the world, and many such areas (Marine Protected Areas, or marine reserves) have already been designated. The reason that this is gaining support as a means to restore fish stocks lies in their impact on reproduction of the stock. The idea is that in the absence of harvest, fish will grow larger and thus produce more eggs in the protected areas compared with the adjacent fished areas. Some of the eggs and larvae produced in the protected areas will drift into the fished areas and settle there to enhance the fished stocks. Although this idea has conceptual merit, many questions remain, and field studies have yet to convincingly show that this result has occurred. A variety of early life-history studies will be required to validate enhanced recruitment to fished stocks from prohibiting fishing in some areas.

Integration of Basic and Applied Studies

As in many areas of science it is sometimes difficult to separate basic from applied studies. Results from what began as a pure scientific study find applications that were not anticipated. The reverse also happens: what starts out as an applied study can lead to basic understanding of processes and events. The following are examples from studies of the early life history of fishes.

As field studies on fish eggs and larvae began, it immediately became evident that plankton nets were not highly selective and caught the eggs and larvae of many species besides the ones being sought. To study the young of the species of interest, they needed to be separated from the young of all other species that might occur in the area that was being sampled. Rearing was the most definitive way (direct method) to accomplish this, but as mentioned before, early on it was not possible to rear larvae of marine fishes beyond the yolk-sac stage. Therefore another method (indirect method) was devised. Early juveniles were sought that had characters sufficient to identify them, but still retained some larval characters. Ever-smaller larvae that only had larval characters were linked to the larger ones so that a developmental series of the species was created. Descriptions of these developmental series were published, which allowed all stages of the species to be identified in plankton samples. It soon became apparent that it would be even more useful to put together descriptive material on all the eggs and larvae of fishes expected to be collected in an area. The first such guide was published for fishes of the Northeast Atlantic by Ehrenbaum (1905–1909). Presently, egg and larval identification guides are available for most marine and several freshwater areas around the world (see Chapter 3).

As larvae of more and more fishes became known, the idea occurred to E. H. Ahlstrom and others that they might be useful in a systematic context: they could be another source of systematic characters to help understand phyletic relationships among taxa. This idea was put to the test in a publication authored by experts on larval and adult taxonomy of various groups of fishes (Moser et al. [eds.] 1984a, www.biodiversitylibrary.org/bibliography/4334). The general conclusion was that larvae were a source of valuable characters for systematic studies, and in some cases they elucidated relationships that were obscure based on adult characters.

Many questions about the ecology of fish larvae require that their age be known. Determination of growth and mortality rates requires knowing the age of fish larvae from field samples. The development of fish eggs is largely temperature dependent, so if the temperature of the water where the eggs were found is known, and their morphological stage of development is determined, their age can be estimated. However, the growth rate of larvae varies with temperature and other environmental factors and food supply. The length and stage of development of larvae generally does not give a good approximation of age. Brothers et al. (1976) discovered that daily rings are laid down in the otoliths of larval fishes, and a count of these rings (increments) could be used to estimate their age. Daily increments on otoliths have since been found on nearly all larval fishes examined.

Beyond their use for age determination, otoliths hold a record of environmental conditions that the larvae experienced. The otolith grows by adding material to its surface. Elements and molecules that are laid down in the otolith remain in place for the life of the fish. Techniques are available to detect minute amounts of chemicals found in small regions of the otolith (representing a few days in the life of the fish). For example, chemical signals associated with individual estuaries where juveniles resided have been found in adult striped bass otoliths (Mulligan et al. 1987).

The Study of Early Life History of Fishes

Scientists at many government fisheries laboratories and at many universities around the world conduct research on the early life history of fishes. Some nongovernmental organizations are also involved, mainly those concerned with pollution and aquaculture. There is no single scientific journal that publishes only papers dealing with early life-history studies. Papers on the early life history of fishes can be found in a wide range of journals including *Fisheries Oceanography, U.S. Fishery Bulletin, ICES Journal of Marine Science, Canadian Journal of Fisheries and Aquatic Sciences, Environmental Biology of Fishes, Bulletin of Marine Science, Marine Ecology Progress Series, Copeia,* and *Transactions of the American Fisheries Society.* Other journals with a narrower focus also publish papers on certain aspects of early life history of fishes such as *Aquaculture, Marine Pollution Bulletin,* and *Estuaries.* Some articles of more general interest also appear in *Ecology, Science,* and *Nature.* Early life history of fishes has been the topic of several large international meetings, beginning with the Lake Arrowhead, California, meeting in 1963 (Lasker 1965), followed by the ICES symposia in Oban, Scotland, in 1973 (Blaxter 1974), the one in Woods Hole, MA, in 1979 (Lasker and Sherman [eds.] 1981), the one in Bergen, Norway, in 1988 (Blaxter et al. [eds.] 1989), and more recently the ones in Baltimore, Maryland, in 1996 (Fogarty [ed.] 2000, 2001) and in Ose, Norway, in 2002 (Browman and Skiftesvik 2003; Govoni [ed.] 2004). Beginning in 1977 Annual Larval Fish Conferences have been held, and since 1980 they have been the annual meeting of the Early Life History Section of the American Fisheries Society. Collections of some papers presented at these larval fish conferences have been published in various journals and special publications of the American Fisheries Society (e.g., Fuiman [ed.] 1993, 1996; Hunter et al. [eds.] 1993; Moser et al. [eds.] 1993; Grant [ed.] 1996; Chambers and Trippel [eds.] 1997; Leis et al. [eds.] 1997 [for a complete listing of publications see http://www2.ncsu.edu/elhs/elhspubs.html]). Fuiman and Werner (2002) provide an introduction to the contributions of early life-history stages of fish to fishery science.

CHAPTER 1

Fish Reproduction

WHAT ARE FISHES?

SEXUALITY
Range of Sexuality
Hermaphroditism
Unisexuality or Parthenogenesis
Bisexuality

REPRODUCTIVE PATTERNS
Live-Bearing
Egg Laying
　Demersal Eggs
　Pelagic Eggs

GONADAL DEVELOPMENT
Gonads
Oogenesis and Spermatogenesis
　Duplication Phase
　Primary Growth Phase
　Maturation Phase
　Follicle Development
　Yolk Vesicle Formation
　Vitellogenesis
　Envelope Formation
　Maturation
　Ovulation

GROSS MATURATION STAGES
Introduction

Gross Anatomical Examination
Example of Maturity Classification
Egg Mass Ratio (EMR) and
　Gonadal-Somatic Index (GSI)

FECUNDITY
General Uses
Definitions
Variations in Fecundity
Methods of Estimating Fecundity
　Mean Ova Diameter
　　Determination
　Histological Examination
　Fecundity Estimation Techniques

SPAWNING
Introduction and Terminology
Factors Triggering Maturation and
　Spawning
　Nutrition of the Female
　Physiological Factors
　Ecological Factors
　　Temperature
　　Photoperiod and Periodicity
　　Tides (Moon Cycles)
　　Latitude and Locality
　　Water Depth
　　Spawning Substrate Type

Salinity	Migrations
Exposure and Temperature	Habitats
Lifetime Spawning Strategies	Behavior
Sites	Secondary Sexual Characters

Each fish species has evolved in response to a unique set of selective pressures, hence species often differ in their life-history strategies; each life-history strategy is a set of developmental adaptations that allows a species to achieve evolutionary success. Each life-history stage (i.e., egg, larval, juvenile, adult) has a number of possible alternative states, but the life history of a given species consists of only one of these states for each life-history period.

Since each fish species evolves under a unique set of ecological conditions, it has a unique reproductive strategy with special adaptations including anatomical adaptations, developmental adaptations, behavioral adaptations, physiological adaptations, and energetic adaptations.

The reproductive process allows species to perpetuate themselves. Almost all fishes reproduce sexually, thus permitting mixing of the genes of the two sexes. The reproductive processes of fishes form the basis for early life-history studies. The great variety of these processes among fishes make their study worthwhile, but also determine how early life-history studies of various fishes can be conducted. For example, fishes reproduce in fresh and marine waters, have external and internal fertilization, have short annual reproductive periods, or produce gametes at regular intervals throughout the year. This chapter summarizes the reproductive patterns of fishes, with an emphasis on how these patterns affect early life-history studies. A more exhaustive account (and extensive reference lists) of modes of reproduction is found in Breder and Rosen (1966) and of developmental biology in Kuntz (2004). There is also a very useful chapter on reproductive ecology by DeMartini and Sikkel (2006).

WHAT ARE FISHES?

The answer to this question is not as simple as it may first appear. This book deals with bony, ray-finned vertebrates, that is, actinopterygians. This includes nearly 24,000 extant species, but excludes such "fishes" as sharks and rays (chondrichthyans) and lungfishes (dipnoians). Sharks and rays have internal fertilization and produce small numbers of large, yolky eggs. Some retain the developing embryos in the parent and give birth to miniature copies of the adults. Others lay the eggs in elaborate egg cases where embryonic development occurs. In either case, this reproductive pattern is quite different from that seen in most bony fishes where larger numbers of small eggs (often ≤ 1 mm) are produced. Even in bony fishes that brood their young (e.g., live-bearers [poecilliids], surfperches [embiotocids]), small eggs with relatively little yolk are produced. Maximum egg sizes in bony fishes are up to 10 mm in salmons and 15 mm in sea catfishes (ariids). Besides the skeletal features that distinguish bony fishes, they generally share the same basic early life-history pattern of producing small unfertilized eggs that undergo indirect larval development before becoming juveniles. Exceptions to this pattern have arisen in several lineages within the bony fishes.

SEXUALITY

The following is a synopsis of sexuality in fishes. A much more detailed account can be found in Breder and Rosen (1966) and DeMartini and Sikkel (2006).

Range of Sexuality

Fishes have a large range of sexuality (they stand out amongst vertebrates), and because of this there have been a number of studies into the questions of sex determination and sex differentiation—are they due to environment or heredity? Fishes may exhibit hermaphroditism, unisexuality (parthenogenesis), bisexuality (gonochorism), or a combination of sexualities.

Hermaphroditism

Hermaphroditism may be genetically programmed or a function of the social surrounding and is particularly prevalent in tropical reef regions.

In **synchronous** or **simultaneous hermaphroditism,** the left gonad is the ovary and the right one is the testis or vice versa, or there may be an ovotestis. Although self-fertilization exists, it is rare because recessive traits become present such as albinism, a physoclistous swim bladder, and others. Synchronous hermaphroditism is usually found in species where potential mates are sparsely dispersed because if the energy costs are the same for a synchronous hermaphrodite as for a male or female, it is clearly advantageous. It is common among bathypelagic and mesopelagic species living in the darkness of the ocean depths in low-population densities. In general, these species are continuously reproductively mature so that a chance encounter is always fruitful, but encounters may only occur once or twice in a lifetime. However, some seabasses (serranids) are also synchronous hermaphrodites, which is perplexing because they are not dispersed in low numbers.

In **asynchronous, consecutive, successive,** or **sequential hermaphrodites,** the fish start functionally as one sex and then switch to another; this strategy is confined to lower percoid groups such as parrotfishes (scarids), wrasses (labrids), and seabasses. A more detailed discussion of this strategy can be found in Warner (1975).

In **protandric hermaphrodites,** the fish are first males, then females; this is more common in porgies (sparids), flatheads (platycephalids), hagfishes (myxinids), and lightfishes (gonostomatids). The theory behind this reproductive method is that it has evolved under conditions where there is a strong female fecundity exponential increase because of volume change and asymmetry of age-specific fecundities between the two sexes. Protandry occurs when there is no advantage to males accruing larger size and, presumably, the gain is considerably greater than the cost of the change.

In **protogyny,** the fish are first female, then male, and there is asymmetrical reproductive success due to greater success by larger males in monopolizing several females. Possibly predation avoidance is also a factor, where males display and females choose, with the male acquiring the female according to their size. It is also an advantageous strategy when a male guards a harem of females since it becomes an advantage to switch from a female to a male because there is greater reproductive success when males are larger.

Unisexuality or Parthenogenesis

Rarer than hermaphroditism, but it is found in some live-bearers (only females). Development of young is without fertilization and females produce only female offspring. There are two ways this is done. In **gynogenesis,** the sperm of a closely related species is used to trigger development of the egg nucleus but does not fuse with it; and in **hybridogenesis,** fusion occurs but only the haploid female genome is transmitted to the developing ovum.

Bisexuality

Also called **gonochorism,** it is by far the most common mode of reproduction practiced by the majority of fishes and is what this book will focus on.

REPRODUCTIVE PATTERNS (FIGURE 1.1, TABLE 1.1)

Live-Bearing

Several groups of fishes have developed a live-bearing (i.e., giving birth to free-living larvae or juveniles, rather than laying eggs) life-history pattern. A prerequisite to this pattern is mating, copulation, and internal fertilization. Viviparous fishes have internal fertilization and are characterized by the embryo developing in close contact with the nourishing maternal tissue— no egg membrane covers the embryo. There are very few truly viviparous species. One of the best known examples is the surfperches (embiotocids) of the coastal waters of the Northeast Pacific, which produce well-developed

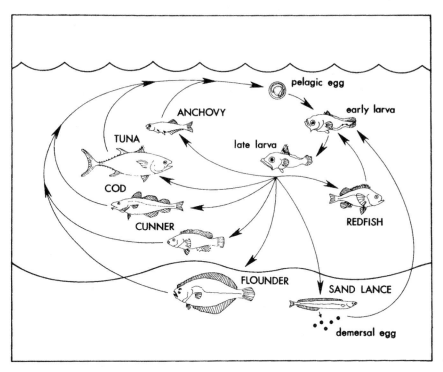

Figure 1.1. Diagram representing some of the variety of reproductive patterns of marine fishes. Objects are not to scale. Most fishes regardless of adult size or habitat spawn pelagic eggs that develop into pelagic larvae (e.g., flounders [pleuronectids], cunner [*Tautogolabrus adspersus*], cod [*Gadus morhua*], tuna [*Thunnus* spp.], anchovy [engraulids]). Other species (e.g., sand lance [*Ammodytes* spp.]) spawn demersal eggs that hatch into pelagic larvae. A few species (e.g., redfish [*Sebastes* spp.]) have internal fertilization and incubation followed by a pelagic larval stage.

young with the males of some species capable of breeding when born. In such fishes parental care of the eggs is maximum since the eggs are within the mother throughout development. Fecundity is low, usually < 100, and nourishment is primarily from the parent, not the egg (eggs small, embryo large), with the weight of the embryo increasing from fertilization to birth. In surfperches, embryos have large medial fins for absorbing nutrients. Once the young are released as juveniles, parental care normally ends. The series of events involving reproduction in viviparous fishes can be somewhat involved. For example, striped seaperch (*Embiotoca lateralis*) in Puget Sound breeds from June to August; after mating, the sperm is stored in a compartment of the ovary until September, when the sperms penetrate the ovary, fertilizing the eggs and causing gestation to start. Ten to 40 young are born, about 50 mm in length, tail first, usually in July of the following year.

TABLE 1.1. Reproductive Patterns of Teleosts

Timing of spawning	Semelparous	Iteroparous		
		Isochronal	Heterochronal	Indeterminate
	(Spawn once, near end of life)	(Spawn one batch of eggs per year)	(Spawn several batches during a spawning season)	(Spawn several times during a protracted spawning season)

Fertilization	External				Internal	
Embryonic nutrition	Oviparous				Lecithotrophic	Viviparous (matrotrophic)
	NO PARENTAL CARE POST-FERTILIZATION					
Release eggs (Spawn)						
Individual pelagic eggs	Anguilla	—	Theragra	Engraulis	—	—
Individual demersal eggs	—	Lepidopsetta, Gadus macrocephalus	—	—	—	—
Pelagic eggs in gelatinous masses	—	Scorpaena, Sebastolobus, Lophius	—	—	Helicolenus percoides	—
Demersal eggs in masses	Onchorhynchus	—	—	—	—	—
					Some cottoids	

	PARENTAL CARE POST-FERTILIZATION			
Release eggs (Spawn)				
Demersal eggs in masses, guard nests	—	*Ophiodon, Scorpaenichthys,* Centrarchids	Pomacentrids	—	—
Release larvae	—	—	—	Hemiramphids	*Sebastes*
Release juveniles	—	—	—	*Poecilia reticulatus*	*Zoarces viviparous,* Embiotocids, Some poeciliids

NOTES: "Isochronal spawn" = total spawn; "Heterochronal spawn" = serial spawn; semelparous, isochronal, and heterochronal are forms of determinate spawning in which all oocytes for a spawning season are matured simultaneously in the ovary.

Since **viviparity** has developed independently several times in fishes, the means of transferring nutrition from the female to the young is quite variable (Wourms 1981). In the case of rockfishes (*Sebastes* spp.), fertilized, relatively undeveloped late yolk sac larvae are produced. The yolk may be supplemented by maternal nourishment as the eggs and larvae develop in the female (Boehlert and Yoklavich 1984). Fecundity of rockfishes does not seem to be reduced significantly in conjunction with the apparent added protection afforded by foregoing the free-living egg stage. Such fishes in the past have been termed "ovoviviparous," and characterized by having internal fertilization. The mother may or may not nourish the embryo, but at any rate the embryo and the maternal tissue are separated by the egg membrane— that is, no "placenta-like" structure is present. Typically, the eggs hatch internally and are released as early larvae. Again, the parental care of eggs is maximum since the eggs are retained internally, but fecundity may be much higher (e.g., in the case of rockfishes tens of thousands of young may be produced per year). As a general example of the type of annual reproductive cycle in rockfishes, we can use golden redfish (*Sebastes norvegicus*), which is an extremely important commercial species in the Atlantic and which has been studied in some detail (Magnuson 1955 in Hempel 1979). Golden redfish mate in the winter when the male is ripe but the female eggs are not yet ripe, so the female stores the sperm until early spring at which time fertilization occurs. Within a single female it has been found that the embryos are all the same size, indicating a single fertilization event. The eggs hatch internally, and then are extruded (born) as early larvae in late spring, which is 2–3 months after fertilization and 6–7 months after mating. At the time of extrusion it is thought that the oxygen requirements of the larvae exceed that provided by the mother, although in the rockfishes it has been demonstrated that in at least some species a special vascular system to the ovary ensures a particularly good gas exchange (Moser 1967).

Egg Laying

Actually, internal fertilization is very rare among teleosts. Eggs of the vast majority of teleosts, especially marine fishes, are released before fertilization— these are the **oviparous** fishes (literally, "egg-laying" fishes). Males and females swim close together so that the eggs are shed into a cloud of spermatozoa. Because mating and courtship occur in many oviparous species, one important aspect is that the activity of sperm depends on small concentrations of Ca or Mg ions, which allows sperm to remain active in salt water for up to an hour or so as opposed to a minute or so in fresh water. The eggs are then fertilized and develop in the environment outside the female. An egg membrane is present, and the embryonic stage is nourished entirely by the yolk. In a few cases (e.g., sea horses and pipefishes [syngnathids]), after being released, the eggs are carried in a pouch in the adult (males in sea horses), or in the mouth of one of the parents (e.g., some catfishes [siluriforms] and cichlids [Cichlidae]). The eggs of most fishes develop either in nests constructed by the adults, or they develop in the environment at large. Fecundity of marine oviparous fishes can be extremely high (> 300 million in the ocean sunfish [*Mola mola*] and commonly 1 million or more). Among the oviparous fishes, there are so-called **demersal spawners** and **pelagic spawners**—actually, the eggs they extrude are either demersal (on the bottom) or pelagic (above the bottom, and often at or near the surface), whereas the juveniles and adults of the species may be either demersal or pelagic in both cases (Table 1.2).

Demersal Eggs. Demersal egg spawners produce eggs that are heavier than the surrounding water and which develop on the bottom. These eggs are either attached to the substrate or float loosely on the bottom and are generally adhesive. Eggs of almost all freshwater fishes are attached to the substratum or are loosely in contact with the bottom, that is, are demersal. This is a reflection of the fact that protein is the main constituent of fish eggs, and protein has a higher specific gravity than fresh water, and demersal eggs have a low water content. Hempel (1979) points out that among the marine groups which spawn demersal eggs are the smelts (osmerids), herring (*Clupea* spp.), greenlings (hexagrammids), and sculpins (cottids). It's interesting that on the Arctic and Antarctic Shelves, fish eggs are mainly demersal with large yolk reserves and long incubation periods (e.g., Greenland cod [*Gadus ogac*] and Arctic flounder [*Liopsetta glacialis*]), whereas most other cods (gadids) and

TABLE 1.2. Summary of Differences Between Demersal and Pelagic Eggs

Characteristic	Demersal	Pelagic
Usual Size	> 1 mm	≤ 1 mm
Specific Gravity	greater	less
Envelope	thicker	thinner
Color	opaque, colored	transparent
Amt. of Yolk	large	small
Period of Dev. (Temperate)	longer (up to 2 mos)	short (~ 1 wk)
Parental Care	common	none
Fecundity	low	great (> 1 million)
Larvae	swim/feed @ once	float/yolk sac
Dispersal	probably low	probably high

right-eyed flounders (pleuronectids) have pelagic eggs. It's possible that spawning demersal eggs may protect the eggs against the risk of freezing and/or also against the low salinity of the surface water (where osmoregulation is difficult) during the melting of the ice.

In demersal spawning fishes there is a more or less continuous line of increasing parental care from deposition of eggs on selected substrate to complete protection of egg masses (Hempel 1979; Gross and Shine 1981). Examples include the eggs of Pacific herring (*Clupea pallasi*), which are plastered on seaweed and no parental care is involved. The male Atlantic spiny lumpsucker (*Eumicrotremus spinosus*) protects the egg mass against a variety of predators and blows water on the egg mass almost continuously for several weeks to increase oxygenation and remove debris. Male lingcod (*Ophiodon elongatus*) protect "nests" of eggs, attacking intruders (including human divers) and picking off snails and other egg-preying invertebrates and small fishes (e.g., sculpins)—normally the only items you find in the stomach of a male lingcod guarding a nest. Some gobies (gobiids) lay their eggs in the shell of a bivalve. Other stages of care include protection by the parent body. Some gunnels (*Pholis* spp.) roll their egg mass into a ball and then one parent coils around the eggs. A well-known example of advanced care is the male sea horses and pipefishes carrying their eggs in the brood pouch. And of course, there is the extreme example of mouth brooding in some freshwater and marine catfishes (siluriforms). Fecundity is generally related to the amount of parental care involved—that is, *the more parental care involved, the lower the fecundity.*

Mating associations in both demersal and pelagic spawners are not related to the type of egg produced, but rather to whether they are schoolers or nonschoolers. Schooling species in which the sexes are notably alike often form large mating associations. Breder and Rosen (1966) say that contrary to general belief, there does not seem to be any evidence of "mass spawning" in the sense of random scattering of the sexual products in a milling school of males and females, although clearly the possibility exists that several males may fertilize the eggs of several females successively.

Bottom-dwelling territorial fishes in which the sexes are notably dimorphic may generally be expected to form breeding pairs and to have demersal eggs. Most of these dimorphic mating pairs with demersal eggs are in fresh water (where nearly all eggs are demersal because of their specific gravity). The fishes in salt water with demersal eggs are found in the nearshore zone and are common from the Arctic (Greenland cod and Arctic flounder) and the Antarctic Shelves (cod icefishes [nototheniids]), and, are also common in the temperate nearshore region, which is unpredictable and changeable. It can be hypothesized that having demersal eggs reduces the risk of getting eggs cast ashore or damaged by abrasion. Fishes with demersal eggs are found not only among families appearing early in the fossil record (i.e., herrings [clupeids], smelts) but also among more recent groups (combtooth blennies [blenniids], sculpins, greenlings, etc.).

Common expressions of parental care are nest building, nest cleaning (removal of dead eggs and decaying matter from the nest), and fanning (behavior for aeration), and guarding (predator removal, e.g., lingcod). In a wide variety of fishes the eggs are laid in nests constructed by the parents. In most cases the male builds the nest and attracts the female to lay her eggs there, where he fertilizes them. In many fishes the eggs are guarded from predators, aerated, and cleaned during development by one or both parents. Care often extends for some time after the eggs hatch. Examples of fishes exhibiting the guarding behavior mode include most sunfishes (centrarchids) and sculpins. However, some fishes that build nests for the eggs or hide them take no further care of them. For example, Pacific salmon (*Oncorhynchus* spp.) lay their eggs in depressions (called redds) they dig in the gravel of the streambed, but the adults die shortly thereafter.

Pelagic Eggs. For the majority of marine fishes that are pelagic spawners, there is little to say about modes of reproduction. As previously mentioned, the male and female swim close together and the eggs and sperm are broadcast into the water. No parental care is involved with the eggs and the spent (spawned-out) adults may resume their prespawning activities. This is the most common spawning mode in the marine environment regardless of whether the habitat is demersal or pelagic, whether it is a coastal or oceanic distribution, whether tropical or boreal ranges, and whatever the systematic affinities are. Out of about 12,000 marine teleosts, about 9,000 (75%) produce pelagic (buoyant) eggs and the eggs are spawned, fertilized, and float individually (although a few species have floating egg masses), usually near the surface. The large number of eggs produced and their rapid dispersal makes it impossible for the adults to show any form of parental care. Sexual dimorphism and dichromism are reduced or absent in pelagic spawners.

A few pelagic spawners lay pelagic eggs in gelatinous masses. Goosefishes (*Lophius* spp.) and thornyheads (*Sebastolobus* spp.) are among these. Thousands to millions of eggs are produced by each female in these cases, and the egg masses can be quite large (several liters). However, the large majority of marine fishes lay pelagic eggs that float individually in the water. The eggs are spawned, fertilized, and develop in the water column as part of the plankton. Rather than caring for the eggs, cannibalism by the adults can be a significant source of mortality in some fishes with planktonic eggs.

GONADAL DEVELOPMENT

Gonads

Female fish have paired ovaries that produce eggs, and male fish have paired testes that produce sperm, which along with seminal fluid is termed *milt*. Although this is the general rule, there are many exceptions. Some fish change sex during their lifetime, some from male to female (protogynous hermaphrodites) and some from female to male (protandrous hermaphrodites), whereas some even produce both sperm and eggs simultaneously (synchronous hermaphrodites). Most fishes are capable of spawning several times during their life (semelparous); however, a few (e.g., Pacific salmon, freshwater eels [anguillids]) spawn once and die (iteroparous).

Oogenesis (Figure 1.2) and Spermatogenesis

Oogenesis is the process that results in the formation of a haploid cell and a mature egg capable of supporting a developing embryo (Foucher and Beamish 1977). The basic stages are the duplication phase, primary growth phase, follicle development, yolk vesicle formation, vitellogenesis and envelope formation, maturation, ovulation, and the spawning-fertilization and egg activation stage.

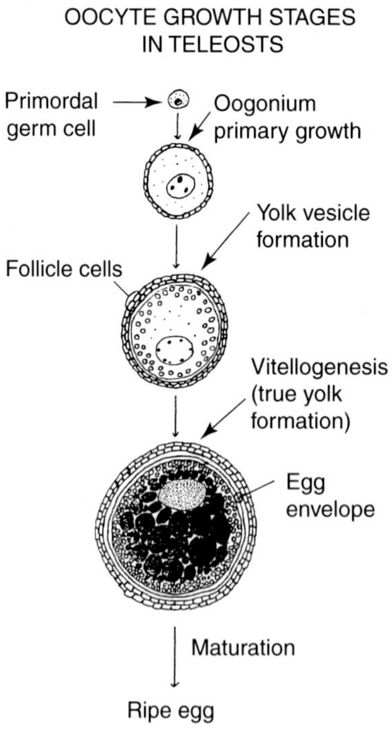

OOCYTE GROWTH STAGES IN TELEOSTS

Primordal germ cell → Oogonium primary growth

Follicle cells

Yolk vesicle formation

Vitellogenesis (true yolk formation)

Egg envelope

Maturation

Ripe egg

Figure 1.2. Oogenesis in teleosts (adapted from Carla Stehr, personal communication).

Duplication Phase. The ovary and oogonia develop from primordial germ cells, which are endodermal in origin. The reserve oogonia undergo repeated mitotic divisions (remaining diploid), and are no longer oogonia after mitosis stops—this is the starting point for oocytes.

Primary Growth Phase. The growth of the primary oocytes which occurs when they are still diploid cells, and ends with a meiotic division forming a secondary oocyte and a polar body each of which are haploid.

Maturation Phase. The secondary oocyte undergoes a second meiotic division resulting in a mature ovum and another polar body, whereas the first polar body may divide into two polar bodies (i.e., results in three polar bodies and one ovum); the polar bodies are eventually resorbed. Follicle development, yolk vesicle formation, vitellogenesis, and envelope formation are occurring during the maturation phase.

Follicle Development. This refers to the structure formed surrounding the developing oocytes, which is formed by two layers of cells that are the outside theca cells and the inside follicle cells (also called the glandular granulosa).

Yolk Vesicle Formation. The formation of yolk vesicles—a misnomer because they are not true yolk, but instead, they contain polysaccharides; they are also known as cortical alveoli. Generally, cortical alveoli are found around the periphery of the oocyte.

Vitellogenesis. This is stimulated by pituitary gonadotropin (Figure 1.3) and is a true yolk formation where the yolk forms and pushes the cortical alveoli to the margins. Microvilli extend from follicle cell (or oocyte) cytoplasm through pore canals providing passageways to transport yolk material and nutrients across membranes to oocytes. Proteins produced in the liver are carried by blood vessels to the follicles, which transfer them to the oocyte cytoplasm where they are assembled into yolk.

Envelope Formation. This occurs during yolk vesicle formation and vitellogenesis. The egg envelope may also be called the chorion, zona pellucida, zona radiata, or vitelline membrane. We prefer the term "envelope" over the more commonly used mammal term "chorion," because "chorion" suggests a cellular layer on the outside, which is not the case in fish eggs. However, chorion is most commonly used in fish literature. The envelope may be composed of as many as three layers, namely: (1) The primary layer, which is produced by the egg, is the innermost layer and is the true "envelope" layer. The primary layer is usually the only envelope present in pelagic eggs for buoyancy reasons, and to allow for the remnants of the pore canals to be seen externally on the egg; (2) the secondary layer, which is produced by the follicle cells (maternal tissue) after rupture, and which usually is found in demersal eggs (if this layer

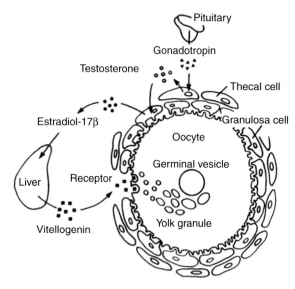

Figure 1.3. Hormonal regulation of vitellogenesis in fishes (from Nagahama et al. 1995).

is present, the remnant pore canals cannot be seen externally); and (3) the tertiary layer, which is produced by the ovary after the egg has ovulated—not common, but present in sharks and rays (elasmobranchs).

Maturation. This is stimulated by steroids, which are produced by follicle cells. At this point the oocyte undergoes the final meiotic division, the oocyte begins to hydrate by taking up ovarian fluid, the egg increases 3–4 times its size, the microvilli withdraw from pore canals which shut off as the egg imbibes body fluid, the yolk coalesces, and the envelope stretches.

Ovulation. This occurs when the follicle cells burst as oocytes hydrate. Not all eggs are ovulated, but there are usually varying amounts of atresia thought to occur due to such things as stress, nutritional factors, and gonadotropin; atretic eggs are reabsorbed.

GROSS MATURATION STAGES

Introduction

A record of the state of maturity of fish is often required to determine the proportion of the stock that is reproductively mature, the size or age at first maturity, the current reproductive status, and the nature of the reproductive cycle for a particular population or species. Some of the general ways of determining the reproductive state are by staging of ovaries using gross anatomical criteria, calculation of Gonadal-Somatic Index (GSI) or Egg Mass Ratio (EMR), by classifying ovaries histologically, and by use of mean egg diameters.

Gross Anatomical Examination

Gross examination can be done because although maturation of the gonads occurs at a cellular level, the maturation state or the spawning state of the gonads can be determined upon gross visual examination of the gonads in the field. This is a particularly useful technique when you are on a trawler or at the fish market and it is not feasible to examine gonads in detail. For this purpose a field key is often handy as an aid in judging the general appearance and size of the gonads. An important aspect of such a classification is that it be standardized. Fortunately, in fishes with one short, clear-cut spawning season per year, there are a number of stages visible by macroscopic inspection. In general, the immature stage (or resting stage) usually covers the major part of the year and is characterized by a very thin ovary similar to that in a juvenile (immature) fish, although the wall of the ovary is usually thicker in repeat spawners than in recruit spawners advancing toward maturity. After the immature stage, there are a number of preparatory phases to spawning which are actually continuous with the immature stage but which investigators have attempted to define for classification purposes. In general terms, there is the "developing" stage during which sperm and eggs are developing, then the "mature" stage characterized by large testes and large ovaries with eggs filled with yellow yolk, which usually are visible through the body wall. Ovulation takes place, the eggs become transparent from the intake of body fluid, and the female is ready to spawn—this is the spawning (or "ripe") stage. (A live fish is referred to as "ripe and running" if pressure on the abdomen results in eggs or sperm being released.) After spawning, the ovary is soft and bloody with some unshed eggs and the fish is referred to as spent (or "spawned out")—the unshed eggs will be resorbed during the recovery phase. In summary, the simplest classification scheme includes the following stages: **Immature, Developing, Mature, Spawning, Spent,** and **Recovery.** Do not dismiss the usefulness of the gross anatomy observations—they are especially valuable when combined with one of the other ways of examining maturity stages (similar to the usefulness of gross pathology observations to histopathology).

Example of Maturity Classification

A more detailed classification system is used by NOAA-Fisheries for walleye pollock (*Theragra chalcogramma*) (Table 1.3).

TABLE 1.3. Maturity Stages (by Code Number) for Walleye Pollock (Sarah Hinckley, personal communication)

Males

1	Immature	Testes thread-like and contained within a transparent membrane.
2	Developing	Testes uniformly ribbon-like. Surface of testes appears smooth and uniformly textured.
3	Mature	Testes large and highly convoluted; sperm cannot be extruded. Body wall incision causes gonads to be expelled from opening.
4	Spawning	Testes milk freely or extrude sperm when compressed.
5	Spent	Testes large, but flaccid, watery, and bloodshot.

Females

1	Immature	Ovaries small, tapered, and transparent. Will not spawn this year. Sex may be difficult to determine.
2	Developing	Early: Ovaries tapered, forming two distinct, transparent lobes with well-developed blood vessels. No or few individual ova present. Late: Developing lobes fill up to half of the body cavity, with distinctly visible opaque, orange eggs.
3	Mature	Ovaries fill more than half of the body cavity and contain distinctly visible eggs. Eggs are not extruded when ovaries are compressed. Most eggs are opaque, but scattered clear (hydrated) eggs may be present. Eggs cannot be easily separated from each other.
4	Spawning	Ovaries large, filling the body cavity. Most eggs are transparent (hydrated) though some opaque eggs may remain. Eggs are extruded from the body under slight pressure or are loose in the ovary and easily separated from each other.
5	Spent	Ovaries are large, but flaccid, watery, and generally reddish. Scattered unspawned eggs can be seen. Ovaries that are "Recovering" will appear red and contain scattered eggs, but will not be as large or quite as flaccid as very recently spawned ovaries, and should be classified as "Early Developing."

NOTE: For both males and females, codes 2 through 5 refer to adult fish.

Egg Mass Ratio (EMR) and Gonadal-Somatic Index (GSI)

The EMR and GSI measure reproductive strain on fishes by measuring general body weight ratios. When males and females are compared, females invest more energy in gonads than males although in pelagic spawners it may be more similar since males need to produce a lot of sperm to ensure fertilization; however, the investment of females is still many times greater than males. These indexes are calculated as follows:

$$\text{EMR} = \frac{\text{Total wt. of all eggs in one season (dry wt. in g or cal)}}{\text{Body wt. (including ovary)}}$$

$$\text{GSI} = \frac{\text{Gonad wt. (g or cal)}}{\text{Body wt. (including gonad)}}$$

FECUNDITY

Fecundity, the number of eggs ripened by female fish during a spawning season, or event, varies from a few dozen in some continuously reproducing livebearing fishes to millions in some species that spawn pelagic eggs on an annual basis. In general, fecundity varies among species inversely with the amount of "care" given to the individual progeny: viviparous fishes have lower fecundity than ovoviviparous fishes, which in turn have lower fecundity than oviparous fishes, and nest builders have lower fecundity than pelagic spawners. Within species fecundity is positively related to size; generally it is close to a function of the cube of fish length. Although fish eggs range from about 0.5 to 20 mm in diameter, the size of the adult limits the fecundity in smaller species. In most species with high fecundity, several batches of eggs are usually produced at intervals of a few days to weeks during the spawning season.

General Uses

Some general reasons for determining fecundity is that it is useful for making total population estimates, it is useful in studies of population dynamics or productivity, and it is useful for characterizing specific populations, subpopulations, and/or stocks of fishes.

Definitions

"Absolute fecundity" is the number of ripe eggs produced by a female in one spawning season or year (this is the usual meaning when the general term "fecundity" is used, although on occasion it might also mean the number of eggs produced in a lifetime). "Relative fecundity" is the number of eggs produced in a season per unit somatic weight of the fish (i.e., eggs/gram), and is useful if it is shown that the fecundity of a fish is proportional to its weight, which is not uncommon. "Population fecundity" is the number of eggs spawned by the population in one season, is the sum of the fecundities of all females, and is usually expressed as the product of the expected fecundity of an average female × the number of breeding females in the population (an example of why classifying maturity stages may be useful).

Absolute fecundity varies with age, length, weight, and type of fish (species and population). For example, sharks and rays have few eggs whereas ocean sunfish have up to 300 million. The highest fecundity is in pelagic spawners, 100,000 to millions (e.g., Atlantic and Pacific cod [*Gadus* spp.] with several million). Intermediate fecundities are found in demersal spawners (1,000 to 10,000, although there are exceptions like the lingcod with 500,000). The lowest fecundity is in live-bearing fishes with few to 100 or so (usually less), although there are quite a few exceptions including rockfishes which often extrude > 100,000 larvae, but which developmentally are closer to the larval stage of pelagic spawners than of other live-bearers, or even of demersal egg spawners. Fecundity is also affected by the nutritional status of the female, and in the population there may also be compensatory mechanisms regulating fecundity.

Many investigators have made scatter plots of fecundity versus length and have come up with the general relationship where the increase in fecundity with length can be described by the power relationship $F = aL^b$, where F = fecundity, a = constant, L = length, and b = nearly 3. A logarithmic transformation gives the straight-line regression of log (or natural log, ln) fecundity on log length, which helps take care of the problem of the larger fish having greater variation in fecundity than the smaller, that is,

log F = log a + b log L. (Equation is of the simple $y = a + bx$ form, i.e., if log F = y, and log a = a, and b = b, and log L = x, then $y = a + bx$.)

When $b \cong 3$, it means F is about proportional to the fish weight because of the surface area-to-volume ratio and allometric growth; that is, as the fish grows in length, the volume (weight or fecundity) gets larger to the cube. However, for Atlantic herring (*Clupea harengus*) stocks, in some situations, values between 3 and 7 have been found, indicating fecundity was increasing faster than body weight (Hempel 1979). For many stocks, in terms of egg production of a population, this implies that if F increases with a power of L higher than 3, a stock of large (and old) fish will produce far more eggs than a spawning stock of the same total weight but consisting of mainly small fish (Figure 1.4). This is an obviously important concept to consider when managing a fishery, but is also an important concept to consider in reproductive ecology and recruitment ecology studies. As fish mature, they have a much higher fecundity because less energy is devoted to growth so more can be given to reproduction.

Variations in Fecundity

Relationships between fecundity and age and weight are not as apparent as with length because there is too much variation in age and weight. This is also a problem with using GSIs to compare fecundities of different fishes from different years, from different localities, and when fish are of different sizes.

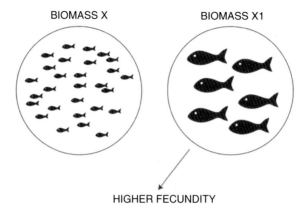

Figure 1.4. Concept of two spawning stocks of the same species and same total weight (biomass), but with one stock (X) composed of mainly small fish and the other stock (X1) of fewer but larger (and older) fish producing far more eggs.

In general, within a population, the size of ripe eggs is not very variable, regardless of the fish size or age. However, there is a relationship between egg size and environmental conditions. For example, there is some evidence that shows a negative correlation between the condition of the mother and egg size—that is, well-fed fish produce more but smaller eggs. Comparisons between species and populations also show that the size of the egg (and of the newly hatched larvae) is negatively correlated with fecundity. Larger eggs have more yolk, which is an extended protection strategy so that when the egg hatches the larva has more yolk although it is now independent of the mother.

One way to think of this is that the production of large eggs with considerable yolk supply and large initial size of the larvae is a kind of extended maternal care for the individual offspring, whereas a high number of small eggs can be considered as a protective measure to ensure survival of the species against high egg and larval mortality, particularly by predation. This argument has been used in the interpretation of fecundity differences between various herring populations in the North Sea, which spawn at different seasons with the result that larvae face different conditions (Hempel 1979). Thus, herring spawning in winter produce large eggs in low numbers, whereas summer spawners produce much smaller eggs in high quantity. The newly hatched larvae of the winter herring meet fewer predators but also less of the very small food organisms that are present for the summer spawners. The general validity of this hypothesis needs much more testing.

These concepts (and other relationships thought to be true) were presented in the 1975 thought-provoking paper by Johnson and Barnett. In their study, "midwater fishes" are real midwater, not herring, but viperfishes (chauliodontids), which live at depths of hundreds of meters. Table 1.4

TABLE 1.4. Larval Survival and Natural Selection
(based on Johnson and Barnett 1975)

Productivity Regime	Average Egg Size	Average Fecundity	Average Larval Size	Number of Meristics
Less productive—low food densities (danger of larval starvation)	Larger	Lower	Larger	Higher
More productive—high predator densities (danger of predation on larvae)	Smaller	Higher	Smaller	Lower

summarizes the authors' findings (how natural selection favors larval survival, based on similar reasoning as in the winter-summer spawning herring concept given previously).

There was definitely a negative correlation found between three measures of productivity and meristic values (anal fin rays, vertebrae, longitudinal photophores). Measures of productivity were phosphate-phosphorus concentrations, net primary production, and zooplankton standing stocks. Meristics were found not to correlate with temperature, salinity, oxygen, or any other physical or chemical factor known to possibly affect meristic variation in fishes.

What these authors are hypothesizing is that the inverse relationship between meristic values and measures of food availability reflects adaptations to low food densities (for larvae) in areas of low productivity and higher predator densities in areas of higher productivity. Their reasoning is that in areas of low food densities, natural selection has favored mechanisms tending to offset the danger of larval starvation (i.e., larger egg size, lower fecundity, larger larval size).

The advantages of being a larger larva in areas of low food density include (1) a longer period of survivorship solely on yolk reserves, (2) increased mobility, (3) wider search volume, and (4) increased diversity of potential prey organisms. Presumably, in areas of high productivity, the danger of starvation is less, but the danger of predation (more potential fish egg and larvae predators) is greater. Here, selection favors smaller average egg size, higher fecundity, and smaller average larval size, the theory being that the increased number of eggs and larvae overwhelms or saturates the predators so that at least some eggs and larvae survive.

Finally, these authors hypothesize that the inverse correlation between productivity and number of meristics is explained by the findings of others that large egg size may result in a longer embryonic and larval duration during which the meristic values are being determined and which will result in higher meristic values.

The caveat is that unfortunately almost no actual data exist for midwater fishes to test their hypothesis—e.g., nothing is known for midwater fishes about age and size at first spawning, number of spawnings per female, fecundity, seasonality of reproduction, course of larval development, or factors actually determining survivorship of larvae. The advantage of using midwater fishes was that reliable productivity measurements were available for the oceanic regimes they are found in, which is not the case with most other groups of fishes, especially nearshore fishes, upon which most experimental meristics and developmental work has been done.

Methods of Estimating Fecundity

Mean Ova Diameter Determination. For many fishes, as part of a fecundity determination, it is important to first do an ova diameter analysis by measuring a representative sample of eggs in the ovary when they are close to being spawned. Plotting of mean ova diameters, for many temperate water fishes, indicates the eggs to be spawned in the present year, and those to be spawned in future years. Laboratory Exercise 1 indicates the general technique followed in doing ova diameter measurements.

Histological Examination. The best method for determining if multiple spawnings are occurring is a histological examination. This must be done if the fish species is a multiple spawner; for example, the northern anchovy (*Engraulis mordax*). The frequency of multiple spawning can be determined by using postovulatory follicles (cannot tell grossly, and often not clear by measuring ova diameter). Ovaries can also be distinguished between immature and postovulatory ovaries by using histology—that is, atresia can easily be seen. For more about this method, see Laboratory Exercise 2.

Fecundity Estimation Techniques. Fecundity is generally estimated by using a volumetric technique, a wet or dry weight technique, or use of a device that is able to actually count individual eggs. The volumetric and weight techniques rely on simple proportionality to estimate the total fecundity from a known number of eggs in a known volume or weight of a subsample, and a known value for the total volume or weight of the sample, and then calculate the total number of eggs in the ovary. The real trick is to make sure that the subsample is truly representative of the whole ovary. Laboratory Exercise 3 demonstrates the fecundity determination technique for the volumetric and wet weight methods.

SPAWNING

Introduction and Terminology

Spawning refers to the release of unfertilized planktonic eggs by female fish, which is the reproductive pattern for most marine fishes. The eggs are fertilized shortly after release by males. Some fishes also deposit unfertilized eggs in nests where they are fertilized and develop. Fishes with internal fertilization release free-swimming larvae, or juveniles. The ripening of eggs and spawning are controlled by hormones, nutrition of the female, and external (ecological) factors (Hempel 1979). Usually maturation and spawning are controlled by a combination of endogenous and exogenous controls and are not governed by any specific factor.

TERMINOLOGY USED IN DISCUSSING SPAWNING

Mating pairing (one-on-one) for the purpose of fertilizing eggs; copulatory organ present.

Spawning release of unfertilized eggs into the environment or release of larvae into the environment; mating and spawning need not occur simultaneously (e.g., surfperches). Spawning can occur without true mating (e.g., herring, which are broadcast spawners).

Fertilization fusion of eggs and sperm (creating diploids from haploids); mating and fertilization need not occur simultaneously (e.g., surfperches and rockfishes).

Incubation time time from egg fertilization to hatching.

Gestation applies only to live-bearing fishes; it is the time young stay within the female.

Hatching when the larva frees itself from the egg.

Breed to produce offspring by hatching (or gestation).

Brood guard and groom eggs until they hatch.

Factors Triggering Maturation and Spawning

There are three primary factors that influence the events leading up to spawning: nutritional state of the female, physiological factors (hormones), and ecological factors.

Nutrition of the Female. The feeding condition of the mother can have an important effect on the final maturation of the eggs. Two examples from Hempel (1979) show that in some of the Atlantic herring populations spawning may occur only every other year if environmental conditions, particularly those affecting food supply, are poor. Also, it has been found in the laboratory that in Atlantic sole (*Solea solea*) no spawning occurred when the flatfish were fed a diet (mussels only) deficient in certain amino acids; however, when the flatfish were force-fed the missing amino acids they spawned, indicating the ovary had been unable to obtain the needed amino acids from maternal tissue when the nutrition of the female had been inadequate (Hempel 1979).

Physiological Factors. Hormones govern migration and timing of reproduction, morphological changes, mobilization of energy reserves, and elicit intricate courtship behavior. The pituitary is the major endocrine gland that produces gonadotropin, which controls gametogenesis, the production of gametes, namely sperm (spermatogenesis) and eggs (oogenesis), by the gonads. The pituitary also controls the production of steroids (steroidogenesis) by the gonads; once the gonads are stimulated by the pituitary they begin producing steroids, which in turn control yolk formation (vitellogenesis) and spawning. The control of spawning by the pituitary is often used in fish farming such as in the production of caviar from sturgeon (*Acipenser* spp.) where spawning is induced by injecting pituitary extract at a late stage of gonadal development, usually in combination with changes in temperature and light periodicity.

Ecological Factors. Often ecological factors are associated with timing so that food availability is optimal for the larvae. Some ecological factors important to spawning are temperature, photoperiod, tides, latitude, water depth, substrate type, salinity, and exposure.

TEMPERATURE. An important factor in determining geographical distributions of fishes. Although little is known about the mechanism by which temperature controls maturation and spawning in fishes, for many marine and freshwater fishes the temperature range in which spawning occurs is rather narrow, so that in higher latitudes the minimum and maximum temperature requirement for spawning is often the limiting factor for geographical distribution and for the successful introduction of a species into a new habitat. For example, Pacific halibut (*Hippoglossus stenolepis*) are found spawning primarily in areas with a 3–8°C temperature on the bottom and therefore do not spawn in Puget Sound, although the adults are caught in the northern areas of Puget Sound. In fact, even in highly migratory tuna, spawning is restricted to water of specific temperature ranges.

PHOTOPERIOD AND PERIODICITY. The daylength (photoperiod), in some cases at least, is thought to influence the thyroid gland and through this the fishes' migratory activity, which is related to gondal development (maturation). In the northern anchovy, by combining the effects of temperature and daylength, continued production of eggs under laboratory conditions was brought about by keeping the fish under constant temperature conditions of 15°C and a light periodicity of less than 5 hours of light per day (Lasker personal communication). In high latitudes, spawning is usually associated with a definite photoperiod (and temperature), which dictates seasonal pulses of primary production in temperate regions to assure survival of larvae. In low latitudes, where there is little variation in daylength, temperature, and food production, other factors may be important such as timing with the monsoons, competition for spawning sites, living space, or food selection.

Reproductive periodicity among fishes varies from having a short annual reproductive period to being almost continuous. There is a tendency for the length of the reproductive period to shorten with increasing latitude. Thus tropical fishes spawn nearly continuously, whereas subarctic fishes spawn predictably during the same few weeks each year. Presumably times of spawning have evolved so larval development will coincide with an abundant food supply. Within spawning seasons, fish may spawn on a daily or monthly tidal cycle or on a diel cycle, or in association with some other environmental cue, such as a change in daylength, temperature, or runoff. A notable instance of spawning periodicity associated with the tidal cycle is the California grunion (*Leuresthes tenuis*), which spawns intertidally at the peak of the spring high tides (Walker 1952). Within species, spawning times may vary with latitude: Generally, in species that spawn as daylength increases, spawning occurs earlier in the year in lower latitudes than at higher latitudes. In species that spawn as daylength decreases, spawning takes place earlier in the year at higher latitudes than at lower latitudes.

TIDES (MOON CYCLES). The dependence of spawning on moon cycles in California grunion spawning on California beaches is an extreme example of external factors controlling reproduction in fishes. Grunion are adapted to spawning on the beach every two weeks in the spring during a new or full moon. Spawning is just after the highest high tide (Figure 1.5); therefore, eggs deposited in the sand are not disturbed by the surf for 10 days to a month later. Eggs will hatch when placed in agitated water (which simulates surf conditions). In Puget Sound, surf smelt (*Hypomesus pretiosus*) spawn year-round, except in March. Surf smelt deposit eggs at high tide in sand and gravel (but not necessarily at the highest tide). On the open ocean shores, spawning occurs at midtidal heights (for different subpopulations) (Dan Pentilla personal communication).

LATITUDE AND LOCALITY (Table 1.5). Pacific herring show a definite relationship between latitude and spawning time. Spawning is early in San Francisco (December, January); later in Washington State (February, March, April, May); and still later in Alaska (April, May, June). These fishes are perhaps of different, distinct subpopulations.

In temperate waters a biomodal distribution of eggs is usually seen, which indicates discontinuous spawners (Figure 1.6). The smaller-sized mode represents resting eggs for a future spawning, and the larger mode represents maturing eggs (oocytes), which will presumably be spawned within the year. Temperate water fishes are also usually deterministic, which means all eggs to be spawned are determined at the start of the year.

A polymodal distribution of eggs is typical of tropical areas and some temperate water fishes, which signifies continuous or serial spawners, and indicates several spawnings. A well-known temperate example would be Pacific sardine (*Sardinops sagax*), which spend 7 months spawning and 2 months developing/maturing (Clark 1934). Batch spawning has been

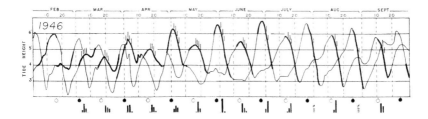

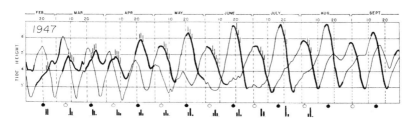

Figure 1.5. California grunion (*Leuresthes tenuis*) runs observed at La Jolla, California, in 1946 and 1947, plotted in relation to variations in observed high-tide heights at La Jolla. Only the heights of high tides have been plotted. The high tides about 24 hours apart have been connected by smooth lines. The two tides of each day yield the two series of curves. Tides occurring during darkness are indicated by the heavier lines. The occurrence of grunion runs is indicated by the short, vertical lines above the tide curves. The moon phases are indicated at the bottom of each graph. A solid circle indicates new moon and a hollow circle indicates full moon. The histograms at the bottom portray the percentage intensity of runs in each series. Seasonal variation in strength of runs is not indicated. All data are based on observations made at Scripps Beach, La Jolla. Data for time and height of tides are from records of the tide-recording machine maintained for the Coast and Geodetic Survey on Scripps Pier (from Walker 1952).

described for northern anchovies (Laroche and Richardson 1983). Tropical spawners are usually nondeterministic, which means the eggs to be spawned are not determined at the start of the year but are produced throughout the year; however, nondeterministic can also represent the spawning potential for successive years, an example being the Atlantic cod (*Gadus morhua*), which will have several years' spawn in the ovary.

In general, older fish usually spawn first and younger fish later, which means that a prolonged spawning period for a population may not be true for individual fish. Once a set of eggs is mature and hydrated, the female may release them all at once or in several batches. An example of releasing several batches is plaice (*Pleuronectes platessa*), where a single

TABLE 1.5. Summary of Spawning Variation with Latitude

	Temperate Latitudes	Tropical Latitudes
Timing	Early winter, spring	Late (spring, summer, or continuous)
Duration	Short (3–4 months)	Long (5–6 months or more)
Frequency (per year)	Once (refers to entire group of eggs to be spawned, not how spawned)	Several times

female two weeks after releasing one batch of eggs releases more eggs, and then three weeks later she releases the remaining eggs. Another example is the Pacific herring, which spawn once a year and females lay about 100 eggs per spawning act, which they repeat several hundred times over a few days (Hourston and Haegele 1980). In the lab, walleye pollock spawned an average of nine times in an average period of 27 days (Sakurai 1983).

It also needs mentioning that a long duration of the spawning season of a population cannot necessarily be taken as an indication of prolonged spawning of the individual fish. The prolonged period may be due to differences in spawning time between age groups since older fish tend to spawn earlier in the season. Furthermore, the coexistence of different spawning subpopulations must be taken into account, since winter and summer spawners may be distinct stocks, although shifts from one seasonal spawning pattern to the other may occur. An example of how unpredictable this can be is that certain Atlantic herring of low fecundity have been found to always spawn in the winter, regardless of whether they originated from winter or summer spawning (Hempel 1979).

WATER DEPTH. Pacific herring spawn along beaches, marine grasses, and algae. Atlantic herring do not spawn along shore but in deeper water up to 200 m (the clearest difference between the Pacific and Atlantic herring,

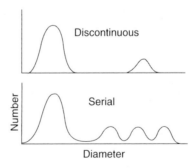

Figure 1.6. Differences in size distribution of oocytes of discontinuous and serial spawners (from Hempel 1979).

which are usually designated distinct species on the basis of genetic analysis). Of course fishes often spawn at one depth but live at different depths during other times of the year. For example, petrale sole (*Eopsetta jordani*), in which spawning occurs in a specific offshore area 300–400 m deep, were found by fishermen and eventually had to be protected with regulations to prevent overfishing (A.C. Delacy personal communication).

SPAWNING SUBSTRATE TYPE. Pacific herring spawn on vegetation whereas Atlantic herring spawn on solid substrate (e.g., gravel). Lingcod spawn on rocks, pilings, and cracks in solid substrate; this species protects the egg mass. Some species such as buffalo sculpin (*Enophrys bison*) and plainfin midshipman (*Porichthys notatus*) spawn intertidally and will stay with the egg mass even when they are exposed at a low tide.

SALINITY. Also a factor affecting spawning. There are varying salinities in many areas of estuaries. Some species will shift spawning sites because of salinity changes. Various degrees of mixing, precipitation, and freshwater runoff may alter spawning habits.

EXPOSURE AND TEMPERATURE. A clear example of shifting spawning sites in response to temperature and exposure is the black prickleback (*Xiphister atropurpureus*), where spawning is shifted from winter in protected areas to spring in exposed areas (Marliave 1975). The complex effects of lower or higher wave action and lower or higher temperatures on courtship, gonadal development, and spawning behavior that result in the spawning site shift.

Lifetime Spawning Strategies

TERMINOLOGY

Semelparous: spawn only once in a lifetime (e.g., many anadromous fishes, the common example is the Pacific salmon).
Iteroparous: spawn many times in a lifetime (e.g., most marine fishes).

Within the same species of fishes one can find two different strategies presumably due to ecological differences, the classic example being American shad (*Alosa sapidissima*), which is entirely semelparous in Florida, whereas those north in the New Brunswick area are ~ 50 to 75% iteroparous, and intermediate populations have intermediate values (T. Quinn personal communication). The explanation is that northern rivers are a more harsh and variable environment for eggs and larvae so iteroparity is a better strategy for this species.

Sites

Many fishes use only a portion of their overall range for reproduction. Many species return to natal areas to reproduce, the Pacific salmon being the best known and most extreme example of this pattern. Adult salmon spend from one to several years in the open Pacific Ocean, and return to their natal streams, which may be hundreds of kilometers from the ocean, to spawn. Even fishes that spend their entire life in the ocean, or a freshwater stream or lake, often select a particular part of their habitat for reproduction.

Migrations. Spawning migrations may require fish to move hundreds of kilometers, or from one depth range to another. The anadromous pattern of salmon and striped bass (*Morone saxatilis*), where the fish move from the marine environment to the estuarine or freshwater environment for spawning, is contrasted to the catadromous pattern of American and European freshwater eels (Anguillidae), which descend rivers and migrate to the Sargasso Sea in the North Atlantic for spawning. Aside from these extremes, most fish move from feeding areas to congregate in spawning areas. Presumably these areas have been selected through evolution to provide a suitable environment for survival of the eggs and larvae.

Habitats. The habitats utilized by fishes for reproduction and development are quite varied. The essentials of the habitat for eggs and larvae are that it remains oxygenated and within temperature and water quality requirements suitable for development. Ecological considerations include protection from predators and microbes, and production of adequate food for the larvae. Most marine fishes produce planktonic eggs and larvae that drift in the upper 200 m of the ocean, although some regularly occur much deeper. Many fishes that occupy much greater depths as adults undergo early development in the epipelagic zone. Various species of fishes build nests or deposit eggs in a wide variety of places. Many fishes dig a nest in the bottom where they deposit and sometimes guard their eggs. Species may have very specific substrate requirements for nest building. Many freshwater and some marine fishes deposit adhesive eggs on the surface of the bottom (gravel or rocks) or plants. The depth chosen for deposition of demersal eggs may be very specific, especially in fishes that spawn intertidally where there is danger of dessication, or exposure to temperature and salinity extremes. Some deposit their eggs in other animals such as clams or crabs in a parasitic relationship.

Behavior

Reproductive behavior of fishes generally involves some sort of courtship, which may aid in species and spawning-readiness recognition. Pairing of individuals probably occurs even when fish appear to spawn in large schools. Communication among potential mates may include visual, olfactory, and auditory cues. As courtship proceeds, the mates eventually swim together with their genital openings touching. Male and female gametes are then released simultaneously. In the case of nesting fishes, the female often deposits a number of eggs, which the male then swims over as he releases sperm (milt). In fishes with internal fertilization the male possesses an intromittent organ to deposit sperm into the female, either directly into the ovary, or into a sperm storage area.

Secondary Sexual Characters

The sexes of species that produce pelagic eggs are generally indistinguishable except for their gonads, and the larger size attained by the females. However, as parental care increases, so do differences between the sexes. Sexual dimorphism includes the intromittent organs of males in species with internal fertilization, and morphological adaptations for nest building and guarding. Sexual dichromatism occurs in some species, apparently as an aid in mate recognition. The ultimate in sexual dimorphism is found in some anglerfishes (ceratioids) in which the males are parasitic on the females.

CHAPTER 2

Development of Eggs and Larvae

EMBRYONIC DEVELOPMENT
   Events Immediately After Spawning
      Fertilization
      Activation
   The Perivitelline Space
   Development
   Rate of Development and Aging of Eggs

LARVAL DEVELOPMENT
   Hatching
   Notochord Flexion and Stages of Larval Development
   Transformation Stage

JUVENILE DEVELOPMENT

Chapter 1 discussed reproduction in fishes, and the development of gametes within the adults. Chapter 2 summarizes information on the development of the eggs and larvae, and introduces juvenile development. Chapter 3 examines the diversity of eggs and larvae, where we find that certainly the eggs, and in most cases the larvae, are quite different in appearance from the adults, but nevertheless they have distinctive characters that allow their identification. These egg and larval characters also include systematic information that has been used to imply relationships among fishes, complementing information provided by the adults and from genetic studies.

In this chapter we focus on the changes that occur as the eggs develop, hatch into larvae, and finally transform into juveniles (**ontogeny**). Most fishes have **indirect development**; that is, as free-living organisms they go through a number of distinct stages or phases of development (egg, larva, juvenile). This is similar to development in amphibians, but opposed to **direct development** seen in reptiles, birds, and mammals in which all the free-living forms appear similar and development is largely adding size to the organism.

Various terminologies have been proposed to define the stages of development in fishes (Figure 2.1) (see Kendall et al. 1984). Here we use **egg** for development from spawning until hatching. This is also considered the embryonic phase of development. At hatching, the fish becomes a **larva.** Most fish have a yolk sac for nourishment at hatching, and until it is absorbed they are termed **yolk-sac larvae.** The larval period lasts until the number of fin rays reach their adult complement when the **juvenile** stage begins. **Squamation** (formation of scales) usually starts at this point in development also. The juvenile stage terminates when the fish first reaches sexual maturity. These dramatic changes in form during development of free-living fishes require similarly different ecologies for each stage; these will be covered in Chapter 4.

Development of fishes has been the subject of many detailed studies, which are summarized by Blaxter (1969, 1988), Browman and Skiftesvik (2003), and in an exhaustive account by Kuntz (2004). Here we give only an overview of fish ontogeny.

EMBRYONIC DEVELOPMENT (FIGURES 2.2.1 AND 2.2.2)

The embryonic development of a variety of fish eggs has been described in detail through the years (e.g., Blood et al. 1994). Here only the basic features of development as they can be discerned by examining whole eggs with a dissecting microscope will be summarized. This level of detail is usually sufficient for early life-history studies dealing with field-caught pelagic eggs. It is also useful in laboratory studies on various environmental effects on development, such as temperature or pollutants.

Events Immediately After Spawning

Two events occur to ripe eggs immediately after spawning: fertilization and activation.

Fertilization. Occurs when the sperm penetrates the envelope (chorion) of the egg and begins the various stages leading to sperm-egg fusion. The ripe egg possesses a micropyle on the surface of the egg envelope at the animal pole, which allows passage of the sperm into the egg. The micropyle is funnel-shaped and only wide enough at its base for the passage of a single sperm. There may be attractants for sperm on the surface of the egg associated with the micropyle. After sperm penetration, a plug forms at the base of the micropyle to prevent polyspermy.

Activation. There are several components to activation, which occurs when the egg is released from the female and first comes into contact with water. When extruded, the yolk occupies all the space inside the envelope of unfertilized ripe eggs.

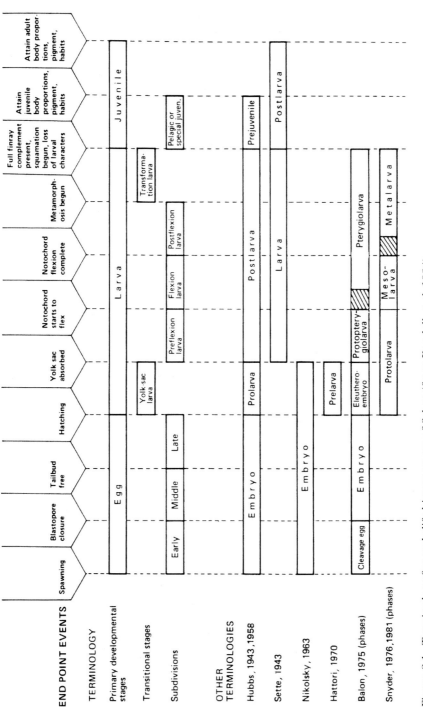

Figure 2.1. Terminology for early life-history stages of fishes (from Kendall et al. 1984).

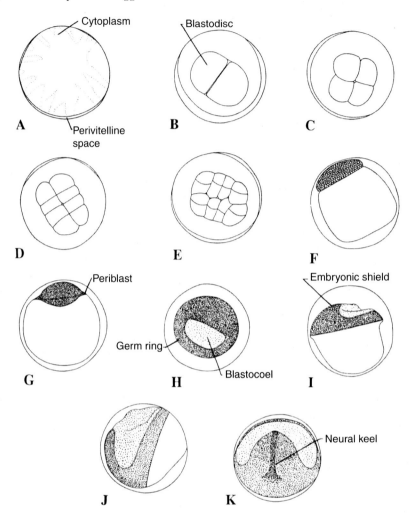

Figure 2.2.1. Stages in the development of a typical fish egg, walleye pollock (*Theragra chalcogramma*) (from Blood et al. 1994). A. Stage 1 (precell); B. Stage 2 (2 cell); C. Stage 3 (4 cell); D. Stage 4 (8 cell); E. Stage 5 (16 cell); F. Stage 6 (32 cell); G. Stage 7 (blastodermal cap); H. Stage 8 (early germ ring); I. Stage 9 (germ ring 1/4, lateral view); J. Stage 10 (germ ring 1/2, lateral view); K. Stage 10 (dorsal view).

The Perivitelline Space

Upon fertilization a perivitelline space forms between the inner edge of the envelope and the egg membrane. It is produced by a cortical reaction, which may take minutes or hours, in which the cortical alveoli are released into the space between the envelope and the cytoplasm. It probably results

Development of Eggs and Larvae 43

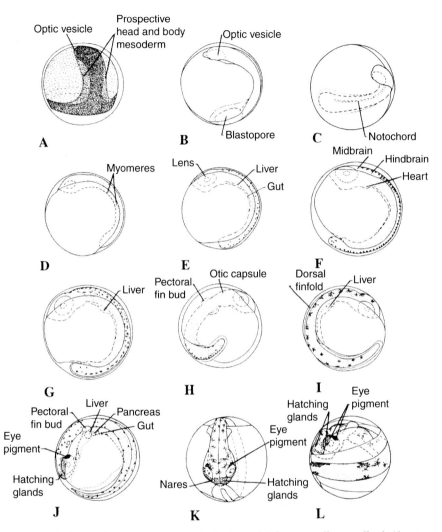

Figure 2.2.2. Stages in the development of a typical fish egg, walleye pollock (from Blood et al. 1994) (continued). A. Stage 11 (germ ring 3/4); B. Stage 12 (blastophore almost closed); C. Stage 13 (early middle); D. Stage 14 (middle middle); E. Stage 15 (late middle); F. Stage 16 (early late); G. Stage 17 (tail 5/8 circle); H. Stage 18 (tail 3/4 circle); I. Stage 19 (tail 7/8 circle); J. Stage 20 (tail full circle, lateral view); K. Stage 20 (dorsal view); L. Stage 21 (tail 1-1/8 circle).

from both an osmotic distension of the envelope and egg shrinkage causing the separation and detachment of the egg membrane. The perivitelline space protects and lubricates the egg, and helps with osmotic regulation. Swelling of the egg, called hydration, also occurs due to absorption of seawater by the colloidal material that was released by the cortical reaction. At the same time the envelope goes through a process called water hardening, which is dependent on Ca^{++} ions being present in the water.

Development

During incubation the embryo develops from a single cell into a complex organism. Embryonic development of fishes can be divided into three major phases: early (from fertilization until blastopore closure), middle (from blastopore closure until the tailbud is free), and late (from free tailbud until hatching). Shortly after fertilization, the cytoplasm becomes thickened at the animal pole of the egg where the nucleus occurs. Early development of fish eggs generally exhibits a meroblastic pattern of cleavage, in that the cells form only at the animal pole of the fertilized egg, and cleavage does not go through the entire yolk (as it does in holoblastic cleavage). The yolk concentrates at the vegetal pole. Cell division at first proceeds in an orderly fashion with a single layer of 2, 4, 8, and 16 uniformly sized cells (blastomeres) forming a blastodisc on the yolk. An acellular layer (the periblast) forms around the blastodisc, which metabolizes the yolk for the developing embryo. At first the periblast is continuous with the marginal blastomeres. The individual cells of the blastodisc decrease in size with each cell division. Beyond the 32-cell stage, it is difficult to count the cells, and cell division is not as synchronous as earlier. Depending on the species and temperature, these early cell divisions can take from less than an hour to several hours. Soon the mass of cells takes on the appearance of a flattened raspberry atop the yolk (the blastodermal cap or morula stage). At this point the individual cells are too small to be distinguished. The cells reach a maximum height of about 10 cells, and the egg usually floats with the blastodermal cap downward. Following this blastodermal cap stage, the blastula stage occurs as the mass of cells begins to flatten and encircle the yolk (epiboly), and the edge of the cell mass becomes thickened (germ ring stage). At the same time, the blastomeres lift up in the center of the mass, creating a central cavity called the blastocoel. By the time the cell mass is about halfway around the yolk, the first indications of the embryo can be seen as a thickened line (neural keel) on top of the yolk perpendicular to the edge of the germ ring, indicating that gastrulation is occurring. During the gastrula stage the single-layered blastoderm becomes a multilayered embryo. As the embryonic shield forms there is a thickening of the caudal margin of the blastodermal cap. In this area cells invaginate to form a gastrula. The outer layer of cells becomes the skin of the adult and the inner layer the gut and mesoderm. The cells continue to grow around the yolk until only a small circle of yolk is not covered (the blastopore). As this happens, the embryo begins to take shape. Along the neural keel the neural tube and notochord are forming. Myomeres gradually form on either side of the notochord, the three main portions of the brain (forebrain, midbrain, and hindbrain) can be seen, and the optic vesicles appear as the blastopore closes (end of gastrulation).

At the beginning of the middle stage of development, internal organs (liver, gut) can be seen and the heart forms and starts to beat when the embryo circles about halfway around the yolk. Auditory organs and the eye lenses appear. Pigment often first appears on the embryo about this time. Myomeres continue to be added both anteriorly and posteriorly to those that first developed midway along the body, and the tailbud margin is defined. Pectoral fin buds and the otic capsules appear as the tailbud lifts off the yolk (beginning of the late stage of development). Following this, the body lengthens, the various organs become better defined, and pigment is added. The embryo begins to move within the egg at this time. Oxygen consumption increases as the embryo starts to move. Hatching occurs at various stages of development, depending on the species. Some fishes hatch when the embryo reaches full circle around the yolk, but other species hatch before or after this.

Rate of Development and Aging of Eggs

The rate of development of fish eggs, and length of time from fertilization until hatching, depends on species, temperature, oxygen, and salinity (Alderdice and Forrester 1971; Laurence and Howell 1981). It is important to know the age of eggs because this reveals the time elapsed since spawning, and it is required to estimate the mortality rate of eggs. The time it takes to reach specific points in development can be determined by rearing eggs under various conditions in the laboratory. The stage of development of wild-caught eggs can then be determined and related to the environmental conditions present when they were collected.

LARVAL DEVELOPMENT (FIGURE 2.3)

Hatching

The larval stage begins at hatching and lasts until complete fin ray counts have been attained and squamation has begun. The yolk-sac larval stage starts at hatching and ends when the yolk sac is absorbed. Hatching occurs once the embryo has reached a certain stage of development. At that point proteolytic enzymes in the perivitelline space begin to soften the envelope. Enzymes produced by the embryo digest proteins to form more soluble compounds. The envelope is digested from inside out, with the embryo using any products that are nutrients. As oxygen demand by the moving embryo exceeds the supply, the embryo moves more, helping it to escape

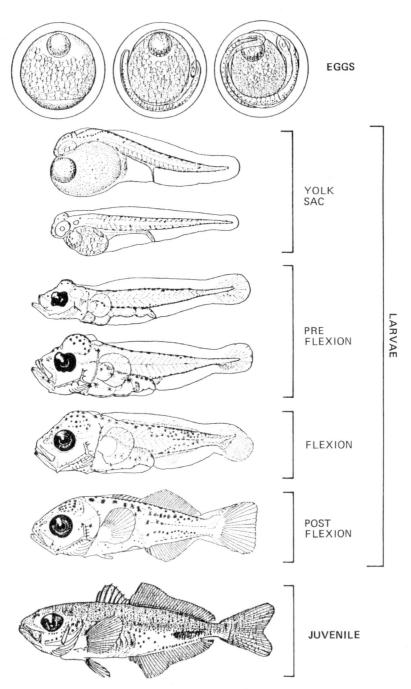

Figure 2.3. Early life-history stages of a typical fish, jack mackerel (*Trachurus symmetricus*) (from Ahlstrom and Ball 1954).

the envelope. In large eggs the embryo tends to emerge tail-first; in small eggs the larva emerges head-first.

Most fishes hatch as yolk-sac larvae, a stage that lasts until the yolk sac is absorbed. During this stage the yolk continues to supply nutrition to the larva (Fukuhara 1990; Johns and Howell 1980). Yolk-sac larvae are about 2–6 mm long, and are generally not well developed. They lack functional mouths and pigmented eyes. They are largely transparent and lack differentiated fins. The large yolk causes the larva to float upside down. Swimming usually consists of short, vertical bursts. The advantages of developing during this stage outside the egg include better oxygen supply and better mechanisms to get rid of metabolites. The larvae can increase in size and development without the restrictions of the egg, while still getting nourishment from the yolk. They can also practice feeding and obtain some nourishment from prey as well as from yolk. Yolk-sac larvae from demersal eggs are usually larger and better developed than those from pelagic eggs (as just described). Some fishes with demersal eggs go through the yolk-sac stage in the egg and are ready to feed at hatching.

Notochord Flexion and Stages of Larval Development

One of the fundamental events in development of most fishes is the flexion of the notochord that accompanies the hypochordal development of the homocercal caudal fin (Figure 2.4). (However, the caudal skeleton in a major group of fishes, codfishes and their relatives [gadiforms], is nearly symmetrical [Figure 2.5].) It is convenient to divide the larval stage on the basis of this feature into "preflexion," "flexion," and "postflexion" stages.

The preflexion stage begins at the end of the yolk-sac larval stage and lasts until the notochord starts to flex. The notochord is straight during the preflexion stage, and the larva is just starting to feed. Its sensory and locomotor powers are developing rapidly since the larva must have operational eyes, a movable jaw, and a functional gut to capture and utilize prey. The median finfold is usually continuous, not divided into separate fins. Larval pectoral fins are well developed and used for positioning. The pelvic fins are usually not developed yet. The larval pigment pattern is becoming established. The retina holds only cones, allowing vision in bright light only. The first red blood cells appear. Respiration is still cutaneous.

The flexion stage in many fishes is accompanied by rapid development of fin rays, change in body shape, change in locomotor ability, and feeding behavior. During the flexion stage the tip of the notochord bends dorsally, and the caudal fin develops ventral to it. This stage ends when notochord flexion is complete. During this stage swimming changes from a burst and stop mode to a more continuous gliding mode. Swim bladder inflation may first occur during this stage, allowing the larva to be neutrally buoyant.

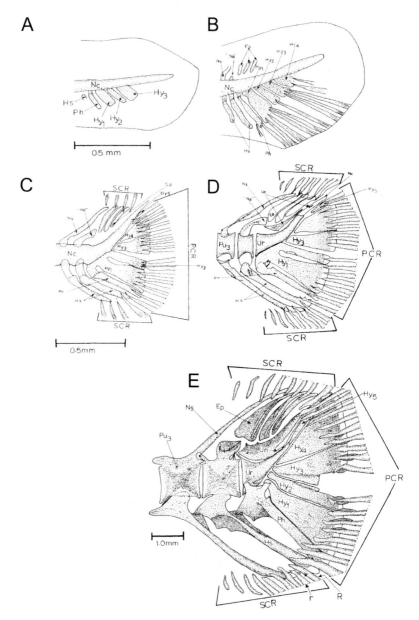

Figure 2.4. Development of the caudal region of a typical fish, longfin escolar (*Scombrolabrax heterolepis*) (from Potthoff et al. 1980). Left lateral views. A. 3.9 mm NL, Indian Ocean. B. 4.4 NL, Atlantic Ocean. Cartilage, stippled. C. 6.6 mm SL, Atlantic Ocean, both cartilage and ossifying cartilage, stippled. D. 9.7 mm SL, Atlantic Ocean, cartilage, white; ossifying, stippled. E. 68.1 mm SL from the Atlantic Ocean, cartilage, white; ossifying, stippled. Ep, epural; Hs, haemal spine; Hy, hypural bone; "Na," specialized neural arch; Nc, notochord; Ns, neural spine; Ph, parhypural; PCR, principal caudal rays; Pu, preural centrum; SCR, secondary caudal rays; Un, uroneural; Ur, urostyle; R, secondary ray with procurrent spur; r, secondary foreshortened ray.

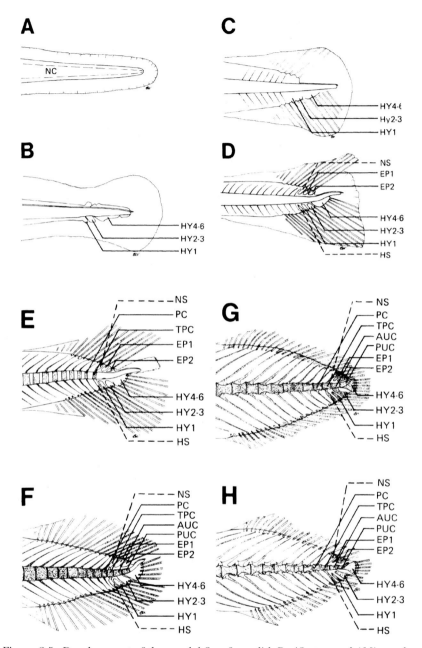

Figure 2.5. Development of the caudal fin of a gadid, Pacific tomcod (*Microgadus proximus*) (from Matarese et al. 1981). A. 5.2 mm SL; B. 7.8 mm SL; C. 9.5 mm SL; D. 10.5 mm SL; E. 11.9 mm SL; F. 15.8 mm SL; G. 25.0 mm SL; H. 41.1 mm SL. AUC, anterior centrum; EP, epural; HS, haemal spine; HY, hypural; NC, notochord; NS, neural spine; PC, preural centrum; PUC, postural centrum; TPC, terminal preural centrum. Ossified elements are stippled.

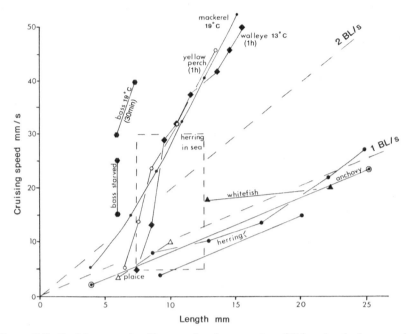

Figure 2.6. Cruising speeds of larvae of various species of fishes (as during searching for food) related to body length (hatched lines show 1 and 2 body lengths per second) (from Blaxter 1986).

Feeding becomes more proficient and the ability to avoid predators increases (Houde and Schekter 1980; Blaxter 1986). The digestive system develops and digestion improves (Govoni et al. 1986; Oozeki and Bailey 1995). Diel vertical migrations and other complex behaviors (e.g., schooling) may commence during the flexion stage. Rods begin to develop in the eyes. Larval pigment patterns are fully developed, and larval specializations such as elongate fin rays and head spines may develop. Fin rays develop rapidly in most fins and they and other skeletal elements start to ossify (Dunn 1984). Cruising and burst swimming speeds of fish larvae increase as the larvae grow (Figures 2.6 and 2.7).

During the postflexion stage larvae continue to develop toward having all of their skeletal elements ossified. The postflexion stage starts when notochord flexion is complete and ends when all of the fin rays have formed, and the juvenile stage begins. Larval specializations become even better developed, and the larva grows in size. Respiration shifts from cutaneous to gills and the stomach differentiates in the gut. Rods further develop in the eyes, allowing vision in lower light.

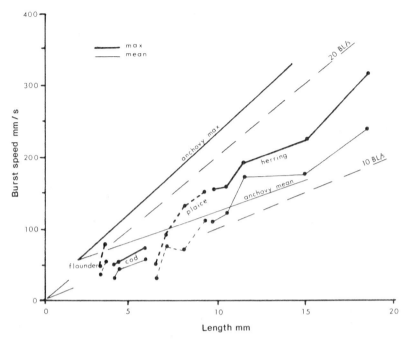

Figure 2.7. Burst swimming speeds of larvae of various species of fishes (as during escape from harmful stimuli) related to body length (hatched lines show 10 and 20 body lengths per second). Maximum speeds are for a very short period after a flight response; mean speeds for a longer period (from Blaxter 1986).

Transformation Stage

Between the larval and juvenile stages, there is a transitional stage, which may be abrupt or prolonged and which, in many fishes, is accompanied by a change from planktonic habits to demersal or schooling pelagic habits. In some fishes migration to a "nursery" ground occurs during or just before this stage. Morphologically the transformation stage is characterized by a change from larval body form and characters to juvenile-adult body form and characters. Metamorphosis occurs during this stage and is considered complete when the fish assumes the general features of the juvenile. Two ontogenetic processes occur during this stage of transition between the larva and juvenile: (1) loss of specialized larval characters, and (2) attainment of juvenile-adult characters. Changes that occur during this stage include pigment pattern, body shape, fin migration (e.g., herrings [clupeids] and anchovies [engraulids]), photophore formation (e.g., lanternfishes [myctophids]), loss of elongate fin rays and head spines (e.g., groupers [epipheline serranids] and squirrelfishes [holocentrids]), eye migration (flatfishes [pleuronectiforms]), and scale formation.

In several groups, where the transformation stage is prolonged, the fish have developed specializations that are distinct from both the larvae and juveniles. This stage has been designated the prejuvenile stage (Hubbs 1943). The specializations generally involve body shape and pigmentation (Figure 2.8). In many, the morph resembles a herring-like fish and is apparently adapted for neustonic life. The dorsal aspect of the fish is dark green or blue and the lateral and ventral is silvery or white. The body tends to be herring shaped and the mouth terminal. Fins are generally unpigmented. Such a stage is present in some codfishes (hakes [*Urophycis* spp.]), squirrelfishes (*Holocentrus* spp.) and relatives (beryciforms), perch-like fishes (perciforms) (e.g., bluefish [*Pomatomus saltatrix*], goatfishes [mullids], and mullets [mugilids]), and mail-cheeked fishes (scorpaeniforms) (e.g., cabezon [*Scorpaenichthys marmoratus*], greenlings [*Hexagrammos* spp.]). In other fishes, such as some lanternfishes and their relatives (myctophiforms) and pearlfishes (carapids), the prolonged transformation stage may have distinctive body and fin shapes.

JUVENILE DEVELOPMENT

The juvenile stage follows the larval stage and begins when fin ray counts are complete and squamation (scale formation) has begun, and ends when the fish enters the adult population or attains sexual maturity. In many fishes the juveniles are miniature adults, in others they are quite distinct morphologically. They may have a distinct body shape as well as pigment. In fact juveniles of some fishes are so different from the adults that they were not recognized when first discovered and were given different names. Presumably juvenile characters are adaptive in the habitats the juveniles occupy. Juveniles are not usually planktonic as are the larvae; some migrate to and occupy a distinct habitat from the adults. Some reside in crevices on the bottom or among floating or attached macroalgae or flotsam and jellyfishes. Some species school as juveniles, even if as adults they do not school.

Figure 2.8 Examples of special juvenile stages of fishes (from Kendall et al. 1984). A. rock greenling (*Hexagrammos lagocephalus*), 28.0 mm. A neustonic or epipelagic form of species that is demersal as an adult; B. Longnose butterflyfish (*Forcipiger longirostris*), 17 mm. A spiny form that lives on tropical reefs as an adult; C. longspine thornyhead (*Sebastolobus altivelis*), 26.8 mm. A barred pelagic form of species that is demersal on the continental slope as an adult; D. coho salmon (*Oncorhynchus kisutch*), 37 mm. The alevin or parr stage of an anadromous salmonid; and E. *Kali macrodon,* 45 mm. The juvenile of a bathypelagic species originally described as *Gargaropteron pterodactylops.*

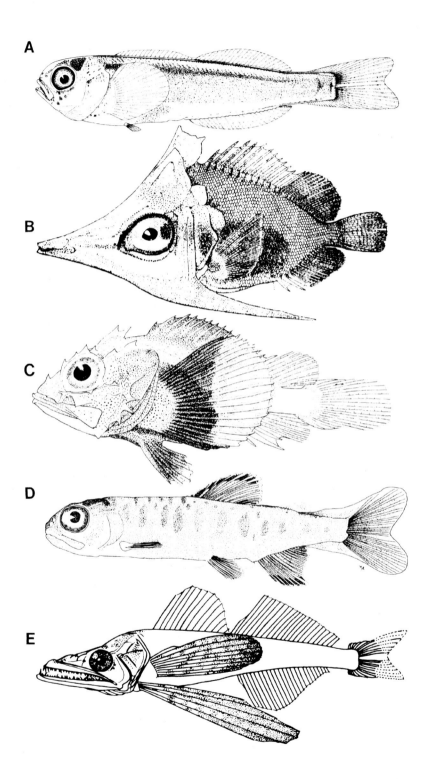

Many flatfishes move to shallow, even intertidal areas as juveniles. A number of species migrate inshore to estuarine nursery areas as juveniles. Juveniles of some species live symbiotically with benthic or pelagic invertebrates. Growth during the juvenile stage in summer can be very rapid, and may be important in determining overwintering survival rates (Sogard 1997). Since most fishes spawn so their larvae can feed on prey produced by the spring bloom, the juvenile stage is reached in early summer. By the end of their first summer, juveniles must have reached a size that will permit them to survive the low food conditions of the winter. During their first winter, juvenile growth is generally quite limited.

CHAPTER 3

Fish Egg and Larval Identification and Systematics

ESTABLISHING THE IDENTITY
OF FISH EGGS AND LARVAE
Morphological
Biochemical Genetics
   Electrophoresis
   Other Genetic Approaches

METHODS AND EQUIPMENT
FOR MORPHOLOGICAL
IDENTIFICATION OF FISH
EGGS AND LARVAE
Knowledge of Adult Fauna and
   Literature on Eggs and Larvae
Laboratory
Microscopes
Tools
Video and Image Analysis
Clearing and Staining
Radiography

IDENTIFICATION AND STAGING
OF FISH EGGS
Characters
   Shape
   Size
   Chorion Texture
   Oil Globules

   Perivitelline Space
   Yolk Characters
   Embryo Characters
Determining Embryonic Stage
   of Development

IDENTIFICATION OF FISH LARVAE
Morphology
Pigment
Meristic Characters
   Myomeres
   Fin Rays
Specialized Larval Characters
Higher-Level Characters

EGG AND LARVAL STAGES
IN SYSTEMATIC STUDIES
Historical Perspective
Theory
Methods
Examples of the Use of Larval Fish
   Characters in Systematics
   The Lightfishes
      (Gonostomatidae)
   The Lanternfishes
      (Myctophidae)
   The Sculpins (Cottidae)

The Seabasses (Serranidae): A Case Study of the Use of Larval Fish Characters	Niphonini
	Epinephelini
	Diploprionini
Serraninae	Liopropomini
Anthiinae	Grammistini
Epinephelinae	

Most ichthyoplankton studies require accurate identifications of eggs and larvae. Generally several kinds of fish eggs and larvae co-occur in plankton samples, so at least the species of interest needs to be separated from the others. It is usually a good practice to attempt to identify all eggs and larvae in the samples. This ensures that the species of interest will be recognized and separated. Also, in some cases other species are found to be important in understanding the species of primary interest; if they are already identified, it makes analysis more efficient.

It is no easy task to identify fish eggs and larvae collected in plankton samples. The larvae, and certainly the eggs (Figure 3.1), do not look like their parents. Different kinds of characters from those used to identify the adults must be examined to identify the eggs and larvae (Ahlstrom and Moser 1976). These identifications are usually done visually, using a dissecting microscope to compare the morphology of specimens with descriptions, illustrations, and characters found in the literature. Matarese and Sandknop (1984) discuss identification of fish eggs, and Powles and Markle (1984) discuss identification of fish larvae.

ESTABLISHING THE IDENTITY OF FISH EGGS AND LARVAE
Morphological

Generally two methods have been used to establish the identity of fish eggs and larvae. The **direct approach** involves rearing larvae from parents of known identity. The larvae are described at various stages of development, allowing wild-caught larvae to be identified by comparing them with the descriptions of the reared larvae. Application of the direct approach is limited, since for most marine fishes it is very difficult to obtain adults and have them produce gametes in captivity, or to collect spawning adults in the sea. The other method, which has been and will probably continue to be more common, is the **indirect approach,** or serial approach (Moser and Ahlstrom 1970), and involves establishing series of various-sized larvae from field collections. The largest larvae in the series should have diagnostic adult characters, while retaining some larval characters. Smaller larvae that do not yet have diagnostic adult characters are identified based on the overlap of their larval characters with those of the larger larvae. Characters of juveniles can

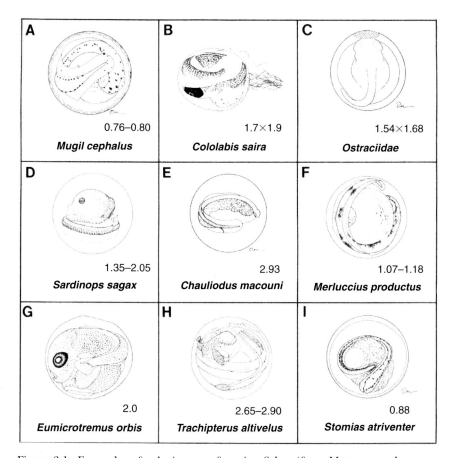

Figure 3.1. Examples of pelagic eggs of marine fishes (from Matarese and Sandknop 1984). A. striped mullet (*Mugil cephalus*): small egg, no oil globules, sculptured chorion, well-developed embryo. B. Pacific saury (*Cololabis saira*): ovoid egg; chorion with filaments; well-developed, heavily pigmented embryo. C. boxfishes (Ostraciidae): slightly ovoid egg, heavily sculptured chorion. D. Pacific sardine (*Sardinops sagax*): wide perivitelline space, embryo coils more than one revolution. E. Pacific viperfish (*Chauliodus macouni*): large egg, wide perivitelline space. F. Pacific hake (*Merluccius productus*): oil globule, pigment bands on embryo, eye of embryo well developed. G. Pacific spiny lumpsucker (*Eumicrotremus orbis*): embryo well developed before hatching. H. king-of-the-salmon (*Trachipterus altivelus*): large egg, ornamented elongate dorsal fin rays of embryo develop in egg. I. black-belly dragonfish (*Stomias atriventer*): small egg, double egg membrane, wide perivitelline space.

be different from those of the larvae and adults, but sometimes help bridge the gap between the two stages. Another method being explored to help identify plankton-caught larvae involves collecting live larvae and rearing them until they develop diagnostic adult characters. Close observations of the larvae as they grow enable developmental series to be obtained from one or more larvae. At intervals during development, the larvae can be photographed or videotaped through a microscope while they are anesthetized, to provide a permanent record of their appearance and to allow various stages to be drawn and described.

Biochemical Genetics

Biochemical genetics can be used to confirm the identity of unknown eggs and larvae. However, the larvae are disintegrated during processing so they are no longer available for microscopic examination. Although not practical for making routine identifications at present, these techniques can be used to confirm the identity of particular types of larvae that cannot otherwise be tied to specific adults.

Electrophoresis. Electrophoresis is a technique that analyzes allozyme patterns and may be helpful in identifying the species of particular series of larvae when morphological criteria lead to ambiguous results. Allozyme patterns of eggs and larvae are similar to those of adults, although not all genes are expressed in young larvae. For example, with rockfish (*Sebastes* spp.) larvae from the Northeast Pacific, several series have been established, but the needed larger larvae with diagnostic adult characters are lacking (see Moser et al. 1977). Electrophoresis may allow such series to be identified. Using electrophoretic patterns of larvae of rockfish to establish their identity produced encouraging results (Seeb and Kendall 1991). This technique is also being explored as a means to identify redfish (*Sebastes* spp.) larvae from the North Atlantic (Nedreaas and Naevdal 1991), and has been used to distinguish fish eggs (Mork et al. 1983) and similar appearing larvae in a number of other cases—for example, white perch/striped bass (*Morone americanus/M. saxatilis*): (Morgan 1975; Sidell et al. 1978) and New Zealand flounders (*Rhombosolea* spp.): (Smith et al. 1980).

Other Genetic Approaches. Mitochondrial DNA (mtDNA) of fish eggs and larvae has been used to determine the identity of species that are morphologically similar, or indistinguishable (e.g., Graves et al. 1989; Rocha-Olivares 1998). Polymerase chain reaction (PCR) is sometimes used to amplify very small amounts of mtDNA, such as available in individual eggs and larvae. Various properties of the mtDNA have been used to compare the mtDNA extracted from eggs or larvae with that from adults that are suspected to be the parent species. Mitochondrial DNA fragments produced by restriction endonucleases were species-specific and allowed identification of eggs and larvae of the three species of sandbasses (*Paralabrax* spp.) that occur off southern California (Graves et al. 1989). This method was applicable to fresh, frozen, or ethanol-preserved material. Another method only amplifies sections of the mtDNA molecule that are exact matches of templates produced from known species (Rocha-Olivares 1998). This is a much more powerful method, but it requires sequencing part of the genome of the suspected parental species.

METHODS AND EQUIPMENT FOR MORPHOLOGICAL IDENTIFICATION OF FISH EGGS AND LARVAE

Knowledge of Adult Fauna and Literature on Eggs and Larvae

Identification of fish eggs and larvae requires a thorough knowledge of the adult fish fauna of the area where the samples were collected. A list of the species present as well as their meristic characters may need to be established, if such is not available for the area being studied. Sometimes adults need to be examined to determine their meristic characters, if they are not available in the literature. Meristic characters of importance include fin rays and vertebrae (which correlate with myomeres seen in larvae before the vertebrae have formed). Since larvae of higher taxonomic groups sometimes share characters, the families and orders of the species should be noted. The life-history pattern of the species should also be noted (e.g., whether they spawn in fresh or marine habitats, whether the eggs are demersal or pelagic, in what season they reproduce). The relative abundance of the various species should be noted, but relative abundance of the eggs and larvae can be very different from that of the adults, depending on the life-history strategy of the species (e.g., the adults of a species may be very abundant in an area, but if the species spawns demersal eggs and you are examining planktonic samples, its eggs would not be expected). Original descriptions of eggs and larvae of fishes in the area of study should be collected, as well as any compilations of descriptions. Illustrations are particularly helpful and may need to be copied from the literature and placed in notebooks and/or on bulletin boards in the laboratory. If descriptions of species in the area of interest are not available, descriptions of other species in the same genus or even family may indicate what the unknown species may look like.

Starting with Ehrenbaum (1905–1909) for fishes of the Northeast Atlantic, a number of guides to the eggs and larvae of fishes of particular marine regions (Figure 3.2) have been published and are indispensable for identifying ichthyoplankton from these regions (e.g., Northeast Atlantic [Russell 1976; Munk and Nielsen 2005], Northwest Atlantic (Mid-Atlantic Bight) [Fahay 1983, 2007a, b], Western Central Atlantic [Richards (ed.) 2006], Japan [Okiyama (ed.) 1988], Indo-Pacific [Leis and Rennis 1983; Leis and

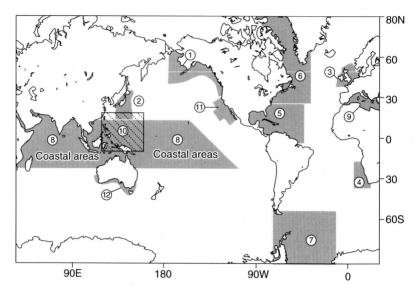

Figure 3.2. Geographic coverage of laboratory guides to eggs and larvae of marine fishes. 1. Matarese et al. (1989); 2. Okiyama (ed. 1988); 3. Russell (1976); 4. Olivar and Fortuño (1991); 5. Richards (1999, ed. 2006); Richards (ed. 2006); 6. Fahay (1983; 2007a; 2007b); 7. Kellermann (1989); 8. Leis and Rennis (1983), Leis and Trnski (1989), and Leis and Carson-Ewart (eds. 2000); 9. Aboussouan (1989); 10. Ozawa (1986); 11. Moser (ed. 1996); 12. Neira et al. (1998).

Trnski 1989; Leis and Carson-Ewart (eds.) 2000; Neira et al. 1998], Antarctic [Kellermann 1989], Northeast Pacific [Matarese et al. 1989], Hawaiian Islands [Miller et al. 1979], Benguela Current [Olivar and Fortuño 1991], California Current [Moser (ed.) 1996]). These guides are also useful in identifying larvae from areas outside the specific region they cover, since they may help to establish the group (e.g., family, genus) of larvae from other areas.

Guides to early life stages of estuarine and freshwater fishes of several areas have been published. These include several volumes for fishes of the Chesapeake Bay region (Fritzsche 1978; Hardy 1978a, 1978b; Johnson 1978; Jones et al. 1978; Martin and Drewry 1978) and publications by Auer (ed. 1982) for fishes of the Great Lakes area of the United States, Wang (1986) for fishes of the Sacramento-San Joaquin estuary, and Pinder (2001) for coarse fishes of the British Isles.

Laboratory

The laboratory for examining ichthyoplankton samples and identifying fish eggs and larvae should be equipped with sturdy tables or benches at desk height for the microscope. Each workstation should be large enough to provide space for recording data, leaving identification manuals open, and having tools and samples to be examined handy. Comfortable chairs are needed to avoid physical stress. Since most ichthyoplankton samples are preserved in formalin, adequate ventilation is needed to remove fumes from the area.

Microscopes

A good-quality binocular dissecting microscope with a clear or frosted glass stage and an adjustable substage mirror is desirable. Powers between about 6 and 100 are useful. Both direct and reflected light are needed, and can be supplied by adjustable fiber-optic light sources. A calibrated ocular micrometer is needed to measure the eggs and larvae. A camera lucida attachment for the microscope is often used to illustrate eggs and larvae.

Tools

Petri or watch dishes (~ 50-mm diameter) with clear bottoms can be used to hold the eggs and larvae for microscopic examination. Specimens are examined as whole mounts in their preserving medium or water. It is often necessary to move the specimens around with probes and fine insect-style forceps, and to change powers as they are examined. Lighting may need to be adjusted, and a piece of white paper inserted between the specimen dish and the stage makes pigment more readily visible. The forceps and pipettes can be used to transfer the specimens from the vials they are stored in to the dishes where they are examined. Larvae, especially undeveloped ones, are very fragile so care must be taken not to damage them while they are being handled. Polarized light can be used to make myomeres and myosepta more readily visible. A steel rule may be needed to measure larger larvae.

Video and Image Analysis

Video systems including videodisc recorders can be useful in documenting larval identifications. When connected with image analysis systems they become powerful tools that aid in describing early life-history stages of fishes by allowing measurements to be made more accurately and efficiently, and the data to be recorded automatically and ready for computer-aided analysis.

Clearing and Staining

Techniques have been developed to stain bone and cartilage of larval fishes and clear the flesh so the stained elements can be seen within the body of the larva (Potthoff 1984; Taylor and Van Dyke 1985; Springer and Johnson 2000). Cartilage can be stained blue and bone red, or the bone alone can be stained. This is particularly useful in tracing the development of the fish skeleton, which is needed for many descriptive and systematic studies. Methods have been developed to stain nerves of fishes (Song and Parenti 1995), but they have not been widely used with larvae.

Radiography

Radiographs (X rays) of later larvae can be used to observe bony elements (Tucker and Laroche 1984). Soft X-ray machines and fine-grained films are needed. Digital X-ray machines are also available and offer advantages over film-based units. Adjustments can be made until the best image is achieved. Images can later be adjusted further and annotated using a photo-editing program. Producing radiographs either on film or digitally is quicker than clearing and staining specimens, and the film (or computer file) provides a permanent record without altering the specimen. However, radiographs are not as easy to interpret as cleared and stained specimens, and cartilage is not distinguished.

IDENTIFICATION AND STAGING OF FISH EGGS

There is a dichotomy in fish eggs that affects their ecology, the way we sample and study them as well as their appearance: Some species spawn eggs that are pelagic and some spawn demersal eggs (see also Chapter 1 for more discussion). Pelagic eggs generally are clearer, smaller, and have a thinner, lighter-colored chorion than demersal eggs. Since most demersal eggs adhere to the bottom or are laid in nests, or in some other protected area, they are not available to be collected by plankton nets. Often adults laying demersal eggs guard the nest, so the identity of the eggs is assured. Winter flounder (*Pseudopleuronectes americanus*), Pacific cod (*Gadus macrocephalus*), striped bass, and white perch are notable exceptions, since they lay demersal eggs that rest singly on or near the bottom. These can be collected with plankton sleds. Since demersal eggs have thick, opaque chorions, it is difficult to see features of development that are used in identifying and staging pelagic eggs.

Characters (Figure 3.3A)

Among all stages in the life history of fishes, the eggs have the least number and range of distinguishing characters. Also there are few egg characters that apply to higher taxonomic groups. Thus the task of identifying eggs is particularly difficult, and it is not unusual to be unable to identify all eggs in a sample. However, eggs do have characters that can be observed microscopically, and that are consistent for species (Box 3.1). Besides characters of the eggs themselves, it is important to consider what eggs could be

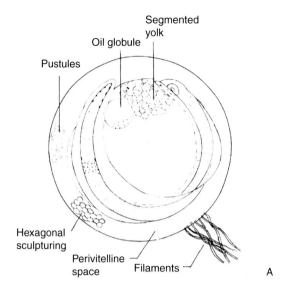
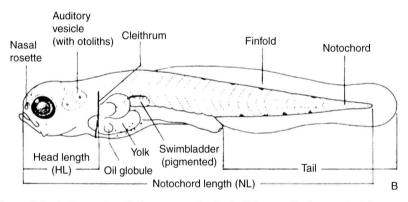

Figure 3.3. A. Examples of characters of pelagic fish eggs. B. Anatomical features and landmarks for measurements of yolk-sac larvae (from Matarese et al. 1989).

expected in the samples at hand. This requires a thorough knowledge of the species present in any given area and their spawning time, location, and depth. It should be noted whether the species produce pelagic or demersal eggs, and if demersal, what the substrate characteristics are. Among demersal egg spawners, the presence of nesting and parental care should be noted. The sampling gear used will determine whether pelagic or demersal eggs can be expected in the samples, although demersal eggs can be collected off the bottom on occasion. In some cases, fish eggs must be reared

> BOX 3.1. Examples of Northeast Pacific Fish Eggs Illustrating Various Characters (See Matarese et al. 1989)
>
> Pacific herring (*Clupea pallasi*): demersal, adhesive, attached to kelp
> round herring (*Etrumeus teres*): segmented yolk
> snubnose blacksmelt (*Bathylagus wesethi*): many oil globules
> northern smoothtongue (*Leuroglossus schmidti*): many oil globules coalesce to 2, unique oil globule movement
> Pacific viperfish (*Chauliodus macouni*): wide perivitelline space (Figure 3.1E)
> walleye pollock (*Theragra chalcogramma*): no oil globule
> jack mackerel (*Trachurus symmetricus*): large size, well-developed embryo with elongated dorsal fin rays
> Pacific saury (*Cololabis saira*): attach to flotsam by filaments, well-developed larva (Figure 3.1B)
> thornyhead (*Sebastolobus* spp.): in masses, pelagic
> ragfish (*Icosteus aenigmaticus*): large size (2.8–3.1-mm diameter)
> sanddab (*Citharichthys* spp.): small size (0.5–0.7-mm diameter)
> witch flounder (*Glyptocephalus zachirus*): well-developed embryo
> flathead sole (*Hippoglossoides elassodon*): large perivitelline space
> Dover sole (*Microstomus pacificus*): large size (2.05–2.68-mm diameter), well-developed embryo
> *Platichthys/Parophrys*: 1-mm right-eyed flatfish (pleuronectid) eggs, no oil globule
> curlfin sole (*Pleuronichthys decurrens*): sculptured chorion
> Alaska plaice (*Pleuronectes quadrituberculatus*): large size (1.7–1.9-mm diameter), no oil globule, wavy pattern on chorion

after collection in order to develop characters that will permit them to be identified (e.g., codfish [gadid] eggs from the Northwest Atlantic).

Shape. Most fish eggs are spherical. However, anchovies (engraulids) have elongated, ellipsoidal eggs. Nonspherical eggs are more common in demersal eggs than in pelagic eggs. Goby (gobiid) eggs are very unusual in shape, being flattened or ovoid. Eggs laid in masses or nests are usually not round because they are deformed by adjacent eggs.

Size. Most fish eggs are about 1 mm in diameter, with the size range of about 0.5 mm to about 8 mm. Pelagic eggs are generally smaller (0.5 mm to 5.5 mm) than demersal eggs (up to 8 mm), and since pelagic eggs are rare in freshwater fishes, eggs of marine fishes are smaller than those of freshwater fishes. Mouth-brooding marine catfishes (ariids) have the largest eggs of any teleosts at 14 to 26 mm.

Chorion Texture. The chorion of most fish eggs is smooth. However, the eggs of some species are ornamented with filaments that are characteristic of particular species. Filaments are more common in demersal eggs than in pelagic eggs; however, as a group the atheriniforms, which includes the halfbeaks (hemiramphins) and flying fishes (exocoetins), have filaments on their eggs (some of which are attached to flotsam). Other fishes have sculpturing of their egg chorions. Similar hexagonal sculpturing seems to have developed independently in several groups including members of the right-eyed flounders (*Pleuronichthys* spp.), the rattails (macrourids; Merrett and Barnes 1996), the pearlsides (*Maurolicus* spp.), and the lizardfishes (synodontids). The chorion of pelagic eggs is thinner than that of demersal eggs.

Oil Globules. Oil globules are characteristic features of most pelagic fish eggs, although their absence is an important character in certain groups (e.g., most right-eyed flounders [pleuronectids]). The size and number of oil globules are specific characteristics for identifying fish eggs. Most eggs possess one oil globule of a specific size, but some eggs have more than 100 oil globules of irregular size. The placement of oil globules within the egg relative to the developing embryo varies, and in some fishes may change during development (Ahlstrom 1969; Kendall and Mearns 1996). The color of the oil globules themselves, as well as any pigmentation on them, is also an important character.

Perivitelline Space. Immediately following spawning during a process called water hardening, a space (perivitelline space) develops between the inner edge of the chorion and the membrane around the cytoplasm of the egg itself. The relative width of the perivitelline space changes little during the rest of embryonic development. The width varies and is characteristic of species. In most species it is fairly narrow (< 0.1 mm), but in some it may be quite wide (e.g., in flathead sole [*Hippoglossoides elassodon*] the yolk is about 1 mm in diameter, whereas the total egg diameter is about 3 mm).

Yolk Characters. The yolk of most fish eggs is homogeneous but it is segmented in some, notably in lower teleosts such as herring-like fishes (clupeiforms), eels (anguilliforms), and salmon and their relatives (salmoniforms), and in some higher teleosts such as jack mackerel (*Trachurus symmetricus*). The yolk of most pelagic eggs is transparent, but in many demersal eggs it is opaque and colored.

Embryo Characters. As the embryo develops, it acquires characters that help identify the egg. Pigment often forms and is seen in characteristic patterns on the embryo, the yolk sac, and oil globules. Myomeres form and reach nearly the number of vertebrae found in the adults during embryonic development. The basic body shape of the larva can be seen in later embryos: whether it will be elongated or deep-bodied, and the relative length of the gut. Some species develop rays in some fins, and some of these can be elongated, pigmented, and ornate. The state of development at hatching is another character that varies among species. In general, larvae from demersal eggs are further along in development than those from pelagic eggs.

Determining Embryonic Stage of Development

Besides identifying the species of fish eggs in samples, for many types of studies it is also important to determine their stage of development. Embryonic development is largely temperature dependent, and along with data from rearing experiments, stage data from field samples can be used to calculate the age of eggs. This information is used to estimate egg mortality, and the number of eggs spawned by the population. Spawning location can also be estimated by knowing the age of the eggs, and the current velocities where they were collected. The number of developmental stages needed depends on the details of the studies involved, and the rate of development of the eggs. For eggs that complete embryonic development in less than one week, three stages (early, middle, late) may be sufficient, whereas with eggs that develop over several weeks, over 20 stages may be needed (Blood et al. 1994). Stages that are recognized and defined based on microscopic examination of the eggs are usually not of equal duration during development, and this must be taken into account when using stage data to calculate mortality. Counts of eggs of some stages, particularly those of short duration, may be combined with others to provide adequate data for mortality studies.

IDENTIFICATION OF FISH LARVAE

Fish larvae possess a larger suite of characters than the eggs, and may undergo dramatic changes with development. In practice, it is often helpful to arrange larvae into groups that look similar before recording specific observations or trying to identify them. Since development is a dynamic process, and appearances can change significantly as the fish grow, larval size (NL or SL), and stage of development (e.g., preflexion, flexion, postflexion, transforming [Kendall et al. 1984]) should be noted (Chapter 2, Figure 2.3). Figures 3.3B and 3.4 illustrate anatomical features and measurements used to describe larvae. The first attempt should be to identify unknown larvae by order or family, based on meristic values, shape, and general appearance. Following this, specific meristic, pigment, and morphological characters are noted, and compared with illustrations and descriptions of larvae occurring in the study area to see if the unknown larva can be identified with certainty. Actual identification is largely a process of eliminating species whose characters do not match the unknown; however, differences between the unknown and described larvae should be noted. If an unknown specimen does not match any available larval

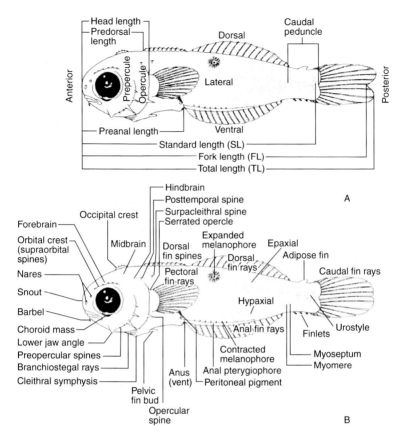

Figure 3.4. A. Landmarks for measurements of postflexion larvae. B. Anatomical features of postflexion larvae (from Matarese et al. 1989).

descriptions, check meristic tables of the most likely taxa to see if the unknown specimen fits a species whose larvae have not yet been described. Keys generally do not work well with fish larvae, because the larvae change so much with development, and the larvae of all species in a study area are rarely known.

Morphology

Larval shape can vary from stout and robust to quite slender and elongated (Table 3.1). The ratio of body depth at the pectoral fin to standard length is usually sufficient to characterize the overall body shape. The head and eye size and shape may also be important. The length of the gut, measured as the ratio of the preanal length to standard length is quite useful. As with other characters, larval shape characters vary with development, so the size and stage of development should be noted when comparing the shape of an unknown larva to illustrations and descriptions of known specimens.

TABLE 3.1. Ordinal/Subordinal Eggs Characters of Northeast Pacific Fishes (modified from Matarese et al. 1989)

Taxon	Families in study area	Development Site			Shape		Chorion	
		Planktonic	Demersal	Internal	Spherical	Elliptical	Smooth	Sculptured
Notacanthiformes/ Anguilliformes	9	X			X		X	
Clupeiformes	2	X	X[a]		X	X	X	
Salmoniformes								
Argentinoidei	5	X			X			X
Salmonoidei	2		X		X		X	
Stomiiformes	6	X			X		X	X
Aulopiformes	7	X			X			X
Myctophiformes	2	X			X		X	X
Gadiformes	5	X	X		X		X	X[b]
Ophidiiformes	3	—	—		—		—	—
Lophiiformes	2	X			X		X	
Gobiesociformes	1		X		X	X	X	
Beloniformes[c]	1	X	X			X		X
Atheriniformes	1		X		X			X
Lampriformes	2	X			X		X	
Beryciformes	5	X			X		X	
Zeiformes	1	X			X			
Gasterosteiformes	3		X	X	X		X	
Scorpaeniformes								
Scorpaenoidei	1			X	X	X	X	
Anoplopomatoidei	1	X			X		X	
Hexagrammoidei	1		X		X		X	
Cottoidei	3[d]		X		X		X	X
Perciformes								
Percoidei	6	X			X		X	
Zoarcoidei	9		X		X		X	
Trachinoidei	1		X		X		X	
Blennioidei	1		X		X			X
Icostoidei	1	X			X		X	
Ammodytoidei	1		X		X		X	
Gobioidei	1		X			X		X
Scombroidei	2	X			X		X	
Acanthuroidei	1	—	—	—	—	—	—	
Stromateoidei	3	X			X		X	
Pleuronectiformes	3	X	X[f]		X		X	X[g]
Tetraodontiformes	1	X			X			X

[a] *Clupea pallasi* eggs demersal.
[b] Macrouridae eggs sculptured.
[c] Nelson (1994) places scomberesocids within the Cyprinodontiformes.
[d] Nelson (1994) considers Psychrolutidae a separate family; we include it in Cottidae.
[e] *Trachurus* yolk segmented.
NOTE: Dash indicates "unknown."

Yolk		Perivitelline Space			Oil Globule			Diameter (mm)		
Segmented	Homogeneous	Wide	Moderate	Narrow	Absent	1 Present	>1	<0.9	1.0–2.0	>2.0
X		X			X	X	X			X
X		X	X	X	X	X			X	
X			X	X		X	X		X	X
X			X				X		X	X
X		X	X	X	X	X		X	X	X
			X		X				X	
X				X	X	X		X		
	X			X	X	X	X		X	X
—	—	—	—	—	—	—	—	—	—	—
X	X			X	X	X		X		
	X			X	X				X	
	X			X	X				X	
	X			X			X		X	X
	X		X		X					X
	X			X		X		X		
	X			X		X			X	
	X			X			X		X	
	X			X		X	X	X	X	
	X			X	X					X
	X			X		X	X		X	X
	X			X		X	X		X	X
X[e]	X			X		X			X	
	X			X		X			X	X
	X			X		X				X
	X			X		X	X			
	X				X	X				X
	X			X		X		X		
	X			X			X	X	X	
	X			X		X			X	
—	—	—	—	—	—	—	—	—	—	—
	X			X		X			X	
	X		X[h]	X	X[i]	X		X	X	X
	X		X	X			X		X	

[f] *Lepidopsetta bilineata* only species with demersal egg.

[g] *Pleuronichthys eggs* have hexagonal sculpturing.

[h] *Hippoglossoides elassodon* and *H. robustus* only species with wide perivitelline space.

[i] Pleuronectidae eggs have no oil globule.

TABLE 3.2. Ordinal/Subordinal Larval Characters of Northeast Pacific Fishes (modified from Matarese et al. 1989)

Taxon	Families in study area	Shape		Preanal Length (% SL)			Gut Shape		
		Elongate	Stocky	<50	50–75	>75	Straight	Coiled	Trailing
Notacanthiformes/ Anguilliformes	9	X			X	X	X		X
Clupeiformes	2	X			X	X	X		
Salmoniformes									
Argentinoidei	5	X			X	X	X		
Salmonoidei	2	X				X	X		
Stomiiformes	6	X	X	X	X	X	X		X
Aulopiformes	7	X		X	X			X	
Myctophiformes	2	X	X	X	X		X		X
Gadiformes	5	X	X	X				X	
Ophidiiformes	3	X		X				X	
Lophiiformes	2		X			X		X	
Gobiesociformes	1	X		X				X	
Beloniformes[a]	1	X		X			X		
Atheriniformes	1	X		X				X	
Lampriformes	2	X	X	X	X			X	
Beryciformes	5	X	X	X	X			X	
Zeiformes	1		X	X				X	
Gasterosteiformes	3	X		X	X		X	X	
Scorpaeniformes									
Scorpaenoidei	1		X	X	X			X	
Anoplopomatoidei	1	X			X		X	X	
Hexagrammoidei	1	X		X	X			X	
Cottoidei	3[b]	X	X	X	X			X	
Perciformes									
Percoidei	6	X	X	X	X			X	
Zoarcoidei	9	X		X	X			X	
Trachinoidei	1	X		X				X	
Blennioidei	1	X		X				X	
Icostoidei	1	X		X				X	
Ammodytoidei	1	X		X			X		
Gobioidei	1	X		X	X		X		
Scombroidei	2	X	X		X			X	
Acanthuroidei	1		X	X				X	
Stromateoidei	3	X	X	X	X		X	X	
Pleuronectiformes	3	X	X	X	X			X	
Tetraodontiformes	1		X	X				X	

[a] Nelson (1994) places scomberesocids within the Cyprinodontiformes.
[b] Nelson (1994) considers Psychrolutidae a separate family; we include it in Cottidae.

Eye Shape		Head Spines		Transformation		Special Juvenile	
Round	Narrowed	Absent	Present	Marked	Gradual	Present	Absent
X	X	X		X			X
X		X		X			X
X	X	X		X	X	X	X
X		X		X		X	
X	X	X		X			X
X	X	X	X		X		X
X	X		X	X	X	X	X
X		X			X		X
X		X			X		X
X		X		X			X
X		X			X		X
X		X			X		X
X		X			X		X
X		X			X		X
X		X	X	X	X	X	X
X			X		X		X
X		X			X		X
X			X		X	X	X
X		X			X	X	
X		X	X		X	X	
X		X	X		X	X	X
X		X	X		X	X	X
X		X			X		X
X			X		X		X
X		X			X		X
X			X		X		X
X		X			X		X
X		X			X		X
X			X		X		X
X			X		X		
X			X		X	X	
X		X	X	X			X
X			X		X		X

TABLE 3.3. Ordinal/Subordinal Fin and Meristic Characters of Northeast Pacific Fishes (modified from Matarese et al. 1989)

		Fin Characters					
				Pelvic Fin			
		Spines		Position			
Taxon	Families in study area	Present	Absent	Absent	Abdominal	Thoracic	Jugular	Formula
Notacanthiformes/ Anguilliformes	9		X	X				
Clupeiformes	2		X		X			6–9
Salmoniformes								
Argentinoidei	5		X		X			6–12
Salmonoidei	2		X		X			8–11
Stomiiformes	6		X		X			6–17
Aulopiformes	7		X		X			8–12
Myctophiformes	2		X		X			7–10
Gadiformes	5	X	X			X		0–17
Ophidiiformes	3	X					X	I,1–2
Lophiiformes	2		X	X				
Gobiesociformes	1	X				X[c]		I,4
Beloniformes[f]	1		X		X			6
Atheriniformes	1	X			X			I,5–6
Lampriformes	2	X	X		X	X		6–7
Berciformes	5	X		X	X	X		0–I,6–8
Zeiformes	1	X			X			I,6
Gasterosteiformes	3	X	X	X	X	X		I,1 or 4
Scorpaeniformes								
Scorpaenoidei	1	X				X		I,5
Anoplopomatoidei	1	X				X		I,5
Hexagrammoidei	1	X				X		I,5
Cottoidei	3[e]	X		XI		X		I,2–5[f]
Perciformes								
Percoidei	6	X				X		I,5
Zoarcoidei	9	X		X		X	X	0–I, 0–5
Trachinoidei	1	X				X		I,5
Blennioidei	1	X				X		I,3
Icostoidei	1	X		X[g]	X			I,4[g]
Ammodytoidei	1		X	X				
Gobioidei	1	X				X		5–6
Scombroidei	2	X				X		I,I,I,5
Acanthuroidei	1	X		X[g]		X		I,4[g]
Stromateoidei	3	X		X[h]		X		I,5
Pleuronectiformes	3		X			X	X	4–6
Tetraodontiformes	1		X	X				

[a] Three dorsal fins in Gadidae.
[b] Macrouridae lack caudal fin.
[c] Pelvic fin modified into a disc.
[d] Dorsal spines isolated from soft-rayed dorsal fin.
[e] Nelson (1994) considers Psychrolutidae a separate family; we include it in Cottidae.
NOTE: Dash indicates "unknown."

	Meristics						
No. Dorsal Fins		Adipose Fin			Principal Caudal Fin Rays		
One	Two	Present	Absent	Vertebrae	Upper	Lower	Total
X			X	68–750	0–9	0–7	0–16
X			X	43–57	10	9	19
X		X	X	34–84	10	9	19
X		X		54–75	10	9	19
X		X	X	35–83	10	9	19
X		X	X	48–87	10	9	19
X		X		29–42	10	9	19
X	X[a]		X	48–64; 84–86	4–6[b]	2–4[b]	6–10[b]
X			X	60–81	4	5	9
X			X	19–21	4	4–5	8–9
X			X	32–36	?	?	14
X			X	62–69	7	8	15
	X		X	44–52	9	8	17
X			X	43–46; 90–94	?	?	?
X			X	23–52	10	9	19
X			X	39	6	7	13
X	X[d]		X	30–64	6–7	6	12–13
X			X	26–31	7	7	14
	X		X	61–66	7	7	14
X			X	36–63	7–8	6–11	13–19
X	X		X	25–71	6	5–7	11–13
X			X	23–41	9	8	17
X	X		X	49–150; 221–250	?	?	12–15
	X		X	44–47	?	?	13
X			X	55–58	9	8	17
X			X	66–68	9	8	17
X			X	65–70	9	8	17
	X		X	26–38	?	?	17
X	X		X	22–23; 30–32; 148–150	9	8	17
	X		X	22–23	?	?	16
X			X	28–31; 52–62	9	8	17
X			X	35–66	9–12[i]	8–11[i]	17–24[i]
X			X	17–18			15 [j]

[f] Pelvic fin absent or modified into a disc in Cyclopteridae.
[g] Pelvic fin present only in larvae.
[h] Pelvic fin absent in *Peprilus*.
[i] Total caudal fin rays.
[j] *Mola mola* has a pseudocaudal fin (see Tyler 1980).

Pigment

Pigmentation available as taxonomic characters on larvae is limited to melanophores, since other pigment cells (e.g., xanthophores) do not retain their color in currently used fixatives and preservatives. Melanophore patterns are very useful for identifying larval fishes. The relative size, position, and sometimes the number of melanophores in series should be noted. In some cases, pigmentation consists of a group of melanophores in a specific area; in others, the pigmentation consists of an individual melanophore. Pigmentation generally changes as larvae develop. Movement of individual melanophores is rather limited, but addition or loss of melanophores is common. Usually preflexion larvae are less pigmented than later larvae, and late in the larval period as transformation occurs, the larval pigment pattern is overgrown by the largely superficial pattern of juveniles. In most fishes, between the preflexion and transformation stages, there is a definite larval pigment pattern, which is relatively stable and unique to species in many cases. Although the position of melanophores is a species characteristic, the degree of contraction seems to be physiologically moderated. Thus, larvae of the same species could have a different overall pigmented appearance, either lighter or darker.

Meristic Characters

Meristic (countable: e.g., vertebrae, fin rays) characters are essential and should be determined to ensure accurate identification of larvae. Larvae that look similar but have substantially different meristic characters cannot represent the same species. Fin elements are gradually added during larval development, so state of development must be considered when comparing counts obtained on larvae with those reported for the adults.

Myomeres. Myomeres are the first meristic character to stabilize, and the number usually reflects the number of adult vertebrae. The number of vertebrae varies from < 20 (ocean sunfishes [molids]) to > 200 (e.g., most eels and relatives [elopomorphs]) (Figure 3.5). Use of polarized light often facilitates counting myomeres. Myosepta are frequently more discernible than the myomeres, and if they are counted, two should be added to the count to account for the myomeres anterior and posterior to the first and last myosepta.

Fin Rays. The developing median fins contain several bits of taxonomic information. Dipping the larva in a potassium hydroxide solution and then in Alizarin Red solution (see Potthoff 1984 for recipes for solutions) for a few seconds, then rinsing and examining it in its preservative allows fin rays and head spines to be seen more clearly, since they stain red. This procedure may interfere with future attempts to clear and stain the larva, however. The principal caudal fin count is often an ordinal character (Table 3.1), and since it generally reaches its adult state shortly after flexion, it is very useful and relatively easy to determine in larvae. The number, position, and order of development of the dorsal and anal fins, and their composition in terms of spines and soft-rays, are important characters (Table 3.1). In several fishes (e.g., rockfishes), the final dorsal and anal spines develop first as soft-rays and transform to spines during transformation. Thus, in late larvae, the median fins may be different from the adults in the composition of spines and soft-rays. Spinous dorsal fins may develop before, concurrently, or after the second dorsal fin. When there is a gap between the spinous and soft-ray portions of the dorsal fin, the spines generally develop after the soft-rays. When two dorsal fins are present and continuous, the second dorsal fin usually develops concurrently with the anal soft-rays. Fin rays that become elongated in larvae often develop precociously.

In the sequence of paired fin development, the pectorals often develop early in the larval period, and the pelvics develop late. The pectoral fins form in the egg as larval pectoral fins, without fin rays. The fin rays generally develop much later in the larval stage. The length and number of rays of the pectoral fin are useful characters. Although the number of pectoral rays may vary within species and among species in a genus, the pelvic fin position and formula is generally stable at a high level of classification (e.g., order; see Table 3.1). The pelvic fin is absent in eels and members of some other groups (e.g., eel pouts [zoarcids], sandlances [ammodytids], and ocean sunfishes). In primitive fishes (e.g., herring-like fishes [clupeiforms], lanternfishes, and relatives [myctophiforms]) it contains no spines, is abdominal, and generally contains more than five soft-rays. In most perch-like fishes (perciforms) and mail-cheeked fishes (scorpaeniforms), the basic pelvic fin count is I, 5, and it is thoracic in position. This count is reduced in some; notably in the north Pacific, sculpins (cottids) often have fewer than five soft-rays. The pelvic fin is modified into a sucking disc in clingfishes (gobiesociforms) and in some snailfishes (cyclopterids). Other meristic characters (e.g., gill rakers, secondary caudal rays, scales) develop too late to be generally useful in identifying larvae, but may be essential when working with pelagic juveniles.

Specialized Larval Characters

Specialized characters of larvae are those that are overgrown or otherwise lost by the end of the juvenile stage. Such characters include elongated fin rays, serrate fin spines, trailing guts, stalked eyes, and pronounced and sometimes serrate head spines. In some larvae, the elongate fin rays are heavily ornamented and pigmented. Head spines, when present, may be more numerous and accentuated in larvae than in adults. Larval head spines are prevalent in

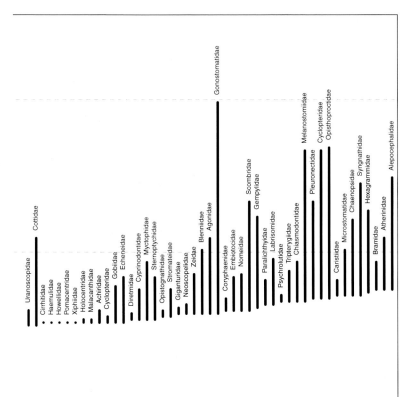

Figure 3.5. Vertebral (myomere) counts of Northeast Pacific fishes (based on Matarese et al. 1989).

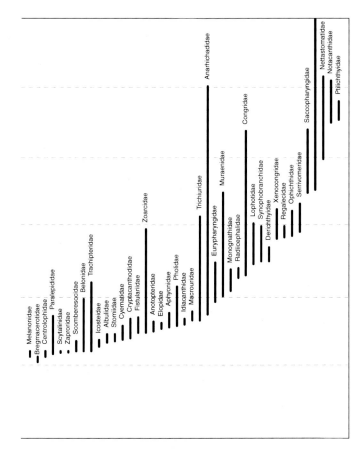

Figure 3.5. Continued.

sculpins and scorpionfishes (scorpaenids) and occur in some members of groups such as squirrelfishes and their relatives (beryciforms), perch-like fishes, and flatfishes (pleuronectiforms) (Table 3.2). Presumably, specialized larval characters have evolved to enhance larval survival in a habitat that may be entirely different from that of the adult. Although the function of such characters is mainly open to speculation, they presumably provide protection from predation by creating visual effects of a larger, or otherwise objectionable, prey organism. The ornamentation and pigment pattern created by some larvae with elongated fins and guts may mimic siphonophores.

Higher-Level Characters

As larvae of more and more fishes started to be recognized, it became apparent that closely related species looked more similar to each other than to more distantly related species. For example, larvae of various species in a genus looked more similar than larvae of other genera. This held true at higher levels of classification also: larval representatives of a family would be more similar to each other than to larvae of different families. This observation provided two insights: (1) larval morphology could be used in a predictive way, that is, unknown larvae of a group (e.g., genus, family) would resemble known larvae of that group, and (2) the larvae possess a suite of characters that can be used to investigate relationships among fishes.

Ahlstrom and Moser (1976) first tabulated larval characters for orders and suborders, and several more recent publications have expanded on their list of characters (Fahay 1983; Matarese et al. 1989; Moser [ed.] 1996; Leis and Carson-Ewart [eds.] 2000). The accompanying box (Box 3.2) and figures that follow this box illustrate some higher-level characters of larvae, primarily those at the order and suborder levels. For example, tarpons (elopiforms), bonefishes (albuliforms), and eels all have a leptocephalus (leaf-like) larval morphology: the larvae are shaped like willow leaves, they are laterally flattened and taper anteriorly and posteriorly. They are lightly pigmented and possess large, sometimes fang-like teeth. Among the orders with leptocephali, the tarpons and bonefishes have forked tails, whereas the eels have pointed tails.

EGG AND LARVAL STAGES IN SYSTEMATIC STUDIES

Historical Perspective

In the early 1900s as larvae of more and more species were described, it became apparent that larvae of closely related species were more similar to each other than to species that were more distantly related. Larvae of most of the codfishes and flatfishes of the Northeast Atlantic were described early on, and all members of these families were found to possess common suites of characters. Codfish larvae generally have a tadpole-shaped body, caudal

BOX 3.2. Higher-Level Characters and Examples of Marine Fish Larvae (Mainly from Fahay [1983], Moser et al. [eds. 1984a], Matarese et al. [1989], and Moser [ed. 1996])

Numbers of orders and families after Nelson 1994; suborders are included for the more complex groups. See figures on pages following this box.

21. Elopiformes: leptocephalus-like, forked tail, 51–92 myomeres, long, straight gut

    63. tarpons (Elopidae): *Elops* sp: Richards 1969 (Figure 3.6A)

23. eels (Anguilliformes): "leptocephalus," some with looped gut, 100–250 myomeres, transparent body, large size (among the largest fish larvae, attaining lengths of 30–184 cm)

    71. false morays (Chlopsidae): *Thalassenchelys coheni:* Matarese et al. 1989 (Figure 3.6B)

25. herring-like fishes (Clupeiformes): elongated, slender body, long, straight gut (48–95% SL)

    88. anchovies (Engraulidae): dorsal and anal fin bases overlap: northern anchovy (*Engraulis mordax*): Kramer and Ahlstrom 1968 (Figure 3.6C)

    91. herrings (Clupeidae): dorsal and anal fin bases do not overlap: Pacific herring (*Clupea pallasi*): Matarese et al. 1989 (Figure 3.6D)

32. Osmeriformes: elongate, slender body, straight gut
    Argentinoidei: "streamer" fin development, stalked eyes in some

    155. deepsea smelts (Bathylagidae): eared blacksmelt (*Bathylagus ochotensis*): Ahlstrom et al. 1984a (Figure 3.6E)
    Osmeroidei: gut pigmented and elongate, 75% SL, adipose fin

    160. smelts (Osmeridae): capelin (*Mallotus villosus*): Fahay 2007a (Figure 3.6F)

34. Stomiiformes: elongated body, medial fins posterior on body in most, photophores in some

    167. lightfishes (Gonostomatidae): noticeable teeth, duck-billed shape, snout in most: Panama lightfish (*Vinciguerria lucetia*): Ahlstrom et al. 1984b (Figure 3.7A)

    170. Idiacanthidae: extremely elongate body, stalked eyes, trailing gut: Pacific blackdragon (*Idiacanthus antrostomus*): Kawaguchi and Moser 1984 (Figure 3.7B)

36. Aulopiformes: elongated body in most, dorsal fin forward on body, series of pigment blotches along gut

(continued)

180. barracudinas (Paralepididae): slender barracudina (*Lestidiops ringens*): Okiyama 1984 (Figure 3.8A)

37. Myctophiformes: round or narrow eyes, some photophores develop in larvae, body shape elongated to moderately stout, generally lightly pigmented

   186. lanternfishes (Myctophidae): pinpoint lampfish (*Lampanyctus regalis*): round eyes. Moser and Ahlstrom 1974 (Figure 3.8B): blue lanternfish (*Tarletonbeania crenularis*): narrow eyes. Moser and Ahlstrom 1970 (Figure 3.8C)

38. Lampriformes: elongated, compressed body, some precocious ornamented and elongate dorsal and pelvic fin rays

   192. ribbonfishes (Trachipteridae): king-of-the-salmon (*Trachipterus altivelus*): Matarese et al. 1989 (Figure 3.8D)

   193. oarfishes (Regalidae): scalloped ribbonfish (*Zu cristatus*): Sparta 1933 (Figure 3.9A)

41. Ophidiiformes: slender body, short, coiled gut, long, medial fin bases, jugular pelvic fins

   198. pearlfishes (Carapidae): first dorsal ray elongated and ornamented: chain pearlfish (*Echiodon drummondi*): Gordon et al. 1984 (Figure 3.9B)

   200. viviparous brotulas (Bythitidae): pigment often as a series of blotches along dorsal and/or ventral body margins: red brotula (*Brosmophycis marginata*?): Gordon et al. 1984 (Figure 3.9C)

42. cod-like fishes (Gadiformes): elongated or stocky body that tapers, caudal fin isocercal or lacking, coiled gut, pigment usually in bands

   205. grenadiers (Macrouridae): bulbous gut, long tapering trunk, pectoral fin on peduncle: Macrourinae (*Nezumia* sp. ?): Fahay and Markle 1984 (Figure 3.10A)

   210. codlets (Bregmacerotidae): elongated but broad trunk, single occipital ray, elongated pelvic rays, long, medial fin bases: spotted codlet (*Bregmaceros macclellandi*): Houde 1984a (Figure 3.10B)

   213. hakes (Merlucciidae): moderately elongated body with large pigment blotches: Pacific hake (*Merluccius productus*): Ahlstrom and Counts 1955 (Figure 3.10C)

   214. codfishes (Gadidae): moderately elongate, tapering body, two or three dorsal fins, pigment often in opposing dorsal and ventral blotches, which sometimes fuse into bands: saffron cod (*Eleginus gracilis*): Dunn and Vinter 1984 (Figure 3.10D)

44. anglerfishes (Lophiiformes): globular body shape, coiled voluminous gut, anterior spines of dorsal fin separate, elongated, and ornamented (forming illicium), low medial fin ray counts, < 25 myomeres

216. goosefishes (Lophiidae): goosefish (*Lophius americanus*): Martin and Drewry 1978 (Figure 3.11A)
46. Atheriniformes: development direct, larvae elongated with preanal finfold, mediolateral pigment
    236. silversides (Atherinidae): jacksmelt (*Atherinopsis californiensis*): Matarese et al. 1989 (Figure 3.11B)
47. Beloniformes: elongated body, preanal finfold, long, straight gut, caudal fin forms before hatching
    243. sauries (Scomberesocidae): Pacific saury (*Cololabis saira*): Matarese et al. 1989 (Figure 3.11C)
49. Stephanoberyciformes: large head, long caudal peduncle, some with dorsal and anal spines
    254. Melamphaidae: highsnout melamphid (*Melamphaes lugubris*): Keene and Tighe 1984 (Figure 3.11D)
50. Beryciformes: head armature in some (mark of Acanthopterygii), slender to stocky body, coiled gut, low number of myomeres (< 35), medial fin spines
    269. squirrelfishes (Holocentridae): pronounced rostral spine: razorfish (*Myripristis* sp.): Jones and Kumaran 1962 (Figure 3.12A)
51. Zeiformes: deep-bodied, heavily pigmented, some with large head spines
    275. boarfishes (Caproidae): boarfish (*Capros aper*): Sanzo 1956 (Figure 3.12B)
54. mail-cheeked fishes (Scorpaeniformes): usually stocky body, coiled gut, head spines (parietal and others), pectoral fins often form early
    291. scorpionfishes (Scorpaenidae): numerous head spines, moderately deep-bodied, < 35 myomeres: darkblotched rockfish (*Sebastes crameri*): Richardson and Laroche 1979 (Figure 3.13A)
    Anoplopomatoidei: elongated body, heavily pigmented long pectoral fins
    301. sablefishes (Anoplopomatidae): sablefish (*Anoplopoma fimbria*): Ahlstrom and Stevens 1976 (Figure 3.13B)
    Hexagrammoidei: elongated body, heavily pigmented, blunt snout
    302. greenlings (Hexagrammidae): rock greenling (*Hexagrammos lagocephalus*): Kendall and Vinter 1984 (Figure 3.13C)
    Cottoidei: head spines, usually stocky body
    304. Rhamphocottidae: grunt sculpin (*Rhamphocottus richardsoni*): Richardson and Washington 1980 (Figure 3.14A)
    306. sculpins (Cottidae): scalyhead sculpin (*Artedius harringtoni*): Richardson and Washington 1980 (Figure 3.14B)

(continued)

309. Hemiptripteridae: sailfin sculpin (*Nautichthys oculofasciatus*): Richardson and Washington 1980 (Figure 3.14C)

310. poachers (Agonidae): fourhorn poacher (*Hypsagonus quadricornis*): Busby 1998 (Figure 3.14D)

311. Psychrolutidae: darkfin sculpin (*Malacocottus zonurus*): Richardson and Bond 1978 (Figure 3.15A)

313. snailfishes (Cyclopteridae): Pacific spiny lumpsucker (*Eumicrotremus orbis*): Matarese et al.1989 (Figure 3.15B)

314. Liparidae: sea tadpole (*Careproctus reinhardti*): Able et al. 1984 (Figure 3.15C)

55. perch-like fishes (Perciformes): most speciose fish order, primarily nearshore marine, variable larval morphology, some with head spines
Percoidei: head spines frequent, stocky to elongated body

350. snappers (Lutjanidae): vermillion snapper (*Rhomboplites aurorubens*): Laroche 1977 (Figure 3.16A)

Labroidei: some lightly pigmented, moderately elongated body, some with elongated eyes

390. wrasses (Labridae): razorfish (*Xyrichthys* sp.): Richards and Leis 1984 (Figure 3.16B)

Zoarcoidei: elongated body, long dorsal and anal fin bases

393. ronquils (Bathymasteridae): northern ronquil (*Ronquilis jordani*): Matarese et al. 1989 (Figure 3.16C)

395. pricklebacks (Stichaeidae): high cockscomb (*Anoplarchus purpurescens*): Matarese et al. 1989 (Figure 3.16D)

396. wrymouths (Cryptacanthodidae): dwarf wrymouth (*Cryptacanthodes aleutensis*): Matarese et al. 1989 (Figure 3.17A)

400. prowfishes (Zaproridae): prowfish (*Zaprora silenus*): Haryu and Nishiyama 1981 (Figure 3.17B)

Notothenoidei: Southern Ocean, some with enlarged fins, beak-like mouth

406. Channichthyidae: *Pagetopsis macropterus*: Stevens et al. 1984a (Figure 3.17C)

Trachinoidei: elongated body, series of pigment spots ventrally along gut

417. sand lances (Ammodytidae): Pacific sand lance (*Ammodytes hexapterus*): Stevens et al. 1984b (Figure 3.17D)

Gobiesocoidei: well-developed at hatching, pelvic fin modified into a disc

427. clingfishes (Gobiesocidae): California clingfish (*Gobiesox rhessodon*): Allen 1979 (Figure 3.17E)

Callionymoidei: small larvae, large head with abruptly tapering trunk, heavily pigmented

428. dragonets (Callionymidae): *Callionymus (Repomucenus) beniteguri*: Houde 1984b (Figure 3.17F)

Gobioidei: moderately elongated body, prominent swim bladder

433. gobies (Gobiidae): green goby (*Microgobius thalassinus*): Ruple 1984 (Figure 3.18A)

Acanthuroidei: highly compressed body, stout fin spines, < 25 myomeres

444. surgeonfishes (Acanthuridae): surgeonfish (*Acanthurus* sp.): Leis and Richards 1984 (Figure 3.18B)

Scombroidei: large heads and mouths, early developing large teeth

449. mackerels (Scombridae): little tunny (*Euthynnus alletteratus*): Collette et al. 1984 (Figure 3.18C)

Stromateoidei: blunt snouts, some deep-bodied, midlateral body pigment

456. butterfishes (Stromateidae): Pacific pompano (*Peprilus simillimus*): D'Vincent et al. 1980 (Figure 3.19A)

56. flatfishes (Pleuronectiformes): perciform derivative, deep, compressed body, long dorsal and anal fin bases, one eye migrates to the other side of the head in late larvae

469. righteye flounders (Pleuronectidae): Dover sole (*Microstomus pacificus*): Matarese et al. 1989 (Figure 3.19B)

57. puffers and filefishes (Tetraodontiformes): perciform derivative, deep, stocky body, some with spines on the head and body, reduced or absent caudal fin, < 20 myomeres

482. molas (Molidae): slender sunfish (*Ranzania laevis*): Leis 1977 (Figure 3.19C)

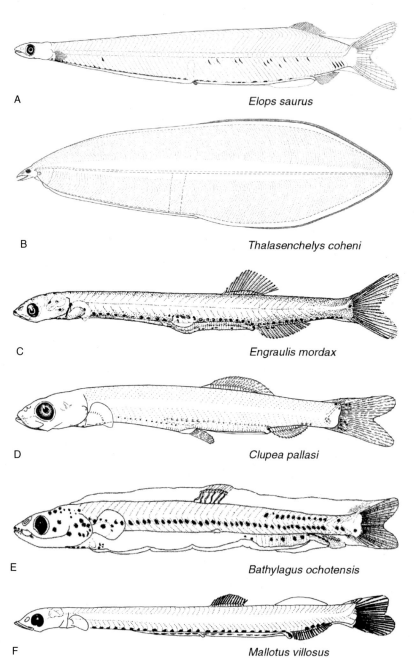

Figure 3.6. Examples of fish larvae. A. ladyfishes (*Elops* sp.): 33.8 mm SL (from Richards 1969). B. *Thalassenchelys coheni*: 190 mm SL (from Matarese et al. 1989). C. northern anchovy (*Engraulis mordax*): 18.4 mm (from Kramer and Ahlstrom 1968). D. Pacific herring (*Clupea pallasi*): 23.8 mm SL (from Matarese et al. 1989). E. eared blacksmelt (*Bathylagus(Lipolagus) ochotensis*): 21.5 mm (from Ahlstrom et al. 1984a). F. capelin (*Mallotus villosus*): 29 mm (from Fahay 2007a).

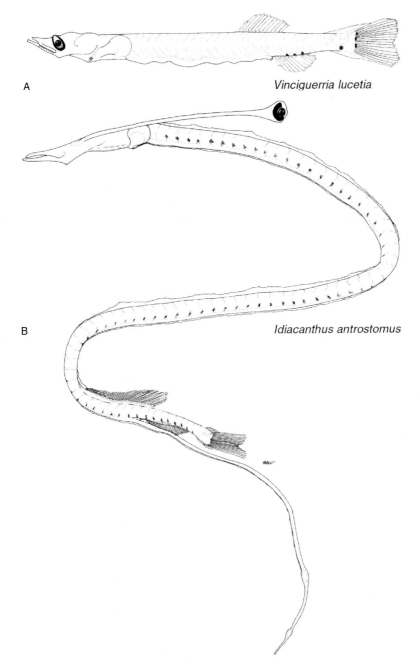

Figure 3.7. Examples of fish larvae. A. Panama lightfish (*Vinciguerria lucetia*): 9.0 mm SL (from Ahlstrom et al. 1984b). B. Pacific blackdragon (*Idiacanthus antrostomus*): 55 mm (from Kawaguchi and Moser 1984).

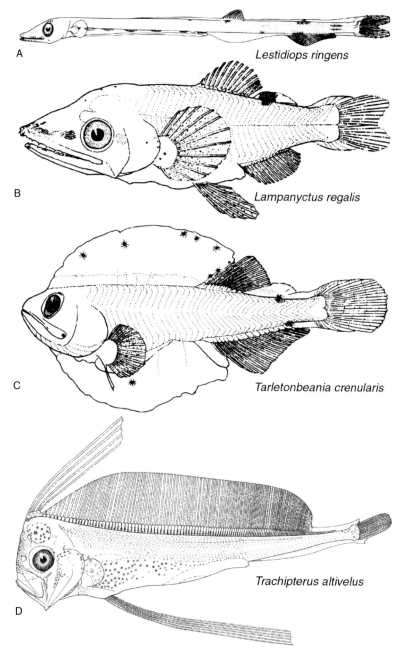

Figure 3.8. Examples of fish larvae. A. slender barracudina (*Lestidiops ringens*): 28.5 mm (from Okiyama 1984). B. pinpoint lampfish (*Lampanyctus regalis*): 9.1 mm SL (from Moser and Ahlstrom 1974). C. blue lanternfish (*Tarletonbeania crenularis*): 18.9 mm SL (from Moser and Ahlstrom 1970). D. king-of-the-salmon (*Trachipterus altivelus*): 24.0 mm SL (from Matarese et al. 1989).

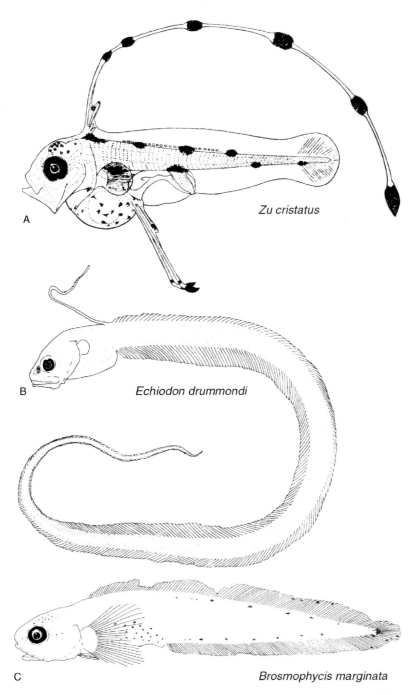

Figure 3.9. Examples of fish larvae. A. scalloped ribbonfish (*Zu cristatus*): 6.5 mm NL (from Sparta 1933). B. pearlfish (*Echiodon drummondi*): 76.5 mm TL (from Gordon et al. 1984). C. red brotula (*Brosmophycis marginata* (?)): 21.9 mm SL (from Gordon et al. 1984).

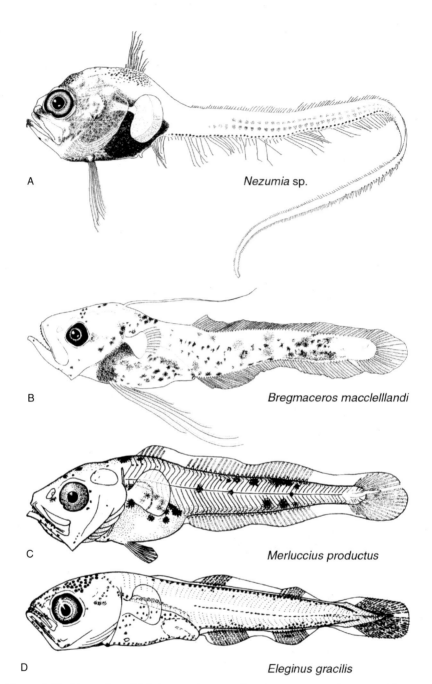

Figure 3.10. Examples of fish larvae. A. grenadier (Macrouronae [*Nezumia*?]): 15 mm TL (from Fahay and Markle 1984). B. spotted codlet (*Bregmaceros macclellandi*): 7.0 mm SL (from Houde 1984a). C. Pacific hake (*Merluccius productus*): 10.1 mm NL (from Ahlstrom and Counts 1955). D. saffron cod (*Eleginus gracilis*): 13.6 mm SL (from Dunn and Vinter 1984).

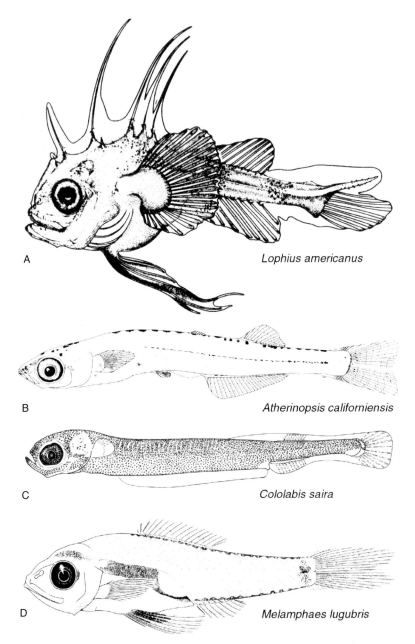

Figure 3.11. Examples of fish larvae. A. goosefish (*Lophius americanus*): 12 mm TL (from Martin and Drewry 1978). B. jacksmelt (*Atherinopsis californiensis*): 18.3 mm SL (from Matarese et al. 1989). C. Pacific saury (*Cololabis saira*): 9.9 mm (from Matarese et al. 1989). D. highsnout melamphid (*Melamphaes lugubris*): 10.4 mm SL (from Keene and Tighe 1984).

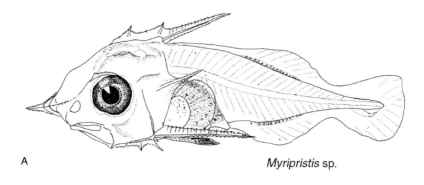

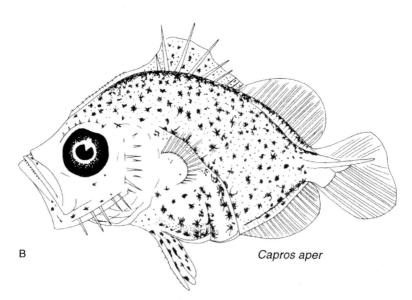

Figure 3.12. Examples of fish larvae. A. soldierfish (*Myripristis* sp.): 4.7 mm NL (from Jones and Kumaran 1962). B. boarfish (*Capros aper*): 5.0 mm NL (from Sanzo 1956).

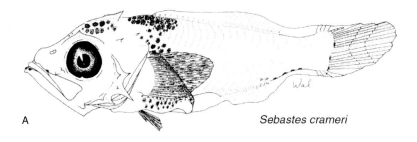
Sebastes crameri

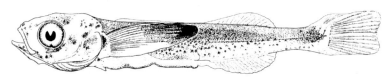
Anoplopoma fimbria

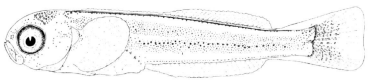
Hexagrammos lagocephalus

Figure 3.13. Examples of fish larvae. A. darkblotched rockfish (*Sebastes crameri*): 9.0 mm SL (from Richardson and Laroche 1979). B. sablefish (*Anoplopoma fimbria*): 18.6 mm SL (from Ahlstrom and Stevens 1976). C. rock greenling (*Hexagrammos lagocephalus*): 16.8 mm SL (from Kendall and Vinter 1984).

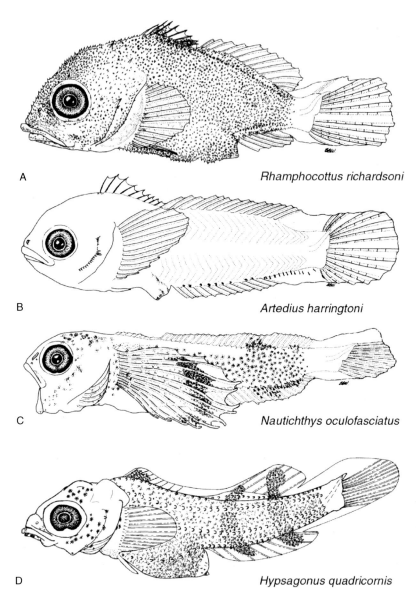

Figure 3.14. Examples of fish larvae. A. grunt sculpin (*Rhamphocottus richardsoni*): 10.6 mm SL (from Richardson and Washington 1980). B. scalyhead sculpin (*Artedius harringtoni*): 9.3 mm SL (from Richardson and Washington 1980). C. sailfin sculpin (*Nautichthys oculofasciatus*): 11.7 mm NL (from Richardson and Washington 1980). D. fourhorn poacher (*Hypsagonus quadricornis*): 10.0 mm SL (from Busby 1998).

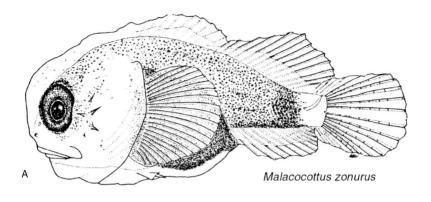

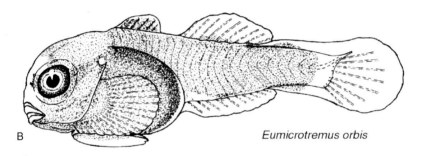

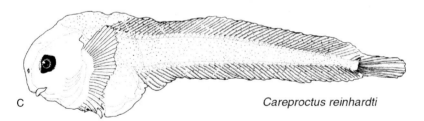

Figure 3.15. Examples of fish larvae. A. darkfin sculpin (*Malacocottus zonurus*): 14.2 mm SL (from Richardson and Bond 1978). B. Pacific spiny lumpsucker (*Eumicrotremus orbis*): 6.4 mm SL (from Matarese et al. 1989). C. sea tadpole (*Careproctus reinhardti*): 12.6 mm SL (from Able et al. 1984).

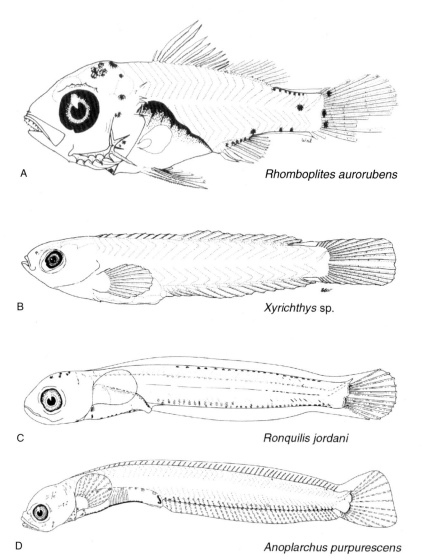

Figure 3.16. Examples of fish larvae. A. vermillion snapper (*Rhomboplites aurorubens*): 6.9 mm SL (from Laroche 1977). B. razorfish (*Xyrichthys* sp.): 10.5 mm SL (from Richards and Leis 1984). C. northern ronquil (*Ronquilis jordani*): 10.4 mm SL (from Matarese et al. 1989). D. high cockscomb (*Anoplarchus purpurescens*): 12.0 mm SL (from Matarese et al. 1989).

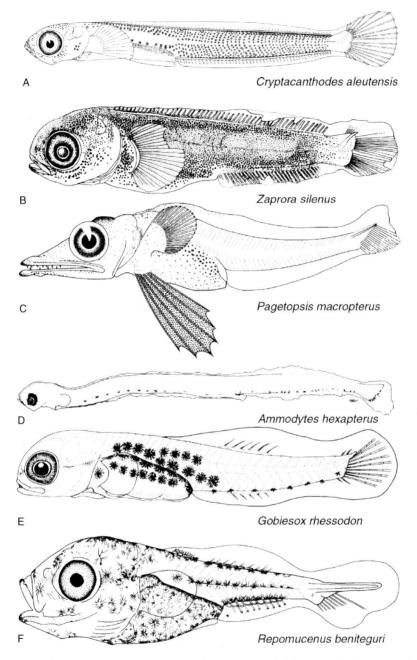

Figure 3.17. Examples of fish larvae. A. dwarf wrymouth (*Cryptacanthodes aleutensis*): 16.0 mm SL (from Matarese et al. 1989). B. prowfish (*Zaprora silenus*): 20.3 mm SL (from Haryu and Nishiyama 1981). C. *Pagetopsis macropterus*: 19 mm (from Stevens et al. 1984a). D. Pacific sand lance (*Ammodytes hexapterus*): 16.0 mm (from Stevens et al. 1984b). E. California clingfish (*Gobiesox rhessodon*): 6.2 mm (from Allen 1979). F. dragonet (*Callionymus* [*Repomucenus*] *beniteguri*): 4.1 mm (from Houde 1984b).

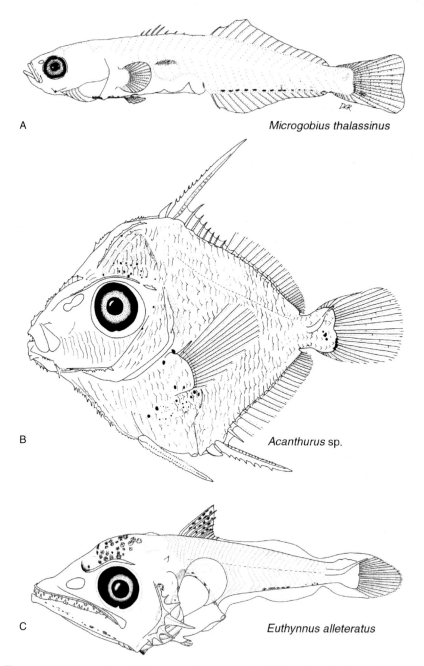

Figure 3.18. Examples of fish larvae. A. green goby (*Microgobius thalassinus*): 8.4 mm SL (from Ruple 1984). B. surgeonfish (*Acanthurus* sp.): 6.0 mm SL (from Leis and Richards 1984). C. little tunny (*Euthynnus alletteratus*): 6.2 mm SL (from Collette et al. 1984).

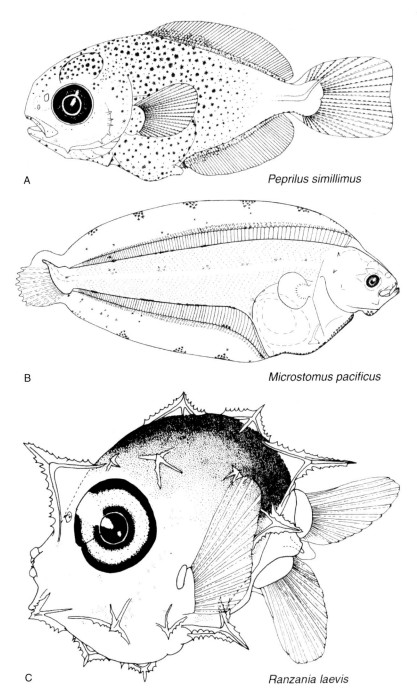

Figure 3.19. Examples of fish larvae. A. Pacific pompano (*Peprilus simillimus*): 10.8 mm (from D'Vincent et al. 1980). B. Dover sole (*Microstomus pacificus*): 15.0 mm SL (from Matarese et al. 1989). C. slender sunfish (*Ranzania laevis*): 3.9 mm (from Leis 1977).

rays developing dorsal and ventral to the notochord tip, and early developing pelvic rays. Flatfish larvae develop asymmetry of the head as metamorphosis approaches, have laterally compressed bodies, and long dorsal and anal fin bases. These early observations led to the consideration of larval characters in classifying fishes. Such studies were enabled by the far-reaching collections of fish larvae gathered during the expeditions of the *Dana*, a research vessel supported by the Carlsberg Foundation to investigate the early life history of the European freshwater eel (*Anguilla anguilla*). Cruises from 1905 to 1930 to all parts of the world ocean provided a wealth of larval specimens to be described (Schmidt 1932; Jesperson and Tåning 1934), and a group of exceedingly competent ichthyologists published many such descriptions over the years as *Dana Reports*. The taxonomy and systematics of many groups of oceanic fishes were elucidated by these works, which usually included larval descriptions and used larval characters in their systematic analyses. A second center of activity in using fish larvae in systematic studies arose as part of the CalCOFI studies of the California Current. Early in these studies, in the late 1940s, it was decided to try to identify all fish larvae caught in the extensive plankton collections that were part of this program. Because of this decision, E. H. Ahlstrom and his colleagues were able to describe larval development of many species, whether or not they had direct commercial importance. As more and more species, particularly of mesopelagic fishes (e.g., deepsea smelts [bathylagids] and lanternfishes [myctophids]), were identified, it again became apparent that they possessed characters that were important for systematics of the groups. Descriptions of groups of species within a family were published that included sections on the application of developmental information to the systematics of the family. At the time of his death, Ahlstrom was planning to write a book summarizing the contributions of larval characters to fish systematics. Shortly thereafter, some of his colleagues decided to enlist world-class ichthyologists to publish the book that he envisaged. The result was a symposium and book (the "Ahlstrom volume") in which invited contributors summarized available knowledge of development of various groups of fishes, and how this contributed to systematics of the group (Moser et al. [eds.] 1984a, www.biodiversitylibrary.org/bibliography/4334). Since then larval characters have continued to be used in a systematic context, although the Ahlstrom volume did not stimulate this type of research to the degree anticipated (see Leis et al. 1997).

Theory

Using development (ontogeny) to help elucidate evolutionary relationships (phylogeny) is a long-recurring theme in biology (Gould 1977; Cohen 1984). In plants as well as animals, characteristics of early stages have figured in classifications. Comparative embryology examines patterns of development among disparate groups of animals for evidence of relationships.

Morphological evidence of ancestral forms was found in early life stages of descendants. This became known as the biogenetic law and was summarized in the phrase "ontogeny recapitulates phylogeny." This idea has passed in and out of favor in biology, but as now formulated provides a basis for understanding development in an evolutionary context. The use of early life-history characters in studies of fish systematics has focused on characters that are evident primarily in the larvae alone. Although embryology has been valuable in understanding relationships among various other animals, it has not received much attention in fishes. Among the vertebrates only the teleost fishes and amphibians have indirect development that includes a larval stage. In both groups the larvae have proven to be a valuable source of characters for systematic analysis (Orton 1953). Although some indications of ancestral conditions may be seen in the development of some fishes (e.g., eye migration in flatfishes, fusion of bones, fin spines developing first as rays, loss of head spines), larval characters themselves have been most useful in studies of relationships among fishes.

Methods

Modern systematics relies on **cladistics** to infer relationships among animals based on shared derived characters (**synapomorphies**). Numerous inherited characters are sought that vary in the group of animals under study (ingroup). Once the characters have been selected they are analyzed to determine their states (primitive [**pleisiomorphic**] or derived [**apomorphic**]) and then used to produce branching diagrams (**cladograms**) to indicate the relationships among the taxa under study.

The characters used for analysis must meet certain criteria. For example, they must be **homologous** (i.e., they have the same ontogenetic origin, regardless of their form in a particular taxon), and they are nonadaptive (i.e., the character states arose by chance, not as an adaptation to a particular situation). **Convergence** (characters that appear the same, but have different ontogenetic origins) is a major problem in determining the suitability of characters for analysis. Studying larval development is useful in determining homology (e.g., Mabee 1988: supraneural bones) and sorting out whether characters are convergent or not. The most robust systematic studies of fishes now use a variety of characters including those based on larval and adult morphology as well as genetics.

Various methods are employed to determine which states are ancestral and which are derived (the **polarity**) for each character. In one method (**outgroup comparison**), a group of animals that is presumed to be primitive to the group under study is selected, and its character states are considered primitive. The states among the species in the group under study are then compared to those in the outgroup to trace evolutionary changes (**transformation series**) in the characters.

The ontogeny of characters can be used in deciding polarity (Mabee 1989). The **ontogenetic criterion** states that in comparing developmental series among a group of animals, the primitive condition is more general (seen in more species), and the derived condition is less general. Also, the condition seen later in ontogeny is considered derived.

The polarity of larval characters can be established based on the outgroup criterion, or on the ontogenetic criterion. In most studies to date, the outgroup method has been used with larval characters, although the use of the ontogenetic criterion has theoretic appeal when dealing with early developmental stages, such as larval fishes.

Once the polarity of the characters has been established, the characters states seen in the taxa under study are evaluated and scored. These data are used to construct trees or branching diagrams (cladograms) representing the relationships among the taxa. Species are grouped within the trees based on their possessing shared derived characters with all other species on a branch of the tree. The trees begin or are rooted with a hypothetical ancestor (as represented by the outgroup) of all of the taxa under study.

Consider a simple example: we have the larvae of three species of fishes and are considering their eye shape (round or oval) in a systematic study. We have an outgroup, based on other characters, and its larvae have round eyes. Among our three species under study, one has round eyes, and two have oval eyes. We have traced eye development in these species and found that the eyes initially develop in the same way in all taxa, and afterwards they elongate similarly in two of the taxa. Based on these observations we would conclude that the taxa with oval eyes form a derived sister group to the species with round eyes.

A very large number of trees (cladograms) can be constructed when numerous characters and taxa are studied. Without any way of knowing which tree represents the actual history of the group, the tree with the smallest number of branching steps is considered the best (i.e., it is the most parsimonious). Computer programs are used to generate trees and then compare trees based on the number of steps required to produce them. The first question to be addressed in such an analysis is whether all taxa under study did have a common ancestor (e.g., are they **monophyletic**). Unless all members of the group under study share some uniquely derived character states that would indicate that they are all part of a single lineage, further analysis is not possible.

Examples of the Use of Larval Fish Characters in Systematics

The "Ahlstrom volume" (Moser et al. [eds.] 1984a, www.biodiversitylibrary.org/bibliography/4334) summarizes information on development of fishes family-by-family, and attempts to use these data in a systematic context. For some groups larval information is inadequate for systematic analysis, for others it is available and quite helpful. Following are a few examples of the use of larval characters in systematic investigations.

The Lightfishes (Gonostomatidae). The lightfishes are a family of about 20 genera of photophore-bearing, midwater marine fishes. This family is closely related to the hatchetfishes (sternoptychids). Ahlstrom (1974) examined photophore development in the larvae of these fishes and concluded that there were three patterns. Several genera in the family exhibited each pattern. One of the patterns was so similar to the hatchetfishes that Ahlstrom (1974) recommended that the two families be combined into one. Ahlstrom et al. (1984b) expanded on this work and considered both adult and larval characters. They questioned the polarity of some of the photophore characters Ahlstrom (1974) used. They concluded that the adult and larval characters indicated similar relationships, but that more research was necessary before a well-founded hypothesis of relationships could be proposed. They emphasized the importance of developmental information in furthering this research.

The Lanternfishes (Myctophidae). The lanternfishes are a very abundant group comprising about 36 genera of photophore-bearing, midwater oceanic fishes. In several papers culminating with Moser et al. (1984b), Moser and Ahlstrom examined relationships among these fishes based primarily on the appearance of their larvae. Larval characters included the shape of eyes, body, gut, and fins as well as the sequence of appearance of photophores. Other characters included larval pigment patterns, size at transformation to juveniles, and specialized larval characters such as elongated eyes, modified fin rays, and chin barbels. The larvae fell into two groups based on eye shape (round or elliptical). These groups were congruent with the subfamilies Lampanyctinae (larvae with round eyes) and Myctophinae (larvae with elliptical eyes) based on adult characters. Within these subfamilies, the genera were recognized by larval characters, and within the genera transformation series of several larval characters could be established. In a systematic analysis of the lanternfishes, Paxton et al. (1984) used both larval and adult characters. Among the 59 characters they used in their analysis, 13 were larval characters. One criterion they used for establishing polarity of larval characters was to compare states of these characters in present genera with those present in a hypothesized generalized ancestor of the group. This analysis separated the lanternfishes into two subfamilies on the basis of three adult characters as well as the shape of the eyes in larvae. Within the subfamilies, larval characters indicated relationships among several genera. For example, the hypothesized ancestral (primitive) larva was assumed not to have elongate lower pectoral fin rays. Four genera shared the derived condition of having elongated lower pectoral fin rays in their larvae. These same four genera also had derived states of two adult characters.

The Sculpins (Cottidae.) The sculpins are a diverse group of mainly shallow-water, Northern Hemisphere, marine mail-checked fishes. Among the 70 or so genera of sculpins, the relationships of three (*Clinocottus, Artedius,* and *Oligocottus*) have been investigated using both adult and larval characters (Strauss 1993). This was primarily a reanalysis of a study by Washington (1986) who used larval characters and a study by Begle (1989) who used adult characters. A total of 47 characters, 10 larval and 37 adult, was used by Strauss (1993). The outgroup for this study was a presumed primitive sculpin, the Irish lords (*Hemilepidotus* spp.). Larval characters included several features of head spines, gut morphology, body shape, and pigment. Using the larval data alone, relationships among several species could not be resolved: monophyly of groups of species could not be established. When combined with the adult data, a much clearer picture of relationships emerged. Four branches were resolved, each representing a genus. Thus, these genera seem to form a monophyletic group within the sculpins, and each of the four genera is monophyletic. Relationships among the species within the genera were well resolved, with uncertain relationships seen in only one group of three species of *Clinocottus*.

The Seabasses (Serranidae): A Case Study of the Use of Larval Fish Characters
Studies of the percoid family containing the seabasses, groupers (e.g., *Epinephelus* spp.), and soapfishes (e.g., *Rypticus* spp.) offer a case study of the application of ontogenetic information in understanding their systematics. The seabasses are primarily tropical to temperate marine fishes that vary in size from < 10 cm to > 300 cm. It is a speciose family with nearly 400 species (Nelson 1994) that has had a history of being hard to characterize and subdivide (e.g., Gosline 1966). The seabasses are continuing objects of taxonomic studies from the species to subfamily levels, and several new species are described each year. Larvae of seabasses show a diversity of characters that permit them to be used in studies of relationships within the group.

The following is a summary of the current status of systematics and larval morphology of the seabassess, and how larval characters have contributed to the systematics of the group. The seabass subfamilies are clearly distinct as larvae. In fact, it is not possible to characterize the seabasses based on larval morphology because no characters unite the subfamilies while separating them from larvae of all other families. Serraninae larvae seem to be the least specialized and are more similar to percoid genera thought to represent the basal stock from which serranids arose (e.g., *Morone, Lateolabrax,* and *Dicentrarchus*). Based on larval and other evidence, it appears that the serranines are the primitive sister group of the anthiines, which in turn are the primitive sister group of the epinephelines (Baldwin and Johnson 1993).

Serraninae. There has been no revision of this primarily Atlanto-American subfamily, and little work on relationships among species in the various genera (Bortone 1977). These are considered the least specialized of the serranids and are mainly united by shared possession of basal percoid characters rather than unique specializations, which would allow a definitive statement about monophyly. They possess the four serranid specializations as mentioned by Johnson (1983), are hermaphroditic or secondarily gonochoristic (see Kendall 1977), share a common predorsal (supraneural) bone pattern (0/0/0/2), and a fairly coherent larval morphology (Figure 3.20). The larvae of *Schultzea, Dules, Acanthistius,* and *Cratinus* are unknown. Knowledge of larval morphology of the serranines is based primarily on Kendall (1977, 1979); more recent contributions are the descriptions of *Paralabrax* (Butler et al. 1982), and the compilation of descriptions of central western Atlantic species in Richards (1999, ed 2006), and that by Watson (1996) for species of the California Current region. Body proportions show rather direct development. There are no elongate spines in the opercular region, rather a series of blunt points. The fin spines are thin and only slightly elongated in some. Most larval pigment consists of melanophores in characteristic positions along the ventral midline. The serranine genera can be distinguished from each other and ordered in a rough progression of divergence from the supposed ancestral larval form (as exemplified by *Morone*) as follows (Figure 3.20): *Serraniculus, Centropristis-Paralabrax, Diplectrum, Serranus,* and *Hypoplectrus* (see Kendall 1979). Characters that lead to this assessment include pigment, body shape, sequence of dorsal spine–soft ray development, and dorsal fin spine elongation.

Anthiinae. This is a cohesive group of fishes that share several specializations in addition to those they hold in common with other serranids. These specializations include large scales, a highly arched lateral line, deep bodies and large heads, mainly 10 + 16 vertebrae, and a predorsal pattern of 0/00/2 or 0/0/2. They are generally small, brightly colored reef fishes. The generic alignments of many species are dubious, and a revision of the group is badly needed. In some cases, larval evidence on relationships is in conflict with that based on adults. Fitch (1982) and Baldwin (1990) showed several incongruencies in generic assignments of larvae in Kendall (1977, 1979), which were corrected in Kendall (1984). Better definitions of the genera must await a worldwide revision that will include information on early life-history stages. Most recent work, however, has focused on describing new species, faunal studies, and some generic revisions. Larvae of 12 of the 24 currently recognized anthiine species from American waters are known to some extent (Figure 3.21). Larvae of a few species from other parts of the world are also known, but have not been studied to the extent that American species have. These deep-bodied larvae have produced spines on several bones in the opercular region, some of which may be serrated. There is a tendency to develop armature on the head, and the interopercular has a characteristic long, posteriorly directed spine that is overlaid by an even larger, similar spine on the preopercular. The pelvic and some dorsal fin spines are strong, serrate in some, but not very elongated. Pigment consists mainly of large blotches and dashes in characteristic positions on the trunk. Anthiine larvae, like the adults, share several characters that unite them, yet they are quite diverse and have proven to be excellent subjects for

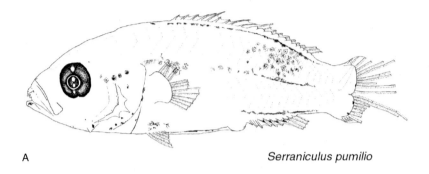

A *Serraniculus pumilio*

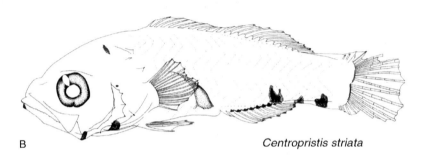

B *Centropristis striata*

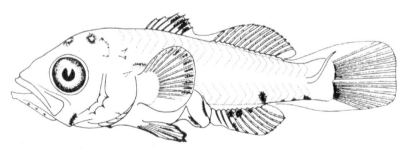

C *Paralabrax clathratus*

Figure 3.20 A-C. Larval representatives of serranine genera. A. pygmy sea bass (*Serraniculus pumilio*): 5.1 mm SL (from Richards 1999). B. black sea bass (*Centropristis striata*): 8.3 mm SL (from Kendall 1979). C. kelp bass (*Paralabrax clathratus*): 7.4 mm SL (from Butler et al. 1982).

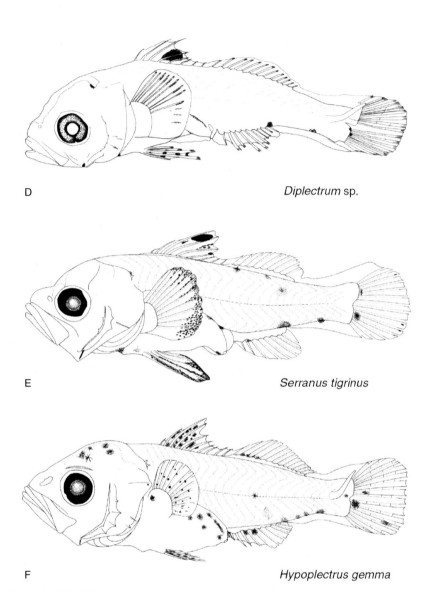

Figure 3.20 D-F. Larval representatives of serranine genera. D. sand perches (*Diplectrum* sp.): 6.1 mm SL (from Kendall 1979). E. harlequin bass (*Serranus tigrinus*): 5.9 mm SL (from Richards 1999). F. blue hamlet (*Hypoplectrus gemma*): 6.4 mm SL (from Richards 1999).

phylogenetic investigation (Baldwin 1990). Within the group, a progression of increasing spinyness and armature is apparent. Among the larvae described to date, armature seems to be added as follows: elongated preopercular and interopercular spines, serrate preopercular and interopercular spines, stout pelvic and first three dorsal spines, supraoccipital spine, serrate dorsal and pelvic spines, serrate head spines on several bones, and spiny scales developing during the larval stage. Baldwin (1990) grouped the known larvae of American anthiines into four groups (Table 3.2). These groups did not align closely with the recognized genera of anthiines. Her cladistic analysis, based on 16 larval and 10 adult characters, indicated that the current genera are not monophyletic (Figure 3.22). Larval group IV, represented by apricot bass (*Plectranthias garrupellus*) (Figure 3.22), is included in "other anthiines" as the primitive sister group of the rest of the American anthiines. Unlike the other anthiines, its larval interopercular spine is smooth rather than serrate. The other three groups of larval anthiines share development of rugosity or serrations on the frontals and parietals. The three representatives of larval group III are all contained in the first derived branch beyond this primitive sister group. They all have a midlateral pigment patch on the trunk, and a simple spine or knob on the supraoccipital. Larval group III represents two genera, neither of which is confined to this branch of the cladogram. Larvae in the other two groups (I and II) have serrate circumorbitals and tabulars. Larval group II contains five species and all are placed in the next branch of the cladogram. Again, larval group II contains representatives of two genera, neither of which is confined to this branch of the cladogram. Within larval group II there is variation in larval scale formation, with two species (roughtongue bass [*Pronotogrammus martinicensis*] and threadfin bass [*Pronotogrammus multifasciatus*]) forming scales with spines on their posterior margins during the larval period, and the rest of the species not forming scales during the larval period. On the basis of this character and two adult characters, Baldwin (1990) proposed that threadfin bass be reassigned from *Holanthias* to *Pronotogrammus*. Larval group I contains three species that are all in the final main branch of the cladogram. These species are also in two genera that contain species found in other parts of the cladogram. Six derived larval characters are possessed by group I: they have three serrate supraoccipital ridges, serrate pterotic, frontal and parietal ridges, serrate articulars, serrate fin spines, and larval scales, which have a spine originating near the center of the scale plate.

Epinephelinae. The epinephelines are mainly genera in the epinepheline-grammistine lineage of Kendall (1976). Johnson (1983) pointed out adult features that characterize this subfamily and the tribes within it. Although he did not provide a detailed analysis of the relationships among the tribes, he proposed that it was a monophyletic lineage (subfamily Epinephelinae) composed of five tribes (Niphonini, Epinephelini, Diploprionini,

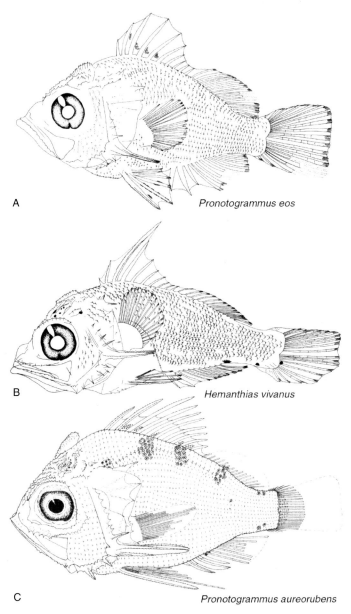

Figure 3.21 A-C. Larvae of anthiines arranged into the four groups proposed by Baldwin (1990). Group I: A. bigeye bass (*Pronotogrammus eos*): 9.3 mm SL (from Kendall 1979, as *Hemanthias peruanus*). B. red barbier (*Hemanthias vivanus*): 6.8 mm SL (from Kendall 1979). C. streamer bass (*Pronotogrammus aureorubens*): 9.8 mm SL (from Kendall 1984). *(See also Figure 3.21 D-K.)*

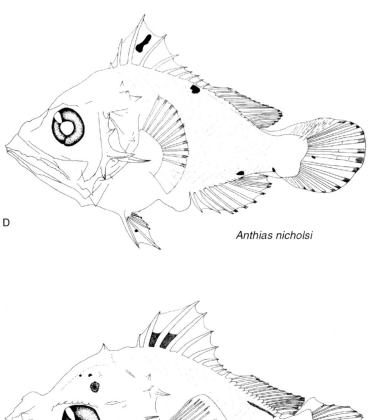

Anthias nicholsi

Pronotogrammus martinicensis

Figure 3.21 D-E. Larvae of anthiines arranged into the four groups proposed by Baldwin (1990). Group II: D. yellowfin bass (*Anthias nicholsi*): 5.3 mm SL (from Kendall 1979, as *Anthias* type 1). E. roughtongue bass (*Pronotogrammus martinicensis*): 5.7 mm SL (from Kendall 1979, as *Anthias* type 2). *(See also Figure 3.21 A-C, F-K.)*

Liopropomini, and Grammistini). Baldwin and Johnson (1993) provided a detailed cladistic analysis of the relationships within the Epinephelinae and between this subfamily and the other two subfamilies of serranids. They used both adult and larval characters in their analysis. Some early

Pronotogrammus multifasciatus

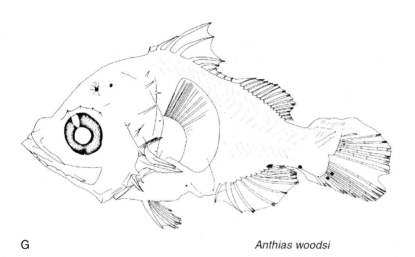

Anthias woodsi

Figure 3.21 F-G. Larvae of anthiines arranged into the four groups proposed by Baldwin (1990). Group II (cont.): F. threadfin bass (*Pronotogrammus multifasciatus*): 6.0 mm SL (from Kendall 1979, as *Anthias gordensis*). G. swallowtail bass (*Anthias woodsi*): 5.1 mm SL (from Kendall 1979, as *Anthias* type 3). *(See also Figure 3.21 A-E, H-K.)*

life-history stages are known for genera in all of the tribes (Figure 3.23). Of the 52 characters used in their analysis, 11 were larval. One of the characters that grouped the Anthiinae and Epinephelinae as the sister group of the Serraninae was a larval character (the presence of a single supraorbital

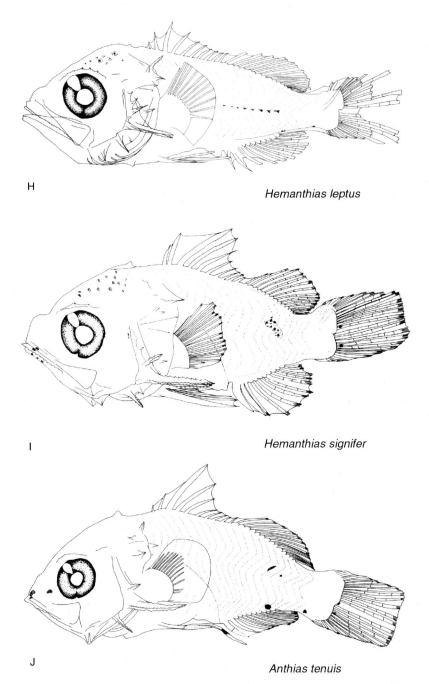

H *Hemanthias leptus*

I *Hemanthias signifer*

J *Anthias tenuis*

Figure 3.21 H-J. Larvae of anthiines arranged into the four groups proposed by Baldwin (1990). Group III: H. longtail bass (*Hemanthias leptus*): 6.0 mm SL (from Kendall 1979, as *Pronotogrammus aureorubens*). I. splittail bass (*Hemanthias signifer*): 7.8 mm SL (from Kendall 1979, as *Pronotogrammus eos*). J. threadnose bass (*Anthias tenuis*): 6.7 mm SL (from Kendall 1979). *(See also Figure 3.21 A-G, K.)*

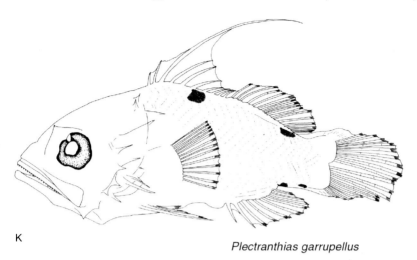

K

Plectranthias garrupellus

Figure 3.21K. Larvae of anthiines arranged into the four groups proposed by Baldwin (1990). Group IV: apricot bass (*Plectranthias garrupellus*): 5.5 mm SL (from Kendall 1979). *(See also Figure 3.21 A-J.)*

spine). One of the two derived characters of the Epinephelinae was larval (the first dorsal pterygiophore serially supports an elongate dorsal spine). Three larval characters were found in the tribe Epinephelini: the second dorsal and pelvic fin spines are robust and serrate, the elongate preopercular spine is serrate, and there is a midventral caudal pigment spot that moves to a midlateral position. Derived states of other larval characters appeared in other parts of the cladogram. For example, in the Niphonini and Epinephelini, the dorsal spines are stout, whereas in the other three tribes they are extremely elongated, flexible, and some have siphonophore-mimicking pigment and shape. In the Niphonini the third dorsal spine is elongated, whereas in the Epinephelini the second dorsal spine is elongated, and in the other tribes the third, and in some, several subsequent spines are elongated. The correspondence of the elongated dorsal spine configuration, and the (presumably) supporting predorsal (supraneural) bones and dorsal fin pterygiophores provided several characters for this analysis.

The following is a summary of what is known of the morphology of early life-history stages of fishes in the epinepheline tribes of Johnson (1983) and Baldwin and Johnson (1993).

NIPHONINI. Larvae of ara (*Niphon spinosus*), the sole member of this tribe, have been described by Johnson (1988), who had correctly speculated (Johnson 1983) on the basis of the predorsal bone patterns, that their third dorsal spine should be elongated (Figure 3.24). Head spines (supraorbital, opercular, preopercular, posttemporal, and supracleithral) and the elongate

TABLE 3.2. Groups of American Anthiines Based on Larval Characters
(based on Baldwin 1990)

Group	Species
I	*Pronotogrammus eos, Hemanthias vivanus, Pronotogrammus aureorubens*
II	*Anthias nicholsi, Pronotogrammus martinicensis, Pronotogrammus multifasciatus, Anthias woodsi*
III	*Hemanthias leptus, Hemanthias signifer, Anthias tenuis*
IV	*Plectranthias garrupellus*

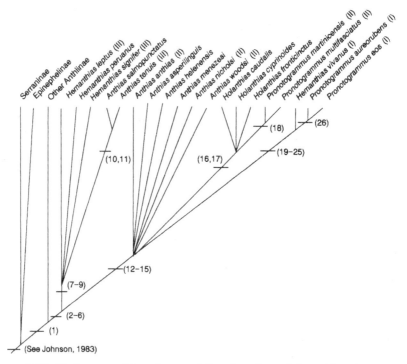

Figure 3.22. Cladogram of American anthiines based on adult and larval characters (after Baldwin 1990).

pelvic and dorsal fin spines are not serrate in ara larvae. The dorsolateral region of the gut is covered with pigment, some pigment develops dorsally on the head, and there is a large spot midventrally on the caudal peduncle where the anal fin will insert. Pigment also develops at the base of the caudal fin.

Baldwin and Johnson (1993) place the Niphonini as the primitive sister of the rest of the Epinephelinae on the basis of larval and adult characters.

EPINEPHELINI. Larvae are known for 6 of the 14 or so genera of Epinephelini (Smith [1971] included five subgenera in American *Epinephelus*, some of which have been considered genera by others). Several species have been reared and their egg and larval development described. Known larvae of this tribe are quite similar, but it is generally possible to assign them to a genus on the basis of larval characters. These are among the most spectacular of fish larvae, with stout, elongated, serrate, and pigmented dorsal and pelvic fin spines (Johnson and Keener 1984). The second dorsal spine is much longer than the others and it, as well as the pelvic spines, is as long as the body. The dorsal spine is often "locked" in an upright position—presumably because of a unique pterygiophore arrangement (Johnson 1983). The first and third dorsal spines and the anal spines are also stout and may bear serrations. The spine at the angle of the preopercular is elongated and serrate; there are two smaller spines dorsal and ventral to the one at the angle, and these may also bear serrations. There is a serrate spine on the supracleithrum. The body is "kite-shaped"; pigment lines the body cavity and there is a large, conspicuous spot on the caudal peduncle that migrates from a ventral midline to a midlateral position during flexion. Leis (1986) developed a cladogram of six genera of the Epinephelini, whose larvae were known, based on 12 characters—11 larval and 1 adult (Figure 3.25). He mainly used the ontogenetic method of polarizing the characters, which Johnson (1988) called into question. However, Johnson concluded that the outgroup method would produce the same polarity in most cases. This analysis indicated that *Plectropomus* was the least derived genus, followed by *Gonioplectrus* and *Cephalopholis* (Figure 3.26). This is followed by the three genera *Epinephelus*, *Mycteroperca*, and *Paranthias*, with the character separating *Epinephelus* from the other two varying among the subgenera of *Epinephelus*.

DIPLOPRIONINI. Until Baldwin et al. (1991) described in detail the development of arrowhead soapfish (*Belonoperca chabanaudi*) (based on field-collected specimens) and barred soapfish (*Diploprion bifasciatum*) (based primarily on reared specimens), all that was known of development of this tribe was a photograph of a transforming larva, a drawing of a juvenile, and a brief description of the juvenile of *Diploprion* showing long, flexible dorsal spines and a rather deep body (Hubbs and Chu 1934). The second and third dorsal spines in *Diploprion* larvae are extremely produced and contained in a heavily pigmented sheath (Figure 3.27). Head spines are small, larval pigment consisting of a band across the body at the insertion of the dorsal and anal fins, and some pigment at the base of the second dorsal fin spine, on the pectoral fin rays, and at the base and on the rays of the pelvic fin. The pelvic fin is not produced. In larvae of *Belonoperca* the second through the sixth dorsal spines are thin and elongated; the

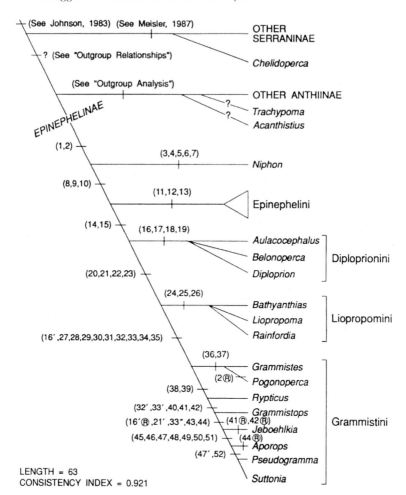

Figure 3.23. Cladogram of epinephelines based on adult and larval characters (after Baldwin and Johnson 1993).

second is longer than the others and is surrounded by a pigmented sheath (its length and the exact nature of the pigment cannot be determined since the spine is broken on all extant specimens). The supraorbital ridge bears two or three simple spines, and spines are present on other head bones, including three on the medial ridge of the preopercle. The larvae are relatively lightly pigmented: some pigment is present on top of the head and on the pectoral and pelvic fin rays, but the bodies of larger larvae are devoid of pigment.

Fish Egg and Larval Identification and Systematics   117

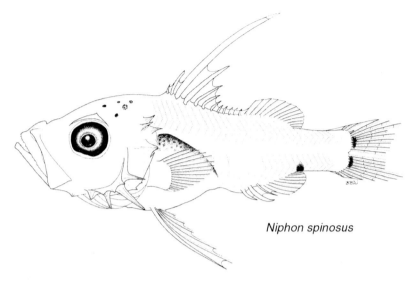

Figure 3.24. Larva of ara (*Niphon spinosus*) (7.0 mm SL), a primitive epinepheline (from Johnson 1988).

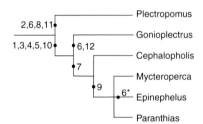

Figure 3.25. Cladogram of genera of Epinephelini (after Leis 1986).

LIOPROPOMINI. Larvae of *Liopropoma* were first described as a new genus, *Flagelloserranus*, by Kotthaus (1970, see Kendall [1977, 1979]) because their larvae are so different from any described fish. They are not as deep bodied as other epinephelines, but are characterized by having the second and third dorsal spines extremely produced and enclosed in fleshy sheaths (Figure 3.28). There are several pigmented swellings on the sheath of the second dorsal spine. The paired fins are not elongated or pigmented. Head spines are poorly developed, and there is little pigment aside from that on the dorsal spines.

GRAMMISTINI. Larvae of five of the eight genera are known, although some are described from single specimens (Baldwin et al. 1991; Baldwin and Johnson 1991). The bodies of the larvae are roughly tubular with a

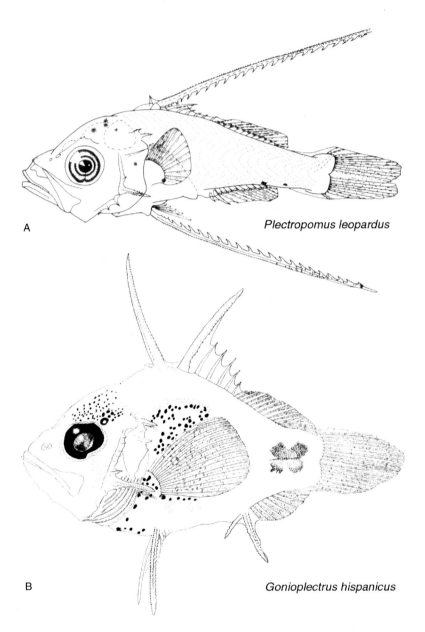

Figure 3.26 A-B. Larvae of representatives of the epipheline tribe Epinephelini.
A. leopard coralgrouper (*Plectropomus leopardus*): 8.1 mm SL (from Leis 1986).
B. Spanish flag (*Gonioplectrus hispanicus*): 13.4 mm SL (from Kendall and Fahay 1979).

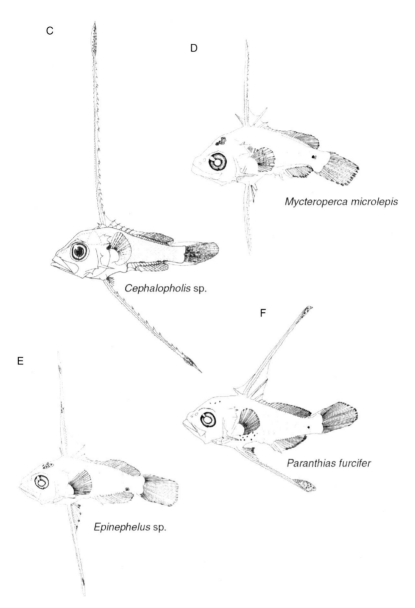

Figure 3.26 C-F. Larvae of representatives of the epinepheline tribe Epinephelini. C. *Cephalopholis* sp.: 5.8 mm SL (from Leis and Rennis 1983). D. gag (*Mycteroperca microlepis*): 7.4 mm SL (from Kendall 1979). E. *Epinephelus* sp.: 8.4 mm SL (from Kendall 1979). F. creolefish (*Paranthias furcifer*): 8.6 mm SL (from Kendall 1979).

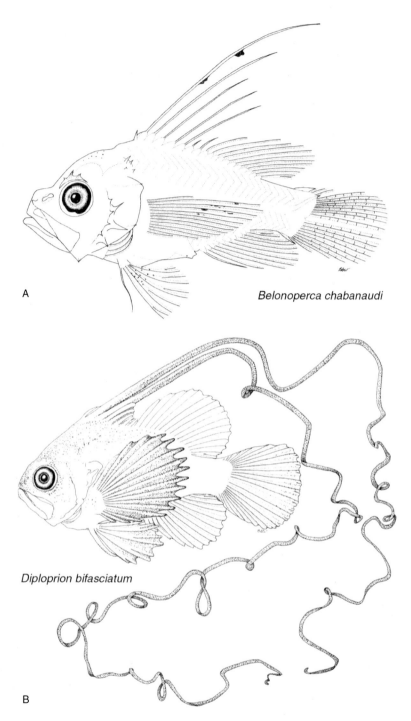

Figure 3.27. Larvae of representatives of the epipheline tribe Diploprionini.
A. arrowhead soapfish (*Belonoperca chabanaudi*): 6.9 mm SL (from Baldwin et al. 1991).
B. barred soapfish (*Diploprion bifasciatum*): 16.2 mm SL (from Baldwin et al. 1991).

Figure 3.28. Larva of a basslet (*Liopropoma* sp.): 11.0 mm SL (from Kendall 1984), a representative of the epipheline tribe Liopropomini.

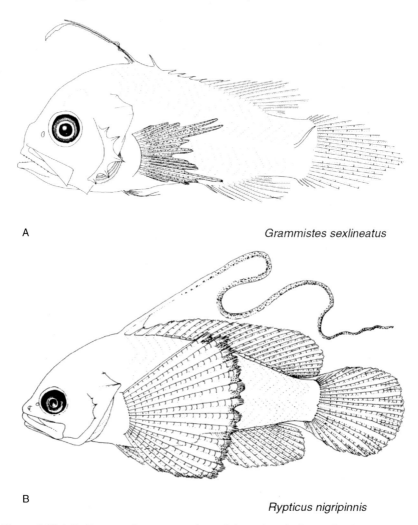

Figure 3.29 A-B. Larvae of representatives of the epinepheline tribe Grammistini. A. sixline soapfish (*Grammistes sexilineatus*): 6.3 mm SL (from Baldwin et al. 1991). B. blackfin soapfish (*Rypticus nigripinis*): 12.3 mm SL (from Watson 1996).

deep, caudal peduncle (Figure 3.29). Among the bones in the opercular series, the preopercular is armed with three (*Rypticus*), five, or six (*Jeboehlkia*) elongated, simple spines. The first (in *Rypticus*) or second dorsal fin spine becomes quite elongated, and is thin and flexible with a pigmented membranous sheath around it. The pelvic fin develops late, but the pectoral fin is large and develops early. Bodies of the larvae are practically devoid of

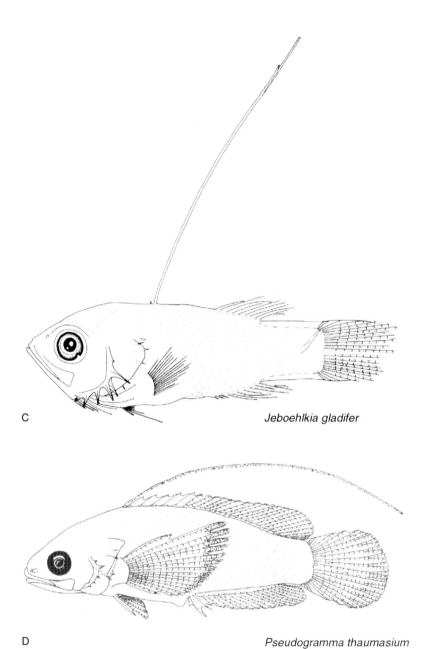

Figure 3.29 C-D. Larvae of representatives of the epinepheline tribe Grammistini. C. *Jeboehlkia gladifer*: 10.2 mm SL (from Baldwin and Johnson 1991). D. Pacific reef bass (*Pseudogramma thaumasium*): 13.3 mm SL (from Watson 1996).

pigment throughout development, although it is present on the pectoral fin and the elongated dorsal fin spines in at least some species (the elongated dorsal spines are broken in many specimens so their intact length and the presence of pigment cannot be determined). Several developmental features of *Jeboehlkia* indicate a relationship with *Pseudogramma* and paedomorphosis in this diminutive species (Baldwin and Johnson 1993).

CHAPTER 4

Ecology of Fish Eggs and Larvae

ECOLOGY DEFINED

THE POSITION OF FISH EGGS
AND LARVAE IN THE
ECOSYSTEM

EGG ECOLOGY
  Vertical Distribution
  Temperature
  Factors Affecting Egg Survival

FUNCTIONAL MORPHOLOGY
  OF LARVAE
  Generalities
  Functional Development

SPRING BLOOM
  Mechanism
  Importance to Larvae

FEEDING AND CONDITION
  The Feeding Mechanism
  Food Organisms
  Environmental Influences on Feeding
  Measuring Condition

GROWTH
  Measuring Growth
    Length Frequency Diagrams
    Otoliths

PREDATION
  Types and Taxa of Predators
  Modes of Predation
  Factors Influencing Rates
    of Predation
  Measuring Predation

ECOLOGY DEFINED

The relationships of plants and animals to their environment are complex. Organisms interact with many facets of their physical environment as well as with other plants and animals. Ecology is the study of these relationships and interactions. An ecosystem is a community of plants and animals that depend on each other and the environment of the area for their existence. Some ecosystems are small and clearly bounded (e.g., caves, hot springs) but some are large and have inexact (leaky) boundaries (e.g., the

Bering Sea). Most species spend their entire lives within the same ecosystem, but may interact differently with other species at various times during their lives.

THE POSITION OF FISH EGGS AND LARVAE IN THE ECOSYSTEM

We generally think of the ecological position of the adults of a species, but in animals, such as fishes with complex life histories, the position of the eggs, larvae, and juveniles may be very different from that of the adults (Figure 4.1). For example, eggs and larvae of most flatfishes are small (eggs 1–3 mm diameter, larvae < 30 mm SL) and planktonic, but the adults may be quite large (>100 cm SL) and are demersal. Larvae of flatfishes (pleuronectiforms) feed in the water column on planktonic crustacea (primarily young stages of copepods), whereas adults feed on a variety of demersal prey such as worms and bivalve molluscs. Likewise, the eggs and larvae of fishes are preyed upon by a different suite of predators than the adults. Eggs and larvae are consumed by larger planktonic crustaceans (e.g., euphausiids, amphipods) and fishes, including their own species in some cases. Although events during the planktonic phase may be very important to the recruitment of fishes, fish eggs and larvae are generally a minor component of the planktonic community. Fish larvae are rarely abundant enough to have an effect on the abundance of their prey, although low prey concentrations may limit larval growth and increase mortality.

EGG ECOLOGY

The characteristics and ecology of pelagic fish eggs are very different from those of demersal eggs. Very few pelagic eggs occur in fresh water. Most marine fishes spawn pelagic eggs. Most freshwater and some marine fishes lay demersal eggs. Pelagic eggs are usually broadcast singly into the water whereas most demersal eggs are laid in some sort of nest and are frequently guarded by one or both parents. Habits of demersal spawners sometimes result in extensive areas of large abundances of eggs produced by many fish spawning in the same suitable area. These areas of high egg abundance can attract predators, and environmental conditions within the areas can be detrimental to egg development. For example, whereas pelagic eggs reside in an oxygen-rich environment, low oxygen levels can be a problem for demersal eggs, especially those in the inner parts of large egg masses produced either by a single spawner or by the spawning of many fish in the same area (Messieh and Rosenthal 1989). Pelagic eggs are generally smaller than demersal eggs and pelagic eggs are usually transparent, or nearly so, whereas demersal eggs are often colored. Demersal eggs have much thicker and more resilient egg envelopes than pelagic eggs. Incubation time, which is temperature

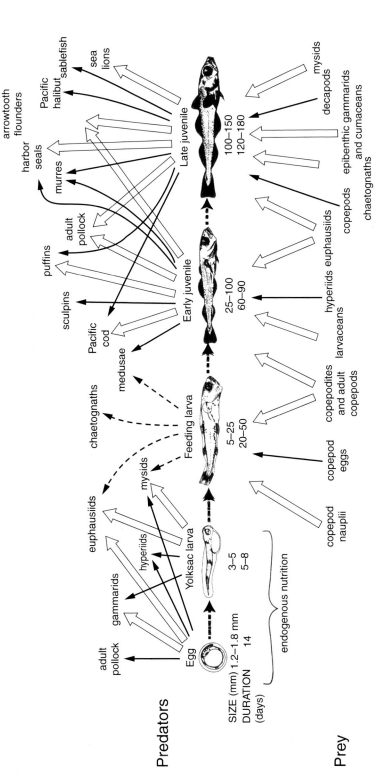

Figure 4.1. Predator-prey relationships of walleye pollock (*Theragra chalcogramma*) eggs, larvae, and juveniles in the Shelikof Strait region (from Kendall et al. 1996). Width of arrows reflects the relative importance of the taxon. Dashed arrows indicate pathways not yet demonstrated in pollock from Shelikof Strait, but shown in other populations.

dependent and related to egg size, is generally longer in demersal eggs than in pelagic eggs. Demersal eggs are subject to a different suite of predators (e.g., benthic invertebrates and fishes) than pelagic eggs (mainly planktonic invertebrates).

Vertical Distribution

Buoyancy of pelagic eggs affects their temperature of development and thus their time to hatch, their rate and direction of drift, the impact of pollutants and light, and the predators they are vulnerable to. Most pelagic eggs rise toward the surface from the depth at which they were spawned and float at a level in the water column where their specific gravity matches that of the surrounding water. (Specific gravity is the ratio of the density of a solid or liquid to the density of an equal volume of distilled water at 4°C. The specific gravity of seawater averages ~ 1.025.) The specific gravity of seawater increases with increasing salinity and decreases with increasing temperature. There are several characteristics of eggs which may affect their buoyancy. Mainly it is the result of the low-salinity water in the yolk and embryo (called the ovoplasm), which is produced by the ovary (Craik and Harvey 1987). This does not change much regardless if the female is held at high or low salinity. Low-density oil globules may play an additional role in flotation. The perivitelline space probably does **not** play a role in buoyancy since the fluid present there is the same density as the surrounding environment. In some species, however, active osmoregulatory mechanisms are capable of adjusting the specific gravity of the egg to the surrounding medium. The thickness of the envelope can also regulate buoyancy; the envelope usually accounts for ⅕ of the dry weight of an egg; this is an important distinction between pelagic and demersal eggs. Membrane elaborations (spines, microstructure, etc.) may also aid in flotation. The specific gravity of eggs increases as the embryo develops so the eggs sink deeper in the water column as they develop (Nissling and Vallin 1996). There are several ecological advantages to this pattern of egg vertical distribution: it puts the larvae into the productive surface waters rich in phytoplankton and zooplankton so that they can feed when they hatch, and surface currents may aid in transport to nursery areas or retention in areas important to the survival of the larvae when they hatch.

Pacific halibut (*Hippoglossus stenolepis*) provides an example of behavioral and species-specific considerations that can affect vertical location of eggs and larvae in the water column (see Thompson and Van Cleve 1936). Mature halibut concentrate on spawning grounds along the edge of the continental shelf at depths from 200 to 500 m from November to March. They are oviparous and the developing eggs are generally found at depths of 100 to 200 m, but can occur as deep as 500 m. The eggs and larvae are heavier than the surface seawater and drift passively in deep-ocean currents. As the larvae grow, their specific gravity decreases and they rise vertically and gradually move toward the continental shelf where surface currents carry them hundreds of kilometers, eventually to shallow inshore waters where they develop (settle as juveniles).

Temperature

Temperature influences incubation time. Within the usual range of temperature in which the eggs of a species develop, there is a linear inverse relationship between developmental time and temperature. *Tagesgrade* (Apstein 1909) is a term that accounts for this influence of temperature on incubation time. Tagesgrade, more commonly known as **Degree-Days,** is useful for rearing fishes at different temperatures (e.g., if you want to rear northern anchovies [*Engraulis mordax*] in Texas), and is commonly used in aquaculture, but seldom in field studies.

Degree-Days = (number of °C above biological zero) × (number of days to reach a given developmental stage).

The term *biological zero* was meant to be the temperature at which embryonic development ceases, but this is a misnomer since for many fishes the biological zero is less than 0°C. The time-temperature relationship is really exponential, but for purposes of the degree-day concept it is a linear relationship within the normal temperature range that eggs encounter in their environment (e.g., in plaice [*Pleuronectes platessa*] degree-days to hatching = 165 degree-days, with biological zero = -2.4°C).

Factors Affecting Egg Survival

Many factors affect egg survival. Lowered salinity usually results in prolongation of incubation in marine fishes, which in turn may increase their mortality. The temperature tolerance range for eggs is smaller than that for larvae and adults. Thus eggs may be the stage in the life history of a fish that is most important in determining the geographic range of a species (Alderdice and Forrester 1971). The temperatures tolerated by eggs of different species decrease in range and absolute values with increasing latitude (i.e., eggs of tropical species tolerate higher and more diverse temperatures than eggs of arctic species). Although adults of temperate species have a great range of temperatures tolerated, eggs do not. The further along the embryonic development, the more temperature tolerant the egg becomes, perhaps in response to the fact that temperatures normally increase as spawning season progresses. As might be expected, in some species fluctuating temperatures yield better hatching rates than constant temperature—this is especially true for shallow-water species and also those subject to tidal effects.

There are also combined effects of temperature and salinity on egg survival. Alderdice and Forrester (1968) did an elegant study on the combined effects of salinity and temperature on English sole (*Parophrys vetulus*) eggs. The optimum combination of temperature and salinity for hatching success is not necessarily the same optimum for larval survival because the early life-history strategy for a subsequent stage may be adapted to an entirely different physical regime.

Oxygen consumption of fish eggs increases about 10 to 20 times after fertilization. This increase is not smooth: it is rapid until gastrulation, then it evens out until heartbeat begins, when the increase is again rapid, and it is even more rapid after the embryo begins to move. Oxygen uptake is temperature dependent. The initial source of oxygen is the yolk and perivitelline fluid. Oxygen diffuses into the eggs from the environment. Reduced oxygen retards development and causes premature hatching (e.g., Blaxter 1969). Oxygen is usually abundant in surface waters unless upwelling brings water low in oxygen to the surface. Low oxygen can be a problem in demersal eggs. Eggs at the bottom of the egg mass may die. Pacific herring (*Clupea pallasi*) lay eggs on algae which moves with the current, increasing oxygen supply. Some nest guarders circulate water on nest. Most demersal eggs are laid in high-current regions.

Mechanical factors may also influence egg mortality. The sea surface is very dynamic during rough weather and pelagic eggs can get jostled (Coombs et al. 1990). Commercial trawling gear can disturb demersal spawning grounds (e.g., capelin [*Mallotus villosus*] and Atlantic herring [*Clupea harengus*] can suffer 5–10% mortality). Two effects of such disturbance are that the envelope can be ruptured (mechanical resistance), and hatching success of the embryos can be reduced (viability). The severity of this damage varies with stage of embryonic development. During early stages eggs are extremely vulnerable. In middle stages mechanical resistance is high, but viability is low. In late stages both mechanical resistance and viability are high. Just prior to hatching there is low mechanical resistance and high viability.

Light and solar radiation can affect egg survival. Most ultraviolet (short wavelength) and infrared (long wavelength) light is absorbed in the first few (about 10 m) meters of water so that eggs occurring deep are not affected by these wavelengths. Pelagic and neustonic eggs are most subject to these wavelengths and they are generally transparent. Their transparency allows infrared light to pass through them. Transparency also protects eggs against predation although many planktivores that eat eggs are not visual feeders (thus protection against predation may be of secondary importance). The primary importance is that transparency may protect the eggs against heat radiation, which will pass through them. Demersal eggs are often sensitive to visible light. They are protected when they are deposited in deeper waters or in crevices where light levels are low. Pigment in the eggs can absorb some wavelengths and reflect others, which may be one reason that demersal eggs are pigmented. Ultraviolet light reduces hatching rates. However, since most ultraviolet light is absorbed within 10 cm of the surface, this effect is only important in eggs residing very near the surface. In trout, visible light destroys lactoflavin (which is involved in the production of respiration enzymes). This in turn retards development and causes early hatching. Blue light was found to be the most destructive. X rays and other shortwave radiation also cause damage.

The effects of disease on fish eggs in the wild are not well studied, although it is obviously important in demersal eggs such as Pacific herring, which when overspawned on either natural or artificial substrate, frequently begin dying of oxygen deprivation and then are entirely engulfed by fungus. Disease is probably very common in the eggs (and larvae) of many fishes, and is probably underestimated substantially as a mortality factor, similar to the way disease has often been badly underestimated as a mortality factor in juvenile fishes (e.g., Angell et al. 1975).

The effects of predation on eggs have received more emphasis recently. There are probably a whole suite of egg predators. However, eggs are often hard to identify in guts because they digest rapidly. Egg predation is probably variable from year to year. Palsson (1984) looked at predation on demersal eggs of Pacific herring using quadrants and predator exclusion cages. He found that daily egg loss rates ranged from 16.9 to 51.8% and differed among cohorts. There was a positive correlation between egg loss rates and cohort density—density decreased exponentially. Based on predator exclusion experiments, Palsson found that 20 to 50% of the egg loss rate was due to bird predation. Predator inclusion experiments were done with snails and amphipods, which consumed 5.6 and 8.0 eggs/day, respectively. *In situ* snail densities combined with potential egg consumption rates equaled or exceeded initial cohort egg density values at some spawning sites. Pelagic eggs are extremely susceptible to nonvisual predators such as ctenophores and medusans, which can be excessive at times.

FUNCTIONAL MORPHOLOGY OF LARVAE

Generalities

The functional morphology of larvae must be considered when thinking about their ecology. Their developing motor and sensory capabilities determine how they can react to their environment. Locomotion of larvae is quite different from larger fish. Small animals, such as fish larvae, cannot glide in the water (Figure 4.2). In effect, the water is much more viscous to these small animals (Müller and Videler 1996). Larvae generally hatch with a continuous median finfold, instead of separate dorsal, caudal, and anal fins. The pectoral fins are usually first to develop, then dorsal and caudal, finally the pelvics and anal fin. The median finfold and pectoral fins do not have fin rays

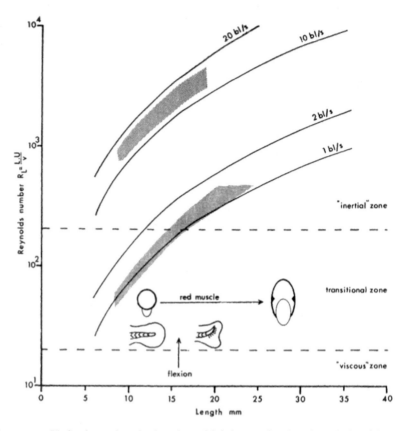

Figure 4.2. Hydrodynamics of swimming of fish larvae showing the relationship between body size and development and the fluid characteristics of water (adapted from Blaxter 1988).

initially—these develop later. A subdermal space is present in pelagic marine larvae, but not in freshwater larvae. The subdermal space separates the skin from the body and is filled with a gelatinous liquid of low specific gravity. It transports gases and yolk matter, functions in osmoregulation, and provides buoyancy and mechanical protection. Originally the gut is suspended under the larval body and is fixed only by mesenteries and the integument within the subdermal space. During larval development metameric muscles gradually enclose the intestine. Also the notochord decreases in size and pigment patterns develop, which are useful in larval identification, but of unknown function for the larvae. Myomeres are present in larvae at hatching and vertebrae develop later on. The number of myomeres closely approximates the number of vertebrae, and counts of both are useful in larval identifications.

Functional Development

Young larvae have no functioning gut, gills, or kidneys so osmoregulation occurs differently than it does in adults (which drink seawater, absorb water and some salts in the gut, and excrete salt through special glands, mainly at the gills and partly in the kidneys). In larvae, osmoregulatory organs occur in the skin (which is otherwise impermeable to salts but not gases), and the subdermal space is also important. Atlantic herring and plaice larvae show a higher salinity tolerance than juvenile and adult fish. Our present knowledge on the water metabolism of fish larvae is based on a very limited number of experiments with very few species.

Buoyancy in marine fish larvae seems to be maintained mainly through the low specific gravity of the fluids of the yolk sac and subdermal space. After the yolk has been absorbed, the subdermal space functions to provide buoyancy because the osmotic pressure in the subdermal space is less than in the environment. The extended dorsal or lateral fins of some species have been interpreted as means for increased buoyancy. However, their protective function against predation seems more likely. Lack of bone reduces specific gravity during early stages. However, in many cases, fish larvae seem to be slightly negatively buoyant after the first days following hatching and have to exert some swimming effort to remain at a given depth in the water. Development of the swimbladder of pelagic larvae starts at an early stage of larval development, but it does not function in Atlantic herring larvae before about 30 mm in total length (when the subdermal space has disappeared). Some flatfish larvae develop a swimbladder whereas it is absent in all adult flatfishes. Dead and sick larvae sink even at early swimbladder stages. This indicates that either swimming activity or osmoregulatory work is required or that the seclusion of the internal body fluid of low specific gravity is maintained by living tissues only.

Although eyes begin developing early in the embryo, in most species, they presumably are nonfunctioning at hatching as they are not pigmented (although even closely related species may differ in the state of visual development at hatching). However, differentiation of the eye and the optic nerve and development of visual pigments is an extremely rapid process. Within two days after hatching, the sardine develops a pigmented retina and visual cells (G. Hempel personal communication). At the third day, visual control of feeding starts, although at this stage the optic nerve has still only 40 individual fibers, thus the image is very coarse, and visual acuity is low compared with adults. In general, the early larval retina contains only cones (i.e., light adapted). In marine fish larvae studied so far, rods (dark adapted) appear in the retina only at metamorphosis, together with the ability of light and dark adaptation. The visual parameters for feeding are (1) acuity, (2) distance of perception, and (3) field of vision. The larval eye is so small and the focal length of the lens so short that even the dense packing of the small cones cannot provide the high acuity found in older fish. Feeding of plaice and Atlantic herring larvae declined in the range of light intensities encountered at the sea surface at dusk and in full moonlight. In this respect the larvae were similar to the adults.

Most larvae have free neuromasts or sensory cells on the skin that are able to sense water vibrations, and help larvae avoid predators before eyes are well developed. The neuromasts later develop into the lateral line system.

Counts of vertebrae, myomeres, scales, gill rakers, and fin rays are often used in subpopulation studies. All of these features, except scales (and possibly gill rakers) are laid down during embryonic and larval development. By the end of the larval period, the fish has all the vertebrae and fin rays it will ever have. The complete set of vertebrae is generally laid down before the complete set of fin rays is. Within limits imposed by genetics, the number of fin rays and vertebrae in a species is influenced by temperature, salinity, and oxygen (but primarily by temperature). At high temperatures fewer meristic elements are laid down than at low temperatures. The temperature effect is greatest if the larva is in the salinity for which it is adapted. Different stages of development are more sensitive than others to environmental influences. For vertebrates, this occurs at gastrulation and just before the last vertebrae are normally laid down. For some species, vertebral counts are completed in the egg, in others in the larva. Fin ray counts may be determined much later (as the complete set is not laid down until transformation). Growth and differentiation are affected by temperature in different ways. If differentiation is slow, more tissue is available to be differentiated. So, cold temperatures may slow the time of differentiation of meristic characters more than that of growth—thus, giving more meristic characters. Conversely, in warm temperatures, the time of differentiation may be earlier, giving fewer meristic characters.

Taning (1952) found the number of vertebrae in brown trout (*Salmo trutta*) to be influenced by incubation temperature during an early period well before any development of the vertebral column. Since the late 1920s, vertebral counts together with other meristic characteristics have been a key element in subpopulation studies in various species. But vertebrae develop relatively late in Atlantic herring (*Clupea harengus*) (e.g., at a stage which is difficult to reach in rearing experiments). However, myomere counts are very closely related to number of vertebrae, and myomeres can be counted fairly reliably already at the yolk sac stage (usually using polarized light).

SPRING BLOOM

Mechanism

The production of food for most larval fishes, except in the tropics, follows an annual cycle that is tied to the spring phytoplankton bloom. The spring bloom is a rapid increase in the growth of phytoplankton that occurs in late winter or early spring. The spring bloom is initiated by the increase in daylength, which causes warming of surface waters. This is accompanied by a decrease in storm activity, which along with the warming, decreases the vertical mixing of the upper water column and eventually leads to the development of a thermocline (an area in the water column where there is a sharp decrease in temperature). Phytoplankton is able to reproduce quickly above the thermocline, in this well-lit, stable upper mixed layer of the water column, which has an abundant supply of nutrients to promote plant growth. Eventually the phytoplankton depletes the nutrients and growth slows.

Importance to Larvae

The spring bloom provides the quantities of the phytoplankton food required for reproduction of copepods and other zooplankton, which in turn are the food for most larval fishes. Thus reproduction of many fishes is tied to the timing of the spring bloom. Interannual variations in the abundance or timing of production of phytoplankton and zooplankton may cause the differences in the amount of food available to larval fishes. This may affect their growth rates and survival.

FEEDING AND CONDITION

Feeding has probably been the most studied facet of the ecology of larval fishes. Since the time of Hjort (1914), it has been thought that food limitation might cause increased mortality of larvae through starvation. Many field studies have examined larval gut contents and compared them to the kinds and amounts of food collected concurrently. Since the early 1960s, laboratory experiments have been performed on food and feeding of larval fishes using a variety of parameters (e.g., food quantity and type, light, size of larvae, previous feeding situation). Many of these studies have relied on technological advancements in sampling, rearing, and analytical procedures. As a result of these studies we know that in some cases larval fish growth in the sea is limited by food, and mortality may be affected either directly by starvation or indirectly by reducing the ability of larvae to avoid predation, or by increasing the time interval that the larvae are vulnerable to predation or to being swept by currents to unfavorable areas.

The Feeding Mechanism

At hatching, most larvae from pelagic eggs still have a large yolk sac and are not yet ready to feed. They live off nourishment provided by the yolk for a few hours to a few days, depending on the species and temperature, after which they must eat to survive. During the yolk sac period, the larva develops the equipment needed to feed: the eyes become pigmented, the mouth opens and the gut forms, and locomotor capabilities increase. The period immediately after the yolk is exhausted and exogenous nutrition needs to commence is fairly brief, and may be a period of increased mortality.

Laboratory experiments have determined that in some species there is a point of no return following the yolk sac period: if the larva does not eat during this period, it cannot survive, even if offered food.

At first feeding the gut of most larval fishes is a straight or coiled tube. Sections gradually develop in the gut: the foregut, midgut, and hindgut (Govoni et al. 1986). Prey are swallowed directly into the foregut, they are digested in the midgut, and material is absorbed in the hindgut. Larval fishes are generally visual feeders (Paul 1983). Thus they usually feed only during the day in the upper layers of the water (Munk et al. 1989). They feed on individual particles, most frequently the naupliar stages of copepods. The feeding mechanism involves swimming and visually searching for and locating a prey organism (Figures 4.3–4.5). Once a prey is located the larva ceases most forward swimming and positions itself near the prey, following the prey and maintaining this position, using its pectoral fins and median finfold. The larva then forms its body into an S shape and quickly strikes forward at the prey (Figure 4.6A, B). Prey are generally swallowed whole.

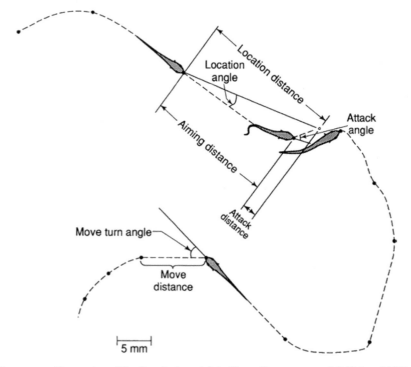

Figure 4.3. Dynamics of feeding in larval fish (from Browman and O'Brien 1992). A typical search path and attack sequence for white crappie (*Pomoxis annularis*) larvae, sketched directly off of a television screen, illustrating the measurements drawn from it. The solid dots along the dashed line (the path of motion) represent stationary pauses; the open dot a prey item.

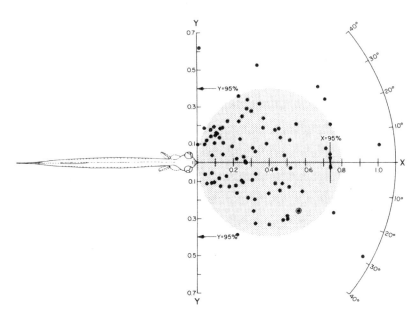

Figure 4.4. Horizontal perceptive field of northern anchovy (*Engraulis mordax*) larvae (from Hunter 1972). Each point represents position of a prey in the horizontal plane at the time larvae first reacted to it. Distances in X and Y axes were divided by the lengths of the larvae and expressed as proportions of larval length. Arrows indicate lines that would enclose 95% of prey sighted in each plane.

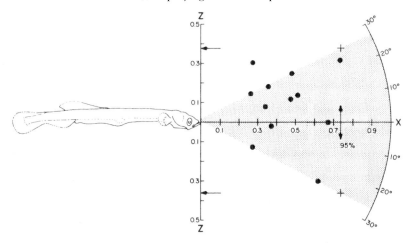

Figure 4.5. Vertical perceptive field of northern anchovy (*Engraulis mordax*) larvae (from Hunter 1972). Distances on X and Z axes are expressed as a proportion of larvae length; points are position of prey when larvae first reacted to them; crosses indicate point of intersection of visual cone with Y = 95% (the 95% limit of prey distribution in Y given in Figure 4.4); and arrows on Z indicate projected values of Z for intersection points.

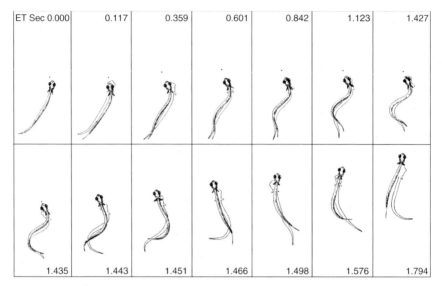

Figure 4.6A. Feeding sequence of a northern anchovy (*Engraulis mordax*) larva. Tracings of selected motion picture frames of an anchovy larva (8.9 mm, 21 days old) and a prey taken at 128 frames per second (from Hunter 1972). Frame lines are fixed such that distance moved by larva is indicated by comparison of tracings; and unshaded image indicates position of larva in the preceding tracing. The following events are illustrated: elapsed time (ET) 0 sec, larva sights prey; ET 0–0.117 sec, larva orients head toward prey and swims toward it; ET 0.117–1.435 sec, larva forms S-shaped strike posture; ET 1.435–1.443 sec, strike begins and prey captured; and ET 1.451–1.794 sec, forward movement continues as tail returns to axis of progression (from Hunter 1972).

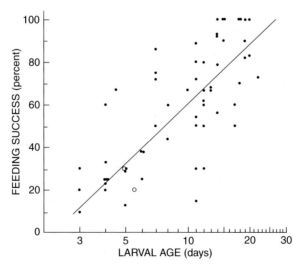

Figure 4.6B. Feeding success (percent of prey captured) of northern anchovy (*Engraulis mordax*) larvae of various ages fed *Brachionus*. Larval age is plotted on log scale, equation for line is percent success = 93.2 (log age) − 33.30. Two open circles, larvae fed *Gymnodinium* (from Hunter 1972).

Food Organisms

Larvae eat a variety of prey organisms (Figure 4.7). The gape of the mouth determines the maximum size of the prey, which increases as the larvae grow (Figure 4.8). In some species the minimum size of prey also increases with the size of the larvae, whereas the larvae of other species continue to eat small prey as they grow. Energetic costs associated with eating small prey may be greater than those of eating larger prey, so even if larvae can eat small prey it may be better for them to eat larger prey. Larvae appear to be selective feeders, and often the prey in the guts is not proportional to the organisms of the same size in the water where the larvae were collected (Checkley 1982). Protozoans and phytoplankton are found in the guts of some small

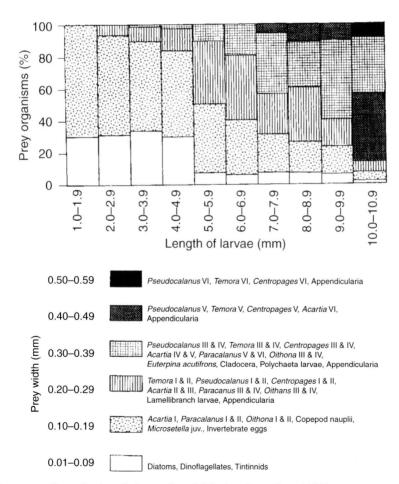

Figure 4.7. Prey size in relation to larval fish size (from Last 1980).

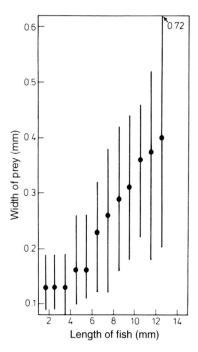

Figure 4.8. Relationship between prey width and larval length (from Kalmer 1992). Based on 8,902 larvae of 20 marine fish species. Points, mean of all species; vertical lines, range between smallest and largest means of single species at any length group (from Table 4 in Last 1980).

larvae, but most commonly planktonic crustacean larvae are consumed (Stoecker and Govoni 1984). Copepod nauplii are the most frequently found food item, and it is becoming possible to identify the species of copepod based on their nauplii. Copepod eggs are another frequent prey item, and although some eggs might be ingested along with egg-bearing adult copepods, eggs of some copepods are shed into the water, and the larvae might eat these directly. Although ingested, copepod eggs may not be nutritious, and may even be viable after passing through the gut (Conway et al. 1994). The consumption of taxa changes somewhat as the larvae grow (Economou 1991). Some of this change may be due to the size of the various taxa of prey (e.g., larvae of euphausiids are larger than larvae of copepods, and they enter the diet of larger larvae).

Environmental Influences on Feeding

Several features of the environment influence feeding of larval fishes. Since larvae are visual feeders, light has an important role in feeding. Larvae are diurnal feeders. The number of food organisms in their gut normally increases in the morning and stays rather constant during the day, but decreases at night. The state of digestion of prey increases during the day, and at night only partially and well-digested prey are present. There is some indication that larvae adjust their vertical position in the water column to reach and stay at light levels that permit feeding. Thus, in the morning as daylight begins they rise in the water column, at midday they sink somewhat, and in the evening as daylight wanes they rise again. On cloudy days they may be higher in the water column than on clear days. The prey organisms of larvae are frequently distributed in patches, and there is evidence that larvae adjust their vertical position and swimming to remain in food patches. Considerable recent research has involved the effects of turbulence on larval feeding (Lough and Mountain 1996). Turbulence in the upper layers is caused by wind and near the thermocline and bottom it can be caused by current shear and tides. Most research has concerned the effects of winds on larval feeding. Since turbulence would dissipate patches of food organisms, it would decrease their density and make it more difficult for larvae to find enough of them to satisfy their needs. However, it has been shown that in moderate turbulence, the encounter rate of food organisms would actually increase. At high levels of turbulence, feeding activity may be disrupted.

Measuring Condition

Since feeding is thought to be important to larval survival, and gut contents only relate to feeding conditions immediately prior to collection, considerable research has been directed at determining larval condition on a longer time scale. Combinations of laboratory experiments to develop and calibrate condition indices under various feeding regimes and application of these indices to field-collected larvae have been performed. Length-weight relationships of larvae have been used, but measurements are tedious and subject to substantial errors. Examining the histology of larvae has also been used and proven to be a very sensitive measure of condition (Theilacker and Porter 1995). Various tissues show characteristic changes during starvation. Based on laboratory experiments, the feeding conditions of field-collected larvae during a few days prior to collection can be surmised. Another measurement that is sensitive to larval feeding condition is the RNA/DNA ratio of the larvae (Buckley 1984; Rooker and Holt 1996). The amount of DNA in each cell is constant, whereas the amount of RNA reflects the amount of protein synthesis that is occurring. Thus, high RNA/DNA ratios indicate good conditions for growth, whereas low RNA/DNA ratios indicate poor growth. A refinement of the RNA/DNA ratio is provided by flow cytometry (Theilacker and Shen 1993). In this technique, cells of certain organs (e.g., brain) are dissociated and the amount of RNA and DNA in each cell is determined. From this the proportion of cells that are actively dividing can be determined, and this is a measure of the feeding conditions experienced by the larvae.

GROWTH

Although the development rate of fish eggs and yolk sac larvae is largely determined by temperature, once the larvae start to feed their growth rate can be variable depending on the amount and kinds of food eaten. There is a rapid growth phase after hatching during the yolk sac stage: nourishment in the form of yolk is plentiful, oxygen is plentiful, and the larva is not constricted by the egg envelope. During this period it is important for the larva to become as large as possible to begin feeding. There is a slow growth phase near completion of yolk resorption since the nutrient supply is becoming exhausted. Larvae may actually shrink after the yolk is resorbed (particularly if food is not immediately available). There is an increase in growth rate after feeding starts. At this point, food supply and temperature are the most important factors controlling the growth rate. The thyroid gland may also be important in controlling growth.

Measuring Growth

In order to measure growth rate, the age of the larvae must be determined. The time to reach the various morphological stages of development varies between species and occurs at different times. Within limits, the stage of development is size dependent, not age dependent. Thus, the stage of development cannot be used to estimate the age of a larva accurately.

Length Frequency Diagrams. Until the mid-1970s, length frequency diagrams were all that were available to estimate larval age (e.g., Graham et al. 1972). Lengths of larvae taken at two or more sampling times were compared. The length frequency diagrams often contained several length modes. The progression of these modes between diagrams from successive sampling times was taken to reflect growth of larvae between the samplings. However, there are a number of problems with this procedure that can lead to erroneous results. For example, it is assumed that the same population is sampled each time, and that there is no movement of larvae in or out of the area sampled. It also assumes that all lengths are completely vulnerable to the sampling gear, but if small larvae are extruded through the nets or larger larvae avoid the gear, biases will be created in the lengths of the larvae collected.

Otoliths. In the mid-1970s, a new technique was discovered to allow direct determination of the age of individual larvae (Brothers et al. 1976).

It was found that daily growth rings were laid down in the otoliths (ear bones) of larvae (Campana and Neilson 1985; Jones 1986). Otoliths are formed in the egg and daily rings are laid down sometime from shortly before hatching to the first feeding stage, depending on the species. The age when the first increment is laid down and the daily nature of deposition should be determined for each species under study through laboratory experiments. By measuring the length of the larvae, and the number of days old they are, a growth rate of the sample can be calculated. By knowing the age of the larva, the date of spawning can be estimated. Given rates of drift, and the age of the larvae, the spawning area can be approximated. In some species (e.g., English sole) otolith increments are not discernible, so growth must be estimated from successive length frequency modes (Shi et al. 1996).

PREDATION

Types and Taxa of Predators

Predation is another source of mortality of fish eggs and larvae, but it has not been studied as thoroughly as nutrition. Demersal eggs, which are usually in clusters and may be guarded by the adults, are subject to a wide variety of predators including invertebrates such as snails and amphipods and other fishes (Frank and Leggett 1982). Even mammals and birds can consume eggs that are in shallow water or exposed to the air at low tides. Planktonic eggs and larvae are preyed upon by a host of planktonic predators, such as jellyfishes (Purcell 1985), chaetognaths, and crustaceans, as well as adult fishes (Daan et al. 1985). Cannibalism is a frequent source of predation for planktonic eggs (Brodeur et al. 1991) and larvae (Brownell 1985; Fortier and Villeneuve 1996) of a number of fishes.

Modes of Predation

Eggs and larvae of fishes are vulnerable to several modes of predation (Figure 4.9). Since the locomotor abilities of eggs are nonexistent, and those of larvae are limited, they have little ability to escape predators, once the predators locate them. Most planktonic fish eggs and larvae are quite transparent, which protects them against predators that locate prey by sight. Startle responses develop early in fish larvae, whereby when they sense nearby water movement, they quickly swim a few body lengths, enabling them to escape predators that only sense prey in the near field or have limited locomotor capabilities themselves (e.g., jellyfishes). Some predators of fish eggs and larvae seek and eat individual particles, however others feed by filtering particles out of the water.

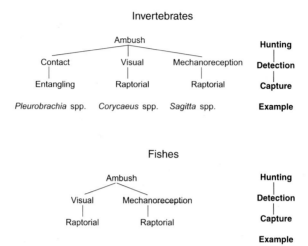

Figure 4.9. Modes of predation on larval fishes (adapted from Bailey and Houde 1989: Figure 2).

Factors Influencing Rates of Predation

Several factors influence rates of predation on fish eggs and larvae. Planktonic fish eggs and larvae are usually rare, so probably few predators rely on them as a primary food source. However, demersal egg masses may provide an attractive food source for predators. As larvae approach the juvenile stage and start to school or settle to the bottom, predation may increase dramatically. It is thought that predation rates are highly size dependent (Bailey 1984), so faster growing larvae are subject to high predation for a shorter period than slow growing larvae. Larvae in poor nutritional state are less able to avoid predation than well-fed larvae. Larvae may avoid predation by seeking levels in the water column that are different from those occupied by their predator. Larvae of some species migrate to deeper levels during the day where there is less light, presumably to reduce vulnerability to predators that locate prey by sight.

Measuring Predation

The study of predation in fish eggs and larvae has been hampered by inability to identify their remains in the guts of potential predators. Many of these predators macerate prey as they ingest it, and the digestive process is rapid. Melanistic pigment in the guts of predators has been used to imply consumption of fish larvae. Otoliths in guts last longer than soft tissue and have been used to imply predation. Biochemical means have been developed to recognize the presence of the remains of fish eggs and larvae in the guts of

Ecology of Fish Eggs and Larvae   145

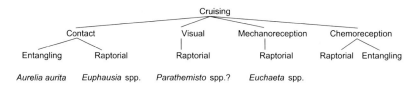

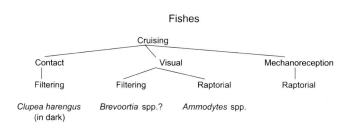

predators (Theilacker et al. 1986). Antibodies are developed against proteins in the yolk or muscle tissue of the fish. Samples of gut contents of potential predators are tested against these antibodies, and positive results indicate that the predator has eaten eggs or larvae of the fish from which the antibodies were developed. It is difficult to determine how many eggs or larvae were consumed, and how long ago they were eaten. Laboratory experiments must be conducted to determine digestion rates and residence time of prey in the guts.

CHAPTER 5

Sampling Fish Eggs and Larvae

PURPOSES
Time and Space Scales
Types of Studies
Uses of Studies

METHODS
Survey Design
Platforms (Ships)
Sampling Gear
  Bongo Nets
  Tucker Plankton Trawls
  Neuston Nets
  Methot Trawl
  Serial Net Samplers
  CalVET
  Continuous Plankton
    Recorders
  Plankton Pumps
  Continuous Underway Fish Egg
    Sampler
  High-Speed Plankton
    Samplers
  Tidal and Channel Nets
  Night Lights
  Light Traps
  Sleds
  Indirect Acoustic and Optical

Ancillary Gear
  Flowmeters
  Depth Determination
  Environmental Data

CHOOSING SAMPLING
METHODS

SHIPBOARD PROCEDURES
Sample Collecting
Sample Preservation
Data Records

SHORESIDE SAMPLE
PROCESSING

DATA PROCESSING
Data Analysis
Variability in Ichthyoplankton
  Data

REPORTING RESULTS
OF SAMPLING
Distribution and Abundance
Vertical Distribution
Spawning Biomass
  Determination
Mortality Estimation
Assemblage Analysis

PURPOSES

Most marine fish eggs and larvae are planktonic, that is they drift with the currents in the ocean with little or no control of their horizontal distribution. Together, these fish eggs and larvae are referred to as **ichthyoplankton.** They can be sampled with gear that has been designed primarily for collecting larger invertebrate zooplankton. This gear is usually a frame covered with mesh to filter the plankton from the water. Since fish eggs and larvae are usually much less abundant than invertebrate zooplankton, relatively large volumes of water need to be filtered to provide adequate samples of ichthyoplankton. Eggs of most freshwater fishes, and some marine fishes, are demersal and some are laid in nests and guarded by one or both parents. The larvae of freshwater fishes may be somewhat planktonic, but freshwater fishes generally do not use this stage for dispersal to the degree that many marine fishes do. Demersal eggs require different sampling than required for pelagic eggs and larvae. Chapter 5 discusses gear and methods for making field collections of planktonic fish eggs and larvae. It also introduces treatment and analysis of ichthyoplankton samples once they are collected.

Several reviews of sampling for fish eggs and larvae have been published (Smith and Richardson 1977; Heath 1992; Gunderson 1993; Kelso and Rutherford 1996) and these should be consulted for additional details. Many articles on sampling fish eggs and larvae in fresh water have been published over the years in *Progressive Fish Culturist* (e.g., Hermes et al. 1984). Also, UNESCO has published two monographs that include sections relevant to sampling fish eggs and larvae (Trantor and Fraser [eds.] 1968; Steedman [ed.] 1976).

Time and Space Scales

Biological processes and variability in numbers and sizes of planktonic organisms, including fish eggs and larvae, occur over a wide range of time and space scales (Figure 5.1). Time scales vary from evolutionary (10^5–10^6 years), to less than a second for biological processes such as individual feeding events. Space scales vary from global (climate variability) to the ambit of individual larvae (a few centimeters). Field sampling of fish eggs and larvae can be used to investigate processes occurring on a subset of the total range of these time and space scales (Figure 5.2). Plankton tows, which are often used in such investigations, generally filter a few hundred cubic meters of water, and require less than an hour to complete. Shipboard cruises to study these processes in the ocean can perform up to about 20 such tows at intervals of about 15 km a day. These factors establish the lower limit of the time and space scale that can be sampled by net tows, and the maximum area that can be sampled synoptically in time, by a single ship. Inshore and freshwater studies are usually conducted at much smaller spatial scales.

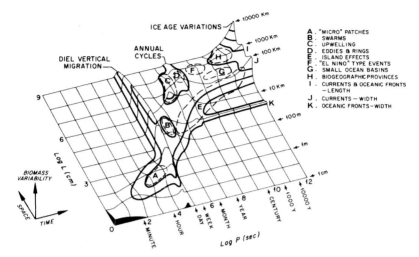

Figure 5.1. Time and space scales of processes in the ocean illustrated by the Stommel Diagram, a conceptual model of the time-space scales of zooplankton biomass variability and the factors contributing to these scales. I, J, and K are bands centered about 1000s, 100s, and several kilometers in space scales, with time variations between weeks and geological time scales (from Haury et al. 1978).

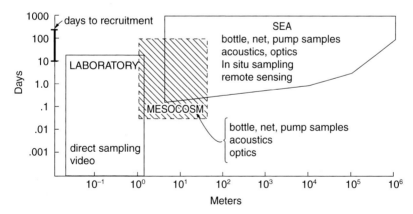

Figure 5.2. Time (ordinate) and space (abscissa) scales of variability within which early life-history research can be pursued in the laboratory, in mesocosm enclosures and through sampling at sea (from Houde 1987). The heavy line along the ordinate represents the possible duration of the recruitment period for teleost fishes. Probable sampling methods in each system are noted.

Types of Studies

Given the temporal and space scales amenable to shipboard plankton sampling for fish eggs and larvae, several types of study are possible (see Lasker 1987; Heath 1992). Time series of egg and larval collections are available in a few parts of the world (e.g., California Current [Moser et al. 1993, 1994] and coastal regions of the Northeast [Berrien and Sibunka 1999] and Northwest Atlantic [Russell 1973]) to investigate decadal scale changes in faunal composition and abundance, which may be related to climatic changes in the environment such as "regime shifts." The spatial scales required to address such problems have been called large marine ecosystems and extend over hundreds of kilometers. Many field studies involving early life-history stages of fishes are related to annual reproductive cycles of particular populations. These studies require sampling on time scales that encompass the spawning and larval period of the population of interest (maybe several months of the year) at space scales that cover the entire distribution of the eggs and larvae (100 to 1,000 km^2). Within the reproductive season of a population, studies on drift, feeding, and mortality may require sampling at time scales from weeks to days, and space scales of less than 10 km. Much variability occurs on a diel basis, and sampling to investigate these processes requires sampling at intervals of a few hours, and at horizontal space scales of a few hundred meters. The vertical scale of such sampling may be less than 10 m and require discrete depth samples from several depths at once.

These types of early life-history studies require collecting samples of pelagic fish eggs and larvae. Pelagic fish eggs and larvae are collectively called ichthyoplankton, and they are usually sampled with nets that filter them out of the water. The nets must be designed and deployed in a manner to minimize the eggs and larvae from passing through the meshes (being extruded), or avoiding the sampler altogether (escapement). Species that have demersal eggs must be sampled with very different gear such as sleds and bottom grabs. Eggs that are laid in nests are particularly difficult to sample. Clearly, the life-history patterns of the species of interest, the purposes of the study, and the habitat studied all influence the sampling plan.

Uses of Studies

Field studies of fish eggs and larvae are conducted to investigate the fauna of an area (Ahlstrom 1972; Russell 1973; Moser et al. 1974), as part of an ecosystem study (Watson et al. 1999), to understand recruitment mechanisms of a species (Houde 1987; Koslow et al.1985), to measure the impact of pollution or other habitat factors (see Fuiman [ed.] 1993; Dempsey 1988), and to assess the abundance of a population (Cunha et al. 1992; Smith and Morse 1993).

Ichthyoplankton sampling is an effective means of determining what species inhabit an area because plankton nets are less selective than most gear used to sample adult fishes. Although fishes vary widely in size, and the habitats that they occupy as adults, most species have planktonic larvae. Some small, cryptic bottom fishes are rarely sampled as adults, but their larvae are common in plankton collections. Many features of the life history of species must be considered when relating the abundance of larvae in an area to the abundance of adults.

As was discussed in more detail in Chapter 4, fish eggs and larvae are part of an ecosystem. They prey on some animals and serve as prey for others. Although fish larvae are usually too rare to influence the abundance of their prey, their time and area of occurrence may coincide with peaks in the production cycle of their prey (match/mismatch theory [Cushing 1990]). Larval survival may in part be dependent on being in an area where their predators are scarce. Thus, sampling programs are sometimes designed to understand the place of the eggs and larvae of particular species in the ecosystem they inhabit. Such programs usually target sampling on a variety of animals besides the fish species of interest. This requires several gears, because the prey and predators are usually not the same size, mobility, and abundance as the fish eggs and larvae.

Most of the uncertainty involving managing economically important fishes relates to differences in levels of recruitment from year to year (Rothschild 1986). In order to manage fisheries conservatively, it must be assumed that recruitment will probably be weak. If it were known in advance that recruitment would be stronger in a particular year, it could be possible to allow more of the stock to be harvested, without endangering the stock in the future. Thus there is considerable economic impetus to develop a better understanding of the recruitment processes of stocks that are heavily fished. Such an understanding requires sampling of eggs and larvae, since changes in their survival are generally thought to be very important in determining year-class strength. Timing, spatial differences, extent, and causes of mortality can be sought through ichthyoplankton collections.

Many fishes spawn and/or their young stages occur in coastal estuarine, or freshwater areas that are subject to degradation by human activities. Also, the young stages of fishes are more susceptible than the adults to the deleterious effect of pollutants and other forms of habitat modification. In order to determine the impact of these factors on survival of fish eggs and larvae, they must be sampled in relation to the source of the problem. Bioassays using fish eggs and larvae may also be required to measure the effect of the particular pollutant.

One of the most common reasons for sampling ichthyoplankton is to provide an independent estimate of population size. Population size of exploited fish populations is estimated from the harvested fish, and from fishery-independent surveys of the adults. Another means of estimating population size involves making quantitative estimates of the numbers of eggs or larvae produced by conducting ichthyoplankton surveys.

The number of eggs collected is then related to the size of the adult population by determining the fecundity, size distribution, and sex ratio of a sample of the adults, and accounting for mortality experienced by the eggs and larvae between the time they were spawned and when they were sampled.

METHODS
Survey Design

When planning a field study of fish eggs and larvae, many factors need to be considered. The purposes of the study, the ship and time available, and the area to be studied will in large part determine the survey design and sampling plan (Kelso and Rutherford 1996; Heath 1992). Compromises between the objectives of the studies and resources available are often required (Cyr et al. 1992). Objectives can range from collecting organisms for live experimental work, or studies requiring preserved eggs and larvae, to taking very detailed quantitative samples to estimate such things as vertical distribution or feeding and mortality rates. The species and life-history stage sought and the habitat to be sampled all determine the gear to use and how to deploy it. The frequency of sampling will be decided by whether the study is to examine diel periods, or longer-term time periods such as seasonal or interannual. The size and speed of the vessel available for the work, and the equipment such as winches will affect the sampling that will be possible. Some studies can use qualitative samples, whereas others require quantitative samples. Qualitative samples can be used to collect fish eggs and larvae for taxonomic, physiological, genetic, bioassay, and embryological studies. In such cases all that is generally required is that the net has fine enough mesh to retain the organisms of interest, and that it is large enough, and is towed in such a manner that they will not be able to avoid it. It must be towed in a region, during a time, and in a depth range where the organisms are expected to occur. For quantitative samples that are required to estimate the density or abundance of fish eggs and larvae, the volume of water filtered by the net must be determined. A flowmeter mounted in the net allows this to be done.

In many studies, a series of quantitative observations using standardized, repeatable procedures is desired. Often tows are standardized for amount of water filtered, so each tow should be at the same speed over the same distance or for the same time. Many sampling plans include making plankton tows and other observations at a series of predetermined geographic points (stations). In cases when ichthyoplankton of an area is to be surveyed and the faunal distribution, and a general idea of abundance by taxa is sought, sampling stations are established in either a grid pattern or along transects or at increasing distances radiating from a point (such as a point source of pollution, or a geographic feature such as an island). Since means are closely correlated with the variance in most plankton sampling, stations should be more closely spaced in areas and at times when high catches are expected. This requires some prior knowledge of the distribution in time and space of the eggs and larvae to be collected. In some cases gradients in fish egg and larval abundance are correlated with geographic or environmental gradients. For example, variability would be expected to be greater across a current than along it. Thus in an area with a strong current, station spacing across the current should be less than along it. Although more research on sampling variability is desired, it seems that station spacing of 15–30 km in coastal areas is often sufficient to derive basic patterns of distribution, and to estimate egg and larval abundance. Generally all of the stations in the established pattern are sampled as quickly as possible, running from one end to the other, and the results are treated as if the sampling were simultaneous. Environmental data such as weather conditions and temperature and salinity with depth are often collected at each station along with the ichthyoplankton. In order to use ichthyoplankton surveys to estimate spawning biomass, several such surveys may need to be conducted over the course of the spawning season, encompassing the entire geographic range of occurrence of the eggs or larvae.

Special process-oriented and hypothesis testing studies require special sampling designs. A frequent study of this type involves investigating the vertical distribution of fish eggs and larvae of a target species, or of the ichthyoplankton assemblage found at a particular place. Sometimes such studies are conducted over a period of 24 to 72 hours, either following a drogued buoy or at a fixed geographic point. One object of such studies is to see how vertical distribution changes with the daily light/dark cycle. Concurrent sampling can measure the abundance and depth distribution of larval fish prey.

Platforms (Ships)

In most cases some form of ship is used for collecting samples. Depending on the area to be sampled, ships for ichthyoplankton work can vary from large, ocean-going research vessels for coastal and offshore areas to outboard motor boats for small rivers and lakes. Many of the requirements for the sampling platform are similar, regardless of its size. The size and seaworthiness of the ship should match the conditions of the sampling area, although at times weather may limit operations even with the best ships. The ship should be capable of traveling between stations quickly, but be reasonably comfortable when at rest, and be able to maintain slow speeds when sampling. Many sampling protocols require precise control of ship speed at very slow speeds. The ship should be equipped with a winch to deploy and retrieve the plankton nets. The speed at which the wire is paid out or retrieved, as well as the amount of wire that has been paid out, needs to be monitored with a mechanical or electronic meter block. The winch should be situated such that the net will not pass through water made turbulent by the passage of the ship. On deck there needs to be room to work with the net as it is deployed, and room to rinse the net and collect the sample after the net has been retrieved. Also aboard the ship there should be a protected area for processing the samples and recording data. Communication either visual or oral or both is required among all who are involved in the sampling operations: the person handling the ship, the winch operator, and the person handling the sampling equipment. Beyond these basic requirements, more sophisticated sampling programs have additional needs that may extend to such things as direct communication links with satellites that provide surface temperature maps of the sampling area and report the location of drifting buoys.

Sampling Gear

Since fish eggs and larvae occupy such a wide variety of habitats, and so many facets of their ecology can be investigated, special sampling devices have been developed to accommodate this variety. Sometimes devices and procedures designed for a particular application can be used as is in other circumstances; sometimes they can be modified for other studies. A tremendous variety in the types of equipment to sample ichthyoplankton has developed over the last 100 years. This resulted from programs with special needs and merely the quest for a "better" sampler. A better sampler was defined in a number of ways by different investigators: one that would collect unbiased samples of a wide range of sizes of eggs and larvae; one that was durable and easy to use; one that minimized avoidance by larger, active larvae; one that could collect discrete samples of parts of the water column. Many studies have been conducted to compare various devices and methods for collecting fish eggs and larvae (e.g., Cada and Loar 1982; Choate et al. 1993; Gallagher and Conner 1983; Gregory and Powles 1988; Marsden et al. 1991).

Most ichthyoplankton studies have used nets that have rigid circular frames to which is attached a cone of fine-meshed, durable nylon netting (Figure 5.3). Mesh sizes vary, depending on the species being studied and other sampling requirements. Generally 0.5- or 0.3-mm mesh is sufficient for most ichthyoplankton. As a rule of thumb, it is assumed that material larger than the diagonal of the mesh size will be retained in the net (e.g., with a 0.5-mm mesh, 0.7 mm and larger eggs would be retained). The net is sufficiently long to allow water to flow through the net and not become clogged. A cylindrical section precedes the conical section in some designs to increase the area of the netting to reduce clogging. The ratio of the open area of the meshes to the mouth opening of the net should be large enough so that water passes through the meshes slowly enough to minimize damage to fragile larvae as they hit the meshes. At the rear of most plankton nets there is a detachable collecting bucket or bag. The bag and the windows in the bucket are made of mesh somewhat smaller than that used in the

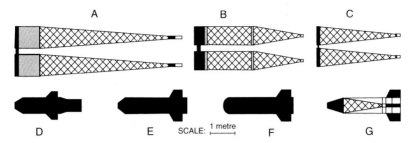

Figure 5.3. Schematic drawings of examples of plankton nets used to collect fish eggs and larvae (from Schanck 1974; Smith and Richardson 1977; and Heath 1992). Some models of Bongo and Gulf III (view from above). A. SIO Bongo (McGowan and Brown 1966): nonfiltering collar. B. BCF Bongo: filtering cylindrical net section. C. Hydrobios Bongo: no cylindrical net section. D. Modified Gulf III (Bridger 1957): aperture 0.2-m diameter, restricted tail unit. E. Dutch Gulf III and Hai: no restricted tail piece. F. Haie with hemispherical nose piece. G. Nackthai: net unencased.

plankton net itself. A rigid collar fits between the end of the net and the collecting bucket. As technology has advanced, the materials available for samplers have changed, and the possibility that samplers could be monitored and controlled by electronic signals sent up and down a conducting wire on the winch has become a reality.

Collecting fish eggs and larvae in estuarine and freshwater habitats often requires specialized samplers to operate effectively in congested areas (Bagenal 1974; Marcy and Dahlberg 1980; Meador and Bulak 1987). Also, eggs and larvae of freshwater fishes are more closely associated with the bottom than most marine fishes, and this necessitates using samplers that can be operated on or near the bottom (Phillips and Mason 1986). Freshwater studies often use much smaller vessels to tow nets than are used in more open waters, and this places limitations on the sampling gear used (Miller 1973).

Sampling juvenile fishes poses significant problems since they occupy a variety of habitats and can actively avoid the samplers used for larvae. Juveniles that occur in midwater can be sampled with small midwater trawls (Isaacs and Kidd 1953; Methot 1986). Small-beam or otter bottom trawls can be used for demersal juveniles, whereas seines can be used in protected shallow-water habitats (Able and Fahay 1998).

Following we briefly describe some of the ichthyoplankton samplers that are presently in use.

Bongo Nets. A variety of paired nets on circular aluminum or fiberglass frames that are connected to each other by a central yoke or axle are called bongo nets (Posgay and Marak 1980). The towing wire and the weight

(22–40 kg) are attached to the axle. Bongo nets are made in a variety of sizes, but those about 60–70 cm in diameter are most commonly used for ichthyoplankton studies. The towing frame is fitted with two cylindrical-conical nets, often of a dark color. The cylindrical section helps prevent clogging, and the dark color reduces net avoidance. The ratio of the sum of all openings in the net to the diameter of the opening is at least 8:1, which increases filtration efficiency and prevents clogging. Flowmeters are mounted in the mouth of each net to provide data on the volume of water filtered. Bongo nets are towed at speeds of 1.5 to 2.0 knots. The advantages of bongo nets are that there is no bridle or other obstruction in front of the net as it is being towed, and it is possible to take two samples simultaneously: these can be used for different processing and analysis, or two mesh sizes can be used to collect different-size fractions of the plankton.

Tucker Plankton Trawls. The Tucker trawl was developed to take a discrete sample within a specified depth range (Davies and Barham 1969). The original design used timing devices to release bars that opened and closed a net (Figure 5.4). Modifications to the original design permit the use of two nets sequentially, which are opened and closed using two messengers sent down the towing wire to activate a tripping mechanism. This allows sampling at specific depths, without contamination from other depths. The nets are usually constructed to have a 1-by-1-m opening when towed. The towing wire attaches to a bar just above the top of the nets, and cables run between the top bar and a weighted bottom bar. The nets are attached to additional top and bottom bars, which slide on the side cables as the tripping mechanism is activated. As with the bongo, there are no obstructions in front of the net as it is being towed. Since the mouth opening of a 1-m Tucker trawl is larger than that of a 60-cm bongo, escapement of larger larvae is reduced.

Neuston Nets. Special nets have been designed to sample the upper few (10–50) centimeters of the water column where the eggs and larvae of a wide variety of fishes reside as either obligate or facultative components of the neuston (e.g., Sameoto and Jaroszynski 1969; Brown and Cheng 1981; Hodson et al. 1981). Neuston nets generally have a rectangular mouth opening and are towed so that they ride with only the lower half of the opening below the surface (Figure 5.5). Some are equipped with flotation, and most have wings that enable them to track at the surface and away from the ship. Flowmeters can be used with them to measure the distance traveled. They are deployed from amidships, as far forward as possible to keep them in undisturbed water out of the wake of the ship. A tag line is frequently used to deploy and retrieve the nets. They are generally towed for a predetermined length of time, at a speed of 2–5 knots. Catchability must be considered in analyzing neuston data, since the occurrence of larvae in the neuston

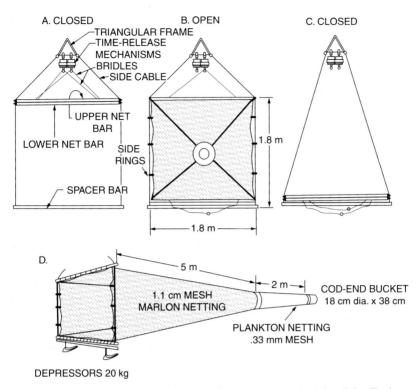

Figure 5.4. A schematic drawing showing the operating principle of the Tucker net and its major dimensions (from Davies and Barham 1969). A. The net is closed and ready to be lowered to sampling depth. B. The bridle from the lower net bar has been released and the net is fishing. C. The upper net bar has been released and the net is closed for recovery. D. Side view of the net while open.

is a function of their behavior. For example, marked diel differences in catches are often seen. A limited size range of larvae of a species may occur in the neuston, reflecting changes in vertical distribution with growth.

Methot Trawl. Several nets have been designed to catch larger planktonic animals (micronekton). Isaacs and Kidd (1953) designed a trawl with a large depressor and graduated meshes that terminated in a fine-meshed codend. This trawl became widely used for collecting larger larval and juvenile fishes in midwater. However, bridles in front of the mouth of the net probably allow such mobile organisms to escape this trawl. Another trawl was designed by Methot (1986) to overcome the problem of the bridle in front of the net. The Methot trawl has a rigid frame (2.2 × 2.2 m) and is

158 Sampling Fish Eggs and Larvae

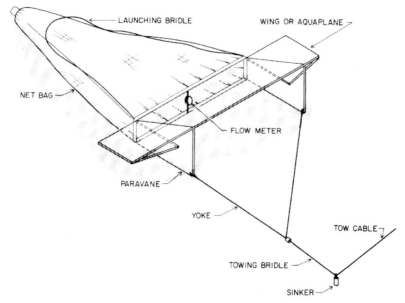

Figure 5.5. A schematic diagram of the Manta net: an example of a neuston sampler (from Brown and Cheng 1981).

made of uniform mesh size, which makes interpreting catch data simpler (Figure 5.6). The Methot trawl has been used in several studies of larger larvae and small pelagic juvenile fishes (e.g., Methot 1986; Munk 1988; Brodeur and Wilson 1996).

Serial Net Samplers. There has long been a desire to collect multiple discrete samples in quick succession. These samples can be used to investigate small-scale vertical and horizontal distribution and patchiness. Early versions of samplers to make such collections changed nets or codends automatically or through mechanical messengers sent down the towing wire to trip releases on the sampler (Davies and Barham 1969; Longhurst et al. 1966; Haury et al. 1976; Pipe et al. 1981). More recently, several electronically controlled discrete depth samplers such as the MOCNESS (*M*ultiple *O*pening and *C*losing *N*et and *E*nvironmental *S*ensing *S*ystem, Weibe et al. 1976) have been developed to obtain vertically discrete plankton samples. These systems consist of a frame with a series of up to 10 nets that are sequentially opened and closed remotely from the ship using electronic signals sent down the towing wire. Most of these systems also telemeter net condition (depth, attitude, flow) and environmental data

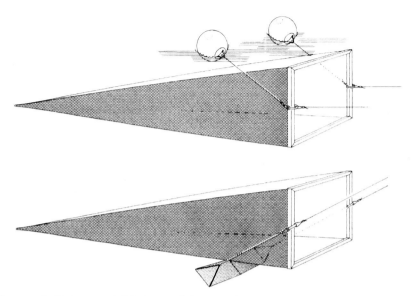

Figure 5.6. Frame trawl (Methot net) for sampling larger larvae and juveniles (from Methot 1986). The trawl in its surface configuration with floats (*above*), and in its subsurface configuration with the Isaacs-Kidd depressor (*below*).

(temperature, conductivity) up the towing wire. MESSIAH, BIONESS, (Sameoto et al. 1977, 1980), and LOCHNESS (Dunn et al. 1993) are other examples of such systems. The MOCNESS is currently most widely used and can be purchased in sizes with openings of 0.25, 1.0, and 10.0 m^2. These systems have greatly improved our ability to investigate the vertical distribution of fish eggs and larvae. They have also been used to obtain replicate samples within a particular depth horizon, and to investigate small-scale variability in distribution and condition of ichthyoplankton. Nets can be opened and closed as they encounter hydrographic features (e.g., thermoclines) of interest and to relate these to concurrently collected fish eggs and larvae.

CalVET. The CalVET (*Cal*COFI *V*ertical *E*gg *T*ow) net was developed to collect samples of Pacific sardine (*Sardinops sagax*) and northern anchovy (*Engraulis mordax*) eggs for estimates of spawning biomass. The 20-cm-diameter net is attached to a hydrographic cable by both the mouth and codend. With the ship at rest, it is lowered vertically to 70 m, and then raised to collect the sample. This procedure routinely takes only about 5 minutes, and allows a large area to be sampled quickly, thus increasing the synopticity of the sampling and decreasing the variance by increasing the number of samples collected. This procedure relies on the fact that the eggs of interest reside in the upper 70 m of the water column, and that they cannot escape the small-mouthed net as it is slowly raised through the water column. Since the net is towed vertically, calculation of volume of water sampled is simplified, and does not require use of a flowmeter.

Continuous Plankton Recorders. Devices that consist of a net in front of an impeller-driven mechanism that rolls plankton onto gauze and preserves it in formalin have been developed to tow behind ships moving at cruising speed. They have been used to monitor seasonal and long-term changes in distribution and abundance of fish eggs and larvae. They can be towed for long distances from high-speed commercial vessels. The Hardy Continuous Plankton Recorder has been used over broad geographical ranges in the North Atlantic. Recent versions of plankton recorders simultaneously collect environmental data.

Plankton Pumps. Large pumps have been used with varying success to collect fish eggs and larvae (see Leithiser et al. 1979; Harris et al. 1986; Solemdal and Ellertsen 1984). They can be used to monitor long-term changes in egg and larval abundance. They may be the most effective means of collecting eggs and larvae of freshwater fishes that occur on or near the bottom in relatively small areas. They can sample unattached benthic eggs such as those of striped bass and Pacific cod. They are usually deployed while the ship is at rest, and they can be used from a pier or other fixed structure. Water is pumped from the desired depth through a flexible hose to the deck of the ship where it is filtered through netting on which the plankton is retained. Sequential samples can be taken by changing the netting. In this way several depths can be sampled with one lowering of the hose. Care must be taken to flush the water from the hose between samples. Since the only "head" the pump must overcome is that between the surface of the water and the pump, samples can be obtained from quite deep in the water column. The advantages of pumps are that the depth and volume of the sample can be known precisely. The disadvantages relate to the sparse numbers of planktonic fish eggs and larvae usually encountered. Rarely are they more concentrated than 1 per m^3. This makes it necessary to pump large volumes of water to obtain enough eggs and larvae to be a meaningful sample. Thus a high-capacity pump and a large-diameter hose must be used, and the ship must have adequate power and space to accommodate such a pumping system. Pumps sample small areas and since fish eggs and larvae are usually patchily distributed, it may be difficult to interpret the results. Another problem associated with pumping is avoidance of the pump inlet by larvae that are able to detect the flow of water and swim away. This avoidance increases with size of the larvae.

Continuous Underway Fish Egg Sampler. The CUFES is a recently developed system that pumps water into a ship while underway and filters it through a collecting device on board (Checkley et al. 1997). The pumping system draws water from a single fixed depth within 3 m of the surface. This device is useful for making rapid surveys of the fish eggs occurring near the surface over large areas. Other measurements can be made on the water stream as it goes through the pumping system (e.g., temperature, salinity, nutrients, fluorescence) or samples can be drawn for later examination (e.g., phytoplankton, microzooplankton). It has been used to assess the abundance of Pacific sardine eggs, and to document the fine-scale distributions of marine fish eggs in protected areas off California (Watson et al. 1999).

High-speed Plankton Samplers. To reduce the length of time required for plankton surveys, nets were developed that could be towed at high speed, allowing samples to be taken more synoptically (Noble 1970). The intent was to use a net that could be towed from the ship traveling at near full speed (about 10 knots) through the sampling area, rather than occupying a series of stations where plankton tows are made with a usual towing speed of about 2 knots. Also, the net could have a smaller mouth opening, but be towed for longer distances to even out some of the small-scale patchiness that is normally observed in plankton distributions. The Miller high-speed sampler (Miller 1961) is an example of such a device that was used in several studies. A series of "Gulf" samplers was developed through International Council for the Exploration of the Sea (ICES) that had the added feature of being constructed entirely of metal, so they were more durable than usual plankton nets with cloth netting (Schanck 1974). The Gulf III, the most commonly used of these samplers, was used in several studies off northern Europe (Gehringer 1962).

Tidal and Channel Nets. Passively sampling water as it moves through inlets with the tides to collect larval fishes that migrate through inlets between the open ocean and inland waters is accomplished with tidal nets (Lewis et al. 1970; Keener et al. 1988). Such nets have been deployed from fixed structures such as bridges, and sampling is designed in relation to the tidal cycle. Nets can be deployed at various depths and positions in the inlets to investigate the dynamics of the exchange of eggs and larvae between offshore and inshore waters. Advantages of tidal nets are their ease and low cost of deployment, and their being less weather dependent than shipboard sampling. The disadvantages include the limited number of sites and species that can be studied using these techniques, and their dependency on tidal currents for collecting samples: they may be so weak that larvae can swim out of the nets, or they may reverse and permit back-flushing of larvae already in the net. Similar nets have been used to collect eggs and larvae in the current of rivers or streams (e.g., Franzin and Harbicht 1992).

Night Lights. Since larvae of many fishes are attracted to artificial lights shown on or in the water at night, lights have been used in connection with dip nets or lift nets to collect larvae (Dennis et al. 1991). The reason for this attraction and how it varies among species and with larval size is unknown. Also, the degree of attraction probably varies with environmental parameters such as water clarity, flow, depth, light level, and choppiness. Once the larvae show up in the orb of the light they are captured with a handheld dip net. Aquarium nets attached to plastic pipe (about 1-inch diameter by 6 feet long) are useful for this. Such collecting can hardly be considered quantitative, but it does provide specimens in excellent condition for descriptive and rearing purposes.

Light Traps. Passive remote samplers have been developed based on the attraction of larvae to light (Doherty 1987; Brogan 1994) (Figure 5.7). The effectiveness of light traps probably depends on the variables mentioned previously under night lights. In contrast to dip nets, they do not depend on the skill of the collector and can be deployed similarly (e.g., for the same length of time) in different situations (e.g., different nights, different locations), to provide at least comparable samples.

Sleds. Devices have been developed to sample the eggs and larvae of some species that dwell on or near the bottom (Phillips and Mason 1986). Essentially, plankton nets have been attached to runners to sample the water just above the bottom (Dovel 1964). Several rectangular frame trawls fitted with plankton nets have been developed (Madenjian and Jude 1985; La Bolle et al. 1985). A Tucker trawl on a sled can be used to obtain a discrete sample of the near-bottom stratum.

Indirect Acoustic and Optical. Indirect methods of assessing ichthyoplankton distribution are being used more frequently. High-frequency hydroacoustics can be used to determine the density of sound reflecting particles in the size range of fish eggs and larvae. However, it is not possible to determine the species of the echosign, and fish larvae are rarely dominant members of the size fraction of the plankton they are part of. When used in conjunction with discrete depth net tows, hydroacoustics can help explain vertical density patterns of the plankton, in relation to the depth of fish larvae. Optical methods hold some promise for sampling fish eggs and larvae at small scales not possible with net tows (Houde et al. 1989). Both photographic and video methods have been developed (Figure 5.8). An optical sensor is usually placed so a stream of water passing through a plankton net passes by it. A silhouette image of material passing by the sensor is recorded on videotape or photographic film. Video images can be transmitted up the wire to the ship to be recorded and viewed in real time.

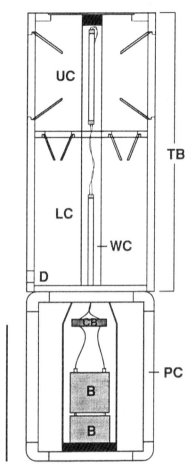

Figure 5.7. Schematic diagram of a light trap for collecting fish larvae (from Brogan 1994). B-batteries; CB-circuit board; D-drain; LC-lower chamber; PC-protective cage; TB-trap body; UC-upper chamber; WC-waterproof core. Scale bar = 30 cm.

Ancillary Gear

Besides needing devices to collect eggs and larvae, other devices are needed to monitor the nets and, in some studies, the environment so that resulting egg and larval data can be interpreted. The volume of water filtered by the net, and the depth of the net is often required.

Flowmeters. To measure the length of the water column through which a plankton net passes, a flowmeter is mounted in the mouth of the net, and

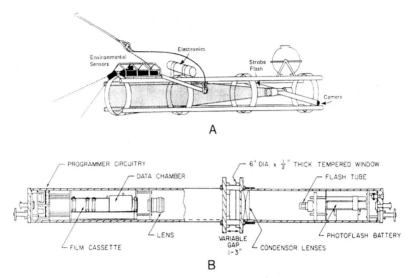

Figure 5.8. Camera-net system to sample plankton (from Houde et al. 1989). A. Diagram of the frame, plankton net, camera, and environmental sensors. B. Diagram of the camera, chamber, and strobe flash apparatus.

the number of revolutions accumulated by the impeller during a tow is recorded. Flowmeters are calibrated to convert revolutions to meters traversed. Calibration can be done in flume tanks, or by towing the nets over known distances. Calibrations should be done at the speeds to be used during actual sampling, since the relationship between revolutions and distance is not always linear. Electronic flowmeters, which telemeter their readings to the ship through a conducting towing cable, are also available.

Depth Determination. For many types of studies it is critical to know the depth stratum that is fished by a net. For vertical plankton net tows, the depth stratum is from the maximum extent of the wire out to the surface, assuming negligible wire angle. For oblique tows, the crudest and least accurate means of estimating net depth is by measuring the amount of wire deployed and the angle of the wire (relative to vertical) as it enters the water. Assuming the wire is straight and the angle at the surface represents the angle of the entire wire, the net depth will be the amount of wire deployed times the cosine of the angle. For example, if 300 m of wire is deployed, and the angle is 45 degrees, the net should be at 210 m. Several mechanical devices have been developed to measure the depth and vertical trajectories of tows. These generally produce a paper plot of depth, as determined by a pressure sensor (a spring), against time (a spring-wound clock). These provide an after-the-fact record of the tow. Recently, compact electronic CTDs (Conductivity-Temperature-Depth), which telemeter data through a conducting tow cable, have been employed with bongo nets on oblique tows to monitor net depth in real time. These also provide temperature and salinity information with depth without requiring additional ship time. Several electronic devices are available that record depth and other parameters as a function of time. These are deployed with the nets and the data downloaded after the tow.

Environmental Data. Concurrent data on environmental conditions are often needed to relate to catches of fish eggs and larvae. Some data, such as time of day and wind and sea state, may help explain differences caused by sampling conditions. Other data such as vertical profiles of temperature and salinity are important in analyzing developmental rates and relating catches to water masses and currents. Objectives of the ichthyoplankton sampling are considered when determining what ancillary environmental data should be collected.

CHOOSING SAMPLING METHODS

Ichthyoplankton sampling methods are as varied as the objectives of the studies and the gear to be used. Many methods involve towing a net in a specified manner at each station. This usually means operating at the same ship speed and towing the net for the same amount of time, through the same depth interval at each station. Some preliminary studies may be needed to determine the appropriate towing scheme for a particular investigation. For example, the depth range of the eggs and larvae of a species may need to be known to efficiently design a survey to determine their geographic range and abundance. A common towing method is a smooth oblique tow of the net between the surface and some specified depth, often 200 m, or the bottom in shallower waters (see Smith and Richardson 1977). A survey using this type of tow with a bongo net equipped with flowmeters and a depth-sensing device at stations spaced about 10 miles (15 km) apart in continental shelf waters generally provides samples that can be used to map the distribution and abundance of most pelagic fish eggs and larvae of the region. Samples from neuston tows taken at the same stations for a specified towing time, say 10 minutes, can be used to determine the distribution of fish eggs and larvae that inhabit near-surface waters. Passive samplers such as light traps and tidal nets are usually deployed at specific sites for predetermined lengths of time. Pumps and nets towed vertically require the ship to be at rest during sampling. Some sampling such as hydroacoustics can be done while the ship is underway, covering transects laid out in the sampling area.

SHIPBOARD PROCEDURES

Sample Collecting

Shipboard sample processing should be done as quickly as possible, since the plankton deteriorates rapidly after being damaged. For accurate measurement of length of larvae and condition factor it is critical that minimal time elapses between collection and fixation (Hay 1981). As soon as the net is out of the water, it should be thoroughly and carefully rinsed with seawater to clean material from the meshes and collect it in the cod-end bucket. Using too much water pressure for this may damage the plankton and render it very difficult to identify and measure.

Sample Preservation

Once the plankton is washed into the cod-end bucket, the bucket is removed and its contents emptied into a jar to which is added concentrated formalin such that when the jar is filled with seawater, the final concentration of formaldehyde will be 5% (50 ml of 37% formaldehyde in 1 liter of seawater). Also, Sodium Borate is added at a rate of 20 g per liter of seawater, for buffering. Such samples are usually sealed and stored for processing ashore.

Data Records

As ichthyoplankton sampling is being planned, it is essential that provisions for keeping accurate and thorough records are considered. Data sheets to record information associated with the tows, and labels for the jars of plankton are required. Ancillary environmental data should be keyed to the plankton samples; using the same station numbers is a convenient way to do this. Ichthyoplankton sampling programs can generate large amounts of data, and adequate data management is essential. The data can be considered a stream that is augmented at each phase of sample collection and processing (e.g., sorting and identification data are added to the stream that began when the samples were collected at sea).

SHORESIDE SAMPLE PROCESSING

Processing ichthyoplankton net samples is a tedious task generally performed in a shoreside laboratory using two or more steps, depending somewhat on the skill of the workers and somewhat on the objectives of the study. The first step is to remove the fish eggs and larvae from the rest of the plankton and detritus in the sample (see Smith and Richardson 1977). The second step is to identify the eggs and larvae, which is covered in Chapter 3. Before the eggs and larvae are removed, the displacement volume of the sample is often determined. In any case, the sample is poured through a mesh cone with a mesh size somewhat smaller than that used to collect the sample and then rinsed to remove most of the preservative. (This, and other operations involving formalin, should be carried out in a well-ventilated area or under a fume hood since formalin is a carcinogen as well as being noxious. Well equipped ichthyoplankton laboratories have flexible tubes at each microscope station connected to a ventilating system to carry the fumes away.) To determine displacement volume after extraneous and nonplankton material (e.g., artifacts) is removed, the plankton material is placed in a 1,000-ml graduated cylinder, and the volume is increased to 1,000 ml by adding water. The liquid and plankton are then poured through a mesh cone that is suspended over another 1,000-ml graduated cylinder. After the liquid has stopped dripping through the meshes (less than one drip per 10 sec), the volume is measured. The difference in volume is the displacement volume of the plankton.

To remove the eggs and larvae from the sample, sequential small amounts of plankton (about 10 ml) are placed in Petri dishes and examined with a dissecting microscope at about 6–10 power. The material is manipulated with fine probes and forceps to ensure that all eggs and larvae can be seen and removed. Once found, eggs and larvae are placed in separate, labeled vials using fine forceps and pipettes. The eggs should be stored in 5% formalin, but the larvae should be placed in 70% ethanol. After each dish of plankton has been thoroughly examined and all eggs and larvae removed, its contents are placed in a container marked "sorted." After the entire sample has been examined and the plankton residue is in the sorted container, some form of quality control can be instituted to ensure that no eggs and larvae remain. Following this, the remaining plankton is usually archived in formalin for further studies.

DATA PROCESSING

Data Analysis

Computer-aided data processing is the norm for ichthyoplankton studies. The specifics of the computers and software used depend on a host of factors including the scope of the study and the skill and resources of the investigators. Spreadsheet and database/programs are frequently used. Most of the commercially available statistical programs can be used to analyze ichthyoplankton data. More advanced studies may require geographical information systems (GIS) if several types of environmental data are to be analyzed along with the ichthyoplankton. Since much of the ichthyoplankton data are related to geography, mapping programs are often needed.

After the data have been entered and edited, the first step in data processing is to standardize the numbers of eggs and larvae of each species collected in each sample to a unit of volume filtered by the net, yielding

168 Sampling Fish Eggs and Larvae

density values (Box 5.1). The number of eggs or larvae is multiplied by the area of the mouth of the net and the distance traveled. The unit of volume frequently used is 1,000 m^3. The density values can be multiplied by the depth of the tow to yield abundance under a standard sea surface area. The unit of sea surface area used is frequently 10 m^2.

Statistical analyses applied to ichthyoplankton data are as varied as the purposes of the studies. In general, the geographic distribution of plankton in the sea is not normal, rather it is frequently lognormal, and the variance is strongly correlated with the mean (see Smith and Richardson 1977).

BOX 5.1. Formulas for Estimating Density and Abundance of Eggs or Larvae in a Plankton Tow

Volume of water strained in the tow (m^3):

  V = R a P
  V = volume of water strained (m^3)
  R = revolutions of the flowmeter
  a = area of net mouth (m^2)
  P = distance (m) traveled per revolution of the flowmeter

Density (eggs or larvae in × m^3):

  D = V/N
  V = volume of water strained (m^3)
  N = number of eggs or larvae caught in the sample

Abundance (eggs or larvae under × m^2):

  A = S × N
  S = standardization factor (Z/V)
  N = number of eggs or larvae caught in the sample
  Z = maximum depth of tow (m)
  V = volume of water strained (m^3)
  S = Z/V (under 1 m^2)
    S = 10 Z/V (under 10 m^2)
    S = 100 Z/V (under 100 m^2)

Example: how many eggs are there beneath 10 m^2

  V = 123 m^3 (volume of water strained during the tow)
  Z = 50 m (maximum depth of the tow)
  N = 32 eggs (number of eggs caught during the tow)
  S = 10 Z/V = 10 × (50 m) / 123 m^3 = 4.065
  Eggs beneath 10 m^2 = 4.065 × 32 = 130 eggs beneath 10 m^2

Note: Make sure the units make sense (i.e., 5.0 eggs beneath 100 m^2 is better than 0.5 eggs beneath 10 m^2).

Thus statistics based on the assumptions of a normal distribution cannot be used unless the data are transformed to approach a normal distribution. Lognormal data can be transformed by using the logarithm of the data and adding 1 to account for zero values; normal statistics can then be applied to these transformed values. As an alternative, nonparametric statistics are often used.

Variability in Ichthyoplankton Data

Variability in ichthyoplankton data occurs because of patchy distributions and limited sampling effort, as well as the biases and inefficiencies of the sampling gear and techniques employed (Jahn 1987; Hauser and Sissenwine 1991). Two main factors are involved in variability associated with the sampling gear: extrusion and escapement through the net, and avoidance of the net (see Morse 1989). Extrusion occurs when fish eggs and larvae are forced through the meshes of the nets by water flow during sampling, and when the nets are cleaned following the tow. Escapement occurs when larvae swim through the meshes of the net. Plankton nets are generally made of woven nylon with square mesh openings that retain material that is larger, in its smallest dimension, than the diagonal of the openings. Selection of the mesh size of plankton nets is important, since it should retain all of the animals of interest, but not be so fine that it retains unwanted smaller organisms, which increases time required for sorting the samples. Pelagic fish eggs range in diameter from about 0.5 mm to 5.0 mm but most are about 1.0 mm. Larvae are mostly larger than 1.0 mm. Commonly used mesh sizes for ichthyoplankton work are 0.333 mm (which will retain 0.47-mm-diameter eggs) and 0.505 mm (which will retain 0.71-mm-diameter eggs). Small larvae are fragile and may disintegrate when they hit the net. Retention of small larvae depends on using small-meshed nets designed to reduce flow through the meshes, and using slow towing speeds (Colton et al. 1980). Nevertheless, in many studies fewer small (yolk sac) larvae are collected than would be expected.

Net design affects filtration efficiency, which is the amount of water presented to a net relative to what goes through the meshes. Filtration efficiency may decrease during tows as organisms accumulate on the net meshes (i.e., clogging). Reduction of clogging is another reason to use mesh that is no smaller than necessary to retain the organisms of interest. High mesh area to mouth opening ratios reduce clogging. Nets used for ichthyoplankton studies usually have 8:1 or 10:1 open mesh area to net mouth opening ratios and produce filtering efficiencies approaching 100%. Filtration efficiency can be determined by comparing readings of flowmeters placed in the mouths of the nets with those towed simultaneously outside of the nets.

Fish larvae are responsive to stimuli produced by plankton nets and can avoid them. Net avoidance increases with larval size in response to increases in both their sensory and locomotor capabilities. Nets should be designed and towed to minimize avoidance. Larvae can respond to disturbance of the water caused by the net. A pressure wave precedes the net as it is towed through the water. The tow wire and bridles as well as the net itself produce this pressure wave. Nets with low filtration efficiencies caused by poor design or clogging produce bow waves. These flow disturbances can be sensed by larvae through their neuromast systems, and they elicit a startle escape response that can lead to net avoidance. Vision can also be used by the larvae to sense and avoid nets. Evidence of this type of avoidance can be seen in lower catch rates and smaller larval lengths in samples taken during the day as opposed to at night (Brander and Thompson 1989). Use of dark-colored nets reduces vision-related net avoidance. Other ways to decrease net avoidance include using nets with larger mouth openings, eliminating bridles and other obstructions in front of the net, and towing the net faster. However, faster towing speeds can increase extrusion and mutilation of the larvae.

Patchiness is a common feature of the distribution of most planktonic organisms including fish eggs and larvae. At several scales the horizontal and vertical distribution of ichthyoplankton is uneven. There are areas and depths where there are less or more organisms than would be expected based solely on physical processes such as advection and diffusion. Several factors can contribute to this uneven distribution. Adults spawning in specific habitats at precise times can cause patchy distribution of eggs. This is quite pronounced when large schools of fishes spawn synchronously. Once spawned, eggs can become more patchily distributed as they sink or rise to a level of water density that matches their own. Other physical phenomena such as eddies and regions of convergence such as fronts can cause accumulation of fish eggs and larvae. Differential mortality of parts of the distribution of eggs and larvae, such as caused by uneven abundance of predators or prey, can also lead to patchiness. Once the larvae can swim and directly influence their location, they can become more patchy, particularly in the vertical. Larvae also can aggregate in areas of high food abundance (Fortier and Leggett 1984). Many studies need to take patchiness directly into account as results are interpreted. For example, larvae may be in better condition when they are in patches where food concentration is elevated. Understanding the cause and scale of the higher food concentrations and their influence on the patchiness of the larvae would be the objects of studies on this observation. Even if understanding patchiness is not the direct objective of studies, it must be accounted for in designing and interpreting sampling. Patchiness will generally result in samples that are highly variable in abundance of eggs and larvae. Thus, many large samples are required to obtain reasonably precise estimates of mean abundance.

REPORTING RESULTS OF SAMPLING

Results of ichthyoplankton sampling can be presented in a variety of ways. For ecological studies the density of eggs and larvae is often needed (e.g., to evaluate the adequacy of prey available). Fish eggs and larvae are usually rather rare components of the total plankton community; their density seldom exceeds 1 per m^3. Densities of fish eggs and larvae are frequently reported as numbers per 100 or 1,000 m^3. Densities are calculated in vertical distribution studies. They are also compared to the density of suitable prey collected concurrently, to determine if feeding conditions are adequate. However, since most plankton tows sample at least tens of cubic meters of water, and larvae can only search 1 m^3 or so, patchiness of the ichthyoplankton and prey limits the usefulness of such samples for these investigations. Finer scale sampling may be required to assess feeding conditions on a scale relevant to larvae.

Distribution and Abundance

For some areas and species where there are limited or no fisheries, collections of fish eggs and larvae can provide a picture of the diversity of the ichthyofauna present. In other cases, fish eggs and larvae can provide an independent estimate of the distribution and relative abundance of commercially important fishes in an area (e.g., Almatar and Houde 1986).

In order to determine the geographic distribution of fish eggs and larvae, the first order of analysis often involves plotting the standardized abundance of eggs and larvae (say, numbers per 10 m^2) of particular species collected at each station on a map. This can be done by hand, although a number of computer programs are now available. Frequently, contours are drawn to illustrate the pattern of abundance of the eggs and larvae. Since the data are usually lognormally distributed, contour intervals based on logarithms are appropriate (i.e., intervals of 0, 1–9, 10–99, 100–999 per 10 m^2, and so forth). Once the distributions of eggs or larvae of various species are plotted, they can be compared with other geographically related data such as spawning areas and water mass distributions. If sequential cruises sampled the same areas, changes in distribution with time can be seen, which may indicate changes in spawning distribution or advection of eggs and larvae by currents.

To determine the abundance of eggs and larvae in a survey area, the results of the catches at each station must be combined in a manner proportional to the area of the sea that each station represents. If the stations were equally spaced, and the survey area had straight perpendicular boundaries, this would be a simple matter of summing the standardized abundances at each station and multiplying this times the survey area. However, stations are frequently not evenly spaced, by prior design or by expediencies during the cruises, and the boundaries of the survey area often include irregular coastlines or other geographic features. To account for this, the area of each station is calculated and these areas, multiplied times the abundances at each station, are summed (Sette-Ahlstrom method; see Smith and Richardson 1977). The polygonal areas of the stations are based on perpendicular bisectors of lines drawn between each pair of stations. The station boundaries are treated somewhat arbitrarily when the edge of the survey pattern approaches the coastline: a line can be drawn halfway between the shoreward-most station and the coast, or along a given minimum depth contour, or at the expected inshore limit of distribution of the eggs and larvae. The total number of eggs or larvae of a particular species can then be computed by summing all the station areas times the abundance at the stations. Means and variances of abundance can also be computed, frequently after transforming the data to logs, as mentioned earlier.

Vertical Distribution

An important facet of any ecological study of fish eggs and larvae, and an important consideration in developing sampling plans, is their vertical distribution (Neilson and Perry 1990). Generally there is an uneven vertical distribution of fish eggs and larvae in the water column. Vertical distribution patterns are species specific and often change with development (Coombs et al. 1985; Kendall et al. 1994). These patterns are based on buoyancy of the eggs and buoyancy and behavioral characteristics of the larvae. Vertical distribution can also vary in relation to environmental characteristics such as light intensity, salinity, temperature, currents, and weather. Frequently, larvae change their vertical distribution on a diel cycle. Although the reverse pattern is occasionally seen, most larvae occur at shallower levels at night than during the day. To determine vertical distribution, specific depths must be sampled without contamination from other depths. Tows should be taken under as many environmental conditions as possible (e.g., day–night, windy–calm, low food–high food) to determine impacts on depth distribution. Discrete-depth tows with opening-closing nets are ideal for these studies. If such nets are not available, a sequence of tows at several depths with standard nets can be used, in which material "expected" to have come from shallower levels is subtracted from the rest of the material.

Spawning Biomass Determination

One of the main purposes of fisheries science is to estimate the abundance of fished populations (Gunderson 1993). Some methods that accomplish this use data from the fisheries themselves whereas others are independent of the fishery. These methods are usually based on collecting samples of the population by using standardized surveys. Although adults are sampled in most cases, when combined with several types of ancillary data, fish egg and larval surveys can be used to estimate the size of the adult population that produced them. Robust estimates of population abundance can be made by combining or synthesizing estimates from several measures of abundance, including egg and larval assessments. There are basically three procedures for estimating population abundance based on egg surveys; and the choice of which to use depends largely on the aspects of the spawning pattern of the population under study.

Historically, the most widely used approach is integrative and is called the annual egg production method (Box 5.2; Saville 1963; Lasker [ed.] 1985; Hunter and Lo 1993; Picquelle and Megrey 1993). This approach is

---

BOX 5.2. Steps in Annual Egg Production Method

1. Conduct plankton surveys to collect eggs.
2. Sort samples and identify eggs in each sample ($_l$).
3. Determine numbers (n) of eggs of each stage ($_g$) in each sample (n_g).
4. Determine numbers of eggs of each stage by surface area sampled at each station ($c_{g,l}$).
5. Determine ages of each stage, using temperature-dependent egg development data, and temperatures measured during egg survey ($c_{a,l}$).
6. Calculate area represented by each survey station (a_l).
7. Estimate abundance of each age group of eggs in survey (sum of abundances of each age group at each station times each station area) ($\Sigma c_{a,l} \cdot a_l = N_s$).
8. Model seasonal spawning curve (shape, peak, length) by regressing abundance of early stage eggs on mean date of each survey.
9. Plot egg abundance by stage from survey nearest peak spawning to estimate egg mortality rate ($N_t = N_0 e^{-zt}$), and then number of eggs spawned on mean survey date.
10. Relate number of eggs spawned on mean survey date (step 9) to that date on spawning curve (step 8).

(continued)

BOX 5.2. Continued

11. Integrate under spawning curve to estimate total seasonal egg production (P_a).
12. Based on samples of prespawning adults, estimate sex ratio (R) and fecundity by weight (E).
13. Divide total seasonal egg production by fecundity by weight times sex ratio to obtain weight of adult population (B).

$B = P_a/ER$
B = spawning biomass
P_a = annual production of eggs in survey area (Determined by steps 1 through 11 above)
R = total weight of fish in sample/weight of females in sample (Determined from samples of prespawning adults)
E = number of eggs/gm of female
$c_{g,l} = n_{g,l} d/ab*10$ (Number of eggs by stage and station per sea surface area)
$n_{g,l}$ = number of eggs by stage ($_g$) in a sample ($_l$)
d = maximum depth of tow (Determined by wire angles or depth sensor on net)
a = mouth area of net (πr^2) (r = radius of mouth of net)
b = length of tow (Determined from flowmeter readings)

appropriate when the population spawns during a rather brief spawning season (Figure 5.9). The abundance of spawned eggs is determined during several plankton surveys covering the entire spawning area during the spawning season. Allowance is made for egg mortality, usually by staging the eggs and using a catch curve to estimate the abundance of eggs spawned. A spawning curve is generated either by noting the progression of abundance of eggs from the plankton surveys or by examining the gonads of the adults. Fecundity and sex ratio of the adults is determined from samples collected just before spawning starts.

Another method for determining spawning biomass from pelagic egg studies is the instantaneous or daily egg production method, which was developed in the 1980s to be used with populations that spawn over a protracted period during the year (Box 5.3; Cunha et al. 1992). In such populations, only a fraction of the adults spawn at a given time. These fish ripen and spawn several batches of eggs during the spawning season. A single plankton survey is conducted, and the gonads of the adults are examined at the same time to determine what fraction of the population produced the eggs collected in the

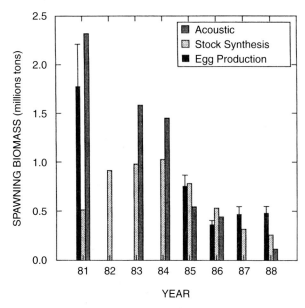

Figure 5.9. Comparison of time series of spawning biomass of walleye pollock (*Theragra chalcogramma*) estimated from three methods: annual egg production method, acoustic survey, and stock synthesis model (from Picquelle and Megrey 1993). There are no egg production estimates for years 1982 to 1984 and acoustic estimates are unavailable for years 1982 and 1987. The egg production biomass estimates have standard error bars. Acoustic and stock synthesis biomass estimates have been modified to include only mature pollock that spawn in Shelikof Strait, Alaska.

plankton survey. By knowing the batch fecundity of the fish, the fraction of the population that produced the eggs in the plankton survey, and the abundance of the eggs in the survey, an estimate of population biomass can be calculated.

A third method combines features of the annual egg production method and the daily egg production method, and is called the daily fecundity reduction method (Box 5.4; Lo et al. 1992). This method is applicable for fishes that have a definite annual spawning cycle (annually ripen a batch of eggs) but do not release all their eggs at once. By making a series of collections of adults and noting the reduction in the numbers of eggs remaining in the ovary as the spawning season progresses, a daily egg production rate for the population can be established. This is compared to the amount of eggs collected in the plankton at specified times during the spawning season to estimate the abundance of spawners.

BOX 5.3. Steps in Daily Egg Production Method

1. Conduct survey to collect eggs and spawning adults.
2. Sort plankton samples and identify eggs in each sample ($_l$).
3. Determine numbers (n) of eggs of each stage ($_g$) in each sample (n_g).
4. Determine numbers of eggs of each stage by surface area sampled at each station ($c_{g,l}$).
5. Determine ages of each stage, using temperature-dependent egg development data, and temperatures measured during egg survey ($c_{a,l}$).
6. Calculate area represented by each survey station (a_l).
7. Estimate abundance of each age group of eggs in survey (sum of abundances of each age group at each station times each station area) ($\Sigma c_{a,l} \cdot a_l = P_a$).
8. Plot egg abundance by stage to estimate egg mortality rate ($P_a = P_d e^{-zt}$) and number of eggs spawned on mean survey date (P_d).
9. Use samples of adults collected during the survey to estimate average female weight (W), sex ratio (R), batch fecundity (F), and the fraction of the females that spawned during the previous day (S).

$B = P_d/(R/W)SF$
B = spawning biomass (Determined by steps 1 through 8 above)
P_d = daily production of eggs in survey area (Determined from samples of spawning adults in steps 1 and 9 above)
R = total weight of fish in sample/weight of females in sample
W = average weight of females in sample
S = fraction of females that spawned during the previous day
F = batch fecundity of females

$c_{g,l} = n_{g,l} d/ab*10$ (Number of eggs by stage and station per sea surface area)
$n_{g,l}$ = number of eggs by stage ($_g$) in a sample ($_l$)
d = maximum depth of tow (Determined by wire angles or depth sensor on net)
a = mouth area of net (πr^2) (r = radius of mouth of net)
b = length of tow (Determined from flowmeter readings)

BOX 5.4. Steps in Daily Fecundity Reduction Method

1. Conduct plankton surveys to collect eggs.
2. Sort samples and identify eggs in each sample ($_l$).
3. Determine numbers (n) of eggs of each stage ($_g$) in each sample (n_g).
4. Determine numbers of eggs of each stage by surface area sampled at each station ($c_{g,l}$).
5. Determine ages of each stage, using temperature-dependent egg development data, and temperatures measured during egg survey ($c_{a,l}$).
6. Calculate area represented by each survey station (a_l).
7. Estimate abundance of each age group of eggs in survey (sum of abundances of each age group at each station times each station area) ($\Sigma c_{a,l} \cdot a_l = N_s$).
8. Plot egg abundance by stage to estimate egg mortality rate.
9. Use mortality estimate with abundance estimate of each stage to estimate numbers of eggs spawned each day ($N_t = N_0 e^{-zt}$).
10. Based on samples of spawning adults, estimate sex ratio (R), average fish weight (W), total fecundity of females (E), and number of females with active ovaries (G).

$B = N_0 / (RW) / D$
B = spawning biomass
N_0 = daily production of eggs in survey area (Determined by steps 1 through 9 above)
R = total weight of fish in sample/weight of females in sample (Determined from samples of spawning adults)
E = number of advanced oocytes in average-weight female
W = average weight of females
G = fraction of females with active ovaries
D = daily fecundity per female (i.e., $d(E*G)/dt$)

$c_{g,l} = n_{g,l} d/ab*10$ (Number of eggs by stage and station per sea surface area)
$n_{g,l}$ = number of eggs by stage ($_g$) in a sample ($_l$)
d = maximum depth of tow (Determined by wire angles or depth sensor on net)
a = mouth area of net (πr^2) (r = radius of mouth of net)
b = length of tow (Determined from flowmeter readings)

Mortality Estimation

Many field studies require that mortality rates of fish eggs and larvae be estimated. In using spawned eggs for biomass estimates, the number of eggs collected at any age needs to be equated to the number of eggs spawned by taking mortality into account. In many ecological studies involving fish eggs and larvae, mortality rates under varying conditions need to be estimated. Mortality rates are generally estimated by developing a catch curve of eggs or larvae of various ages. To determine the ages of eggs, age/stage of development relationships at various temperatures must be developed for the species under study. For larvae, determining age by otoliths is usually applied. The decreases in numbers of subsequent ages are then interpreted as mortality. This can be presented as a mortality rate per day, or the proportion of eggs or larvae that die each day. Caution must be taken to ensure that the same population of eggs and larvae are represented in each sampling. If eggs or larvae are recruited to the sampling area, or leave it through advection, the changes in numbers represent other processes besides mortality. Sampling biases can also affect computed mortality rates. For example, if larger larvae escape the net, the decrease in their numbers would inflate the apparent mortality rate.

Another method for estimating mortality rate from ichthyoplankton samples is to follow the reduction in the abundance of cohorts of eggs and larvae. This is particularly effective when spawning occurs during a short interval on a diel cycle, and when egg development lasts only a few days. Otherwise, it is difficult to assign eggs and larvae in samples to cohorts. Population mortality rates can be estimated this way, if the sampling represents the eggs and larvae produced by the entire population.

Assemblage Analysis

For many ecosystem studies that include ichthyoplankton, it is useful to determine which species frequently occur together. This can lead to investigations of the environmental conditions that characterize certain groups (assemblages) of eggs and larvae (Moser and Smith 1993). Concordant or disconcordant changes in abundance of various members of assemblages may signal changes in adult abundance or changes in survival of the eggs and larvae. Such changes may be due to changes in the ecosystem of which the eggs and larvae are components. Various methods have been used to analyze assemblages of ichthyoplankton. A common method, which uses only presence and absence data, is called REGROUP (e.g., Moser and Watson 2006). Both groups of species and groups of stations can be derived using this method. Results can be mapped and compared with geographic and environmental information. Often, groups of species that are associated with various seasons and offshore, nearshore, northern, and southern parts of the sampling area are seen. The distributions of these groups are frequently associated with certain ranges of environmental parameters such as temperature and salinity. More sophisticated methods that take into account abundance as well as occurrence include various forms of cluster analysis and principal component analysis (e.g., Doyle et al. 2002). Environmental data can be included in these analyses as well as egg and larval abundance.

CHAPTER 6

Population Dynamics and Recruitment

POPULATION DYNAMICS
   Population Characteristics
      Fecundity
      Longevity
      Individual Size
      Growth Rate
      Population Size
      Specificity of Habitat Requirements
      Spawning Frequency
      Parental Investment in Individual Offspring
      Early Life-History Traits

POPULATION FLUCTUATIONS
   Scales of Population Fluctuations
   Ranges of Population Fluctuations
   Concordant and Disconcordant Fluctuations
   Causes of Population Fluctuations

SPAWNER-RECRUIT RELATIONSHIP
   General Theory
   Density Dependence
   Environmental Influences
      Density-Independent Effects
   Empirical Relationships
   Recruitment Theories

      Critical Period Theory
         Survival Curves for Natural Populations of Fishes
         Evidence for Starvation of Larvae at Sea
         Evidence for the Sensitivity of Larval Fish to Lack of Food
      Offshore Transport-Retention Theory
      Growth-Mortality Theory

FACTORS AFFECTING SURVIVAL OF EGGS AND LARVAE
   Physical Factors
      Temperature
      Salinity
      Storminess and Turbulence
      Currents and Flow Features
   Biological Factors
      Feeding
      Predators

IMPORTANCE OF JUVENILE STAGE
   When Is Recruitment Set?
   Factors Affecting Survival of Juveniles
      Food
      Predators

Currents
Juvenile Habitat Requirements
(Nursery Grounds)

RECRUITMENT STUDIES
History of Recruitment Studies in the United States
(From Kendall and Duker 1998)
Recent Recruitment Studies
Fisheries-Oceanography Coordinated Investigations (FOCI)
South Atlantic Bight Recruitment Experiment (SABRE)

Global Ocean Ecosystem Dynamics (GLOBEC)
ICES: Cod and Climate Change
U.S. GLOBEC: Georges Bank
Future Directions of Recruitment Studies
Timescales of Variability and Biological Responses
A New Paradigm for Recruitment Research
Timescales
Multiple Trophic Levels
Multispecies Focus

POPULATION DYNAMICS

In order to manage harvested fish populations, the sources and magnitude of changes in population size must be known. How much of the variation in population size is due to natural causes and how much is due to fishing? How much of the population can be harvested without affecting its ability to reproduce itself? When considering the impact of pollution or other anthropogenic habitat changes, it is essential to know how fish populations are affected; are changes due to these factors, or can such changes be expected without these environmental changes? Population dynamics is the study of fluctuations in abundance of populations. It considers how much populations vary, and over what timescales, and seeks to find the causes of these variations. These are complex issues with most fish populations having high fecundities, complex life cycles, and multiple year-classes in the mature population (Fogarty et al. 1991).

Much of the study of population dynamics in fishes has focused on recruitment, since variations in annual reproductive output of populations seem to drive population abundance in most fishes. However, the term *recruitment* has several meanings in fishery literature. It can mean recruitment to the adult population, to the fished population, to the spawning stock, or to the juvenile stage. In this discussion the latter meaning is used; that is, recruitment is the number of juveniles in a population that annually survives the egg and larval stages.

Populations of most nontropical fishes are composed of several distinct year-classes (cohorts) (Figure 6.1). The variable contribution of annual recruitment to the population produces year-classes of varying abundance (Figure 6.2). Once a year-class has recruited to the population (i.e., reached the juvenile stage), there is limited interannual variation in its natural mortality rate,

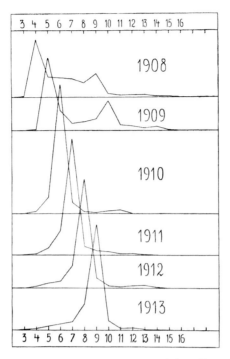

Figure 6.1. Example of year-class phenomenon in fishes (from Hjort 1914). Composition in point of age of spring Atlantic herring (*Clupea harengus*), 1908–1913. 4 = 4 years old (scales showing 4 winter rings).

however it may also be subjected to fishing mortality. In each successive year, the abundance of a year-class is diminished, however the relative abundance of each year-class maintains itself from year to year in the population. Strong and weak year-classes can be seen as peaks and valleys in age frequency diagrams of the population. Year-classes can be seen moving through the population in successive annual age frequency diagrams of the population.

A common feature of population dynamics of fishes is that the number of progeny recruiting to the population is often not proportional to the number of adults (there is little if any relationship between stock and recruitment [Koslow 1992]). However, the range of variations in population fecundity is related to the range of variations in recruitment in some fishes (Serebryakov 1990). In the long term each adult must reproduce itself for population stability. However, with fishes that produce on the order of 10^5 to 10^6 eggs per year, most must die or the population will explode. Very small changes in the mortality of early stages will have dramatic impacts on recruitment to the adult population. Given this basic life-history pattern,

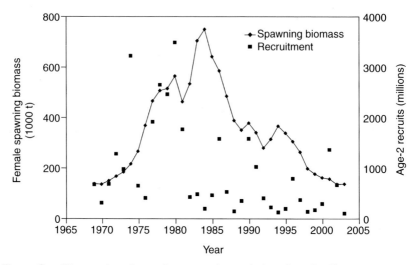

Figure 6.2. Time series of recruitment and population size of walleye pollock (*Theragra chalcogramma*) from Shelikof Strait, Gulf of Alaska (based on Dorn et al. 2003).

the most remarkable feature of population dynamics of fishes may be their relative stability rather than their variability (see Heath 1992). Here we briefly consider some population characteristics of fishes that affect their dynamics, with emphasis on those components most directly related to early life-history studies. We then discuss some of the factors that are involved in population fluctuations in fishes.

Population Characteristics

Several interrelated characteristics affect population dynamics. These vary considerably among fishes. Each species has evolved ranges of values of these characteristics that assure its continuation. Feedback (compensation or density dependence) may be observed in some of these characteristics as population size tends to be beyond limits for the species.

Fecundity. The number of eggs spawned is termed *fecundity*. In fishes, individual fecundity is usually related directly to female size, after the female reaches first maturity. The number of eggs produced is usually closely related to the cube of female length. Some species develop all the eggs for a spawning season at once (synchronous spawners), whereas others develop several batches during the spawning season (serial spawners). The number of eggs produced by the population (population fecundity) is a product of the number of eggs produced by the females in the population and the individual fecundity. Since individual fecundity is a function of female age and size, population fecundity depends not only on the number of individuals in the population but also their age, size, and sex ratio. Because individual fecundity is a function of weight once maturity is reached, population fecundity is frequently expressed as a function of population biomass. Errors in stock-recruitment relationships can occur from using spawning-stock biomass as a measure of egg production to compare with recruitment (Rothschild and Fogarty 1989).

Specific fecundity (the number of eggs per gram of female weight) may change in relation to condition of the female. Condition of adults might be related to changes in their food supply due to changes in the environment. Density-dependent changes in fecundity have been observed in some species. At low population sizes, age at first maturity can be reduced, and specific fecundity can be increased. At high population sizes food may be limiting, and specific fecundity may be reduced. However, most harvested species are probably at densities far below the carrying capacity of their environment, so density-dependent reduction in fecundity would not be expected.

Longevity. The maximum age of fishes varies widely among species. A few tropical fishes are annual, in that they regularly live for only one year. At the other extreme, several species of rockfishes (*Sebastes* spp.) live for over 100 years; the oldest rockfish found to date was 209 years old (Love et al. 2002). Many fishes have longevities of 10–20 years. Clearly once maturity is reached in iteroparous species, the number of times an individual spawns is directly related to its longevity. Conversely, the number of offspring that need to survive from each spawning to maintain the species is inversely related to the number of times the individual spawns. Also, the importance of each year-class to the population abundance is inversely related to the number of year-classes present in the population.

In many fishes harvest has changed the age frequency distribution by reducing or eliminating the older age classes in the population. This reduces the population fecundity, but also frequently causes other changes associated with reproduction (Longhurst 2002). In some species older (larger) fish spawn earlier or later in the spawning season than other members of the population. This can have an effect on the match/mismatch of the larvae and their food supply. Older spawners often have a more prolonged spawning season, allowing more resiliency to unfavorable environmental conditions. Also, the quality of eggs varies with the age of the spawners: the eggs of younger fish are not as fit as those of older fish.

Individual Size. Maximum length among fishes ranges from about 15 mm (*Schlinderia*) to 11 m (oarfish [*Regalecus glesne*]). Since fecundity is related to fish size, species that are smaller in size cannot produce as many eggs per individual as species that are larger. Also, the range of egg sizes is not as great as the range of adult sizes. Regardless of adult size, eggs of most species are about 1–3 mm. Even species with very large adults, such as tunas (*Thunnus* spp.), have eggs that are about 1 mm in diameter. The size of the eggs in fishes has more to do with early life-history traits than it does with the size of the adults.

Growth Rate. Most fishes show indeterminate growth; that is, they continue to grow throughout their lifetime. Roughly, growth is linearly related to length, and related to the weight cubed. Growth rate may vary considerably during a fish's life. Growth from the time the fish becomes a juvenile until it approaches first maturity may be quite rapid, but then energy is put into gonad development and somatic growth slows. After first maturity, growth varies according to the annual cycle of spawning, recovery, and feeding. Some fishes continue to grow at a fairly steady annual rate, however, in others growth slows considerably in older fish (e.g., sablefish [*Anoplopoma fimbria*]). Outside of the tropics, growth varies seasonally, being slow or negligible during colder seasons but more rapid during warmer seasons. Growth rate can increase in the presence of abundant food and decrease when food is less abundant. Intraspecific competition for food can result in decreased growth at large population sizes. Since fecundity is related to fish size, it can be impacted by changes in growth rate.

Population Size. Population sizes of fishes vary tremendously (Figure 6.3). Some species of fishes are the most abundant vertebrates on Earth, whereas other species are very restricted in distribution, and their population sizes number only in the hundreds or thousands. Desert pupfishes (*Cyprinodon* spp.) of the arid Southwest United States and cavefishes (amblyopsids) are examples of such limited populations. These are small, short-lived species whose fecundity is relatively low. Clearly species like these are at risk of extirpation from recruitment failures or environmental problems. Special measures such as preservation of critical habitat or artificial culture and breeding programs are sometimes required to maintain these populations. The journal *Environmental Biology of Fishes* includes a series of articles on endangered species of fishes, which highlight problems associated with preventing their extinction (cf Cambray 2000).

At the other extreme, numerous species of coastal marine fishes have population sizes that number in the billions (10^9). For example, walleye pollock (*Theragra chalcogramma*) ranges throughout the North Pacific Rim from Japan to Washington State. There are several populations within this extensive range, the largest of which (about 20 million tonnes) occurs in the eastern Bering Sea. Herrings (clupeids), anchovies (engraulids), mackerel (*Scomber* pp.), and

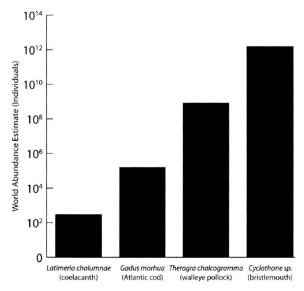

Figure 6.3. Examples of variability in world population abundance of fishes. Coelacanth (*Latimeria* sp.) from Hissmann et al. 1998, cod (*Gadus* spp.) and walleye pollock (*Theragra chalcogramma*) from Myers et al. 1995, and bristlemouths (*Cyclothone* spp.) estimated from information in McClain et al. 2001.

some other codfishes (gadids) also have large populations. Most of these fishes are heavily exploited. Many of these populations comprise several year-classes, and their fecundity is relatively high. Annual population egg production is often on the order of 10^{14}. Although these populations can fluctuate widely in abundance interannually and on longer timescales, their large abundances make them at much lower risk of extinction than species with smaller overall population sizes. These species can weather several consecutive years of poor recruitment. Nevertheless, overharvest combined with poor recruitment has recently reduced the population sizes of some of these species to the point that fishing has been severely restricted or closed in an effort to rebuild the stocks.

There is much more resiliency to recruitment failures in fishes with large population sizes than in those with small population sizes. This is true because of their sheer numbers, the fact that the population comprises several year-classes, and because with the large numbers there is more genetic diversity so gene frequencies can change and thus the population can adapt more quickly to environmental changes. For example, Myers and Pepin (1994) found more variability in recruitment among populations that occupied isolated banks than those of the same species that occupied larger shelf areas.

Specificity of Habitat Requirements. Each species has habitat requirements for each life stage. The life-history strategy of many species includes migration of prespawning adults to specific spawning areas, and drift of eggs and larvae to juvenile settling or nursery grounds, followed by gradual or rapid return of adults to the spawning area. Some species inhabit large geographic areas, and presumably are tolerant of a range of environmental conditions. Other species have very restricted ranges and more stringent habitat requirements. The habitat requirements for some life-history stages may be more restricted than for other stages. For example, eggs of some species are deposited on certain vegetation (Pacific herring [*Clupea pallasi*] on eelgrass or kelp), or in other animals such as bivalves or sponges. To complete their life cycles these species need to live where these other organisms are present. Some species have restricted physical or geographic requirements, such as salmon that spawn in freshwater streams but spend most of their lives in offshore oceanic waters. Many species of flatfishes (pleuronectiforms) have pelagic eggs and larvae, settle as juveniles in shallow waters with sandy bottoms, and as adults live on the bottom in deeper waters. Juveniles of coral reef fishes often require specific reef environments for successful settlement. Some fishes that occur offshore as adults occupy estuaries as young-of-the-year juveniles. The amount of such habitat for the juveniles can limit population size. Density-dependent effects are seen in such populations when juvenile habitat is reduced, or population numbers are high. Population size can be affected by problems with the habitat needed by any life stage. Various species are more or less vulnerable to these problems depending on how much of their critical habitat is available and how specific their requirements are.

Spawning Frequency. Aside from the tropics, most fish spawn once annually, although some spawn several batches of eggs over a period of days or weeks. Some temperate species, such as northern anchovy (*Engraulis mordax*), spawn batches of eggs every few days over a spawning season of several months. At lower latitudes, where seasonal environmental signals are less intense, fish spawn more frequently, up to nightly in some reef fishes. Once maturity is reached, spawning usually continues at the established frequency for the species for the lifetime of the individual (iteroparity). In a few fishes (e.g., Pacific salmon [*Oncorhynchus* spp.], and freshwater eels [anguillids]), spawning occurs only once near the end of the life of the individual (semelparity).

Parental Investment in Individual Offspring. In all species, choosing the right time and place to release young is important to the survival of the species (Hinckley et al. 2001). Beyond that, fishes vary tremendously in how much care they give their offspring. At one end of the spectrum are the surfperches (embiotocids), which carry and nourish their young internally in the females until they are released as fully transformed juveniles. The males of some species of surfperches are sexually mature when they are released. In other fishes such as the live-bearers (poecilliids) the females provide nourishment, but release young that grow considerably before reaching maturity. Rockfishes demonstrate a lesser degree of maternal care in that the young are released at the preflexion stage with little yolk left. Other fishes (e.g., sunfishes [centrarchids]) spawn unfertilized and undeveloped eggs, but protect the eggs during development in nests that one or both parents guard. Salmon and other fishes place their eggs in nests but do not guard them. Most fishes, particularly marine species, spawn unfertilized eggs into the open water where the male releases sperm, and afford the young no additional care. In fact, egg cannibalism is a regular part of the life history in some fishes (e.g., northern anchovy).

Generally, fecundity is inversely related to the amount of care that the parents afford to the young (see Barlow 1981). Thus, fecundity of live-bearers that release well-developed young is usually less than 100, whereas fecundity of broadcast spawners can be several million. The importance of mortality during early life in population dynamics is greatest in fishes with high fecundities and no parental care.

Early Life-History Traits. Concomitant with the wide variation in early life-history traits among fishes is the importance of this portion of their life history in variation in population sizes. In fishes with rather direct development (e.g., live-bearers) most variation occurs because of factors affecting fecundity or adult mortality. In fishes that protect the eggs in nests, variations in population size are probably due to changes in mortality of later stages: larvae and juveniles. However, in highly fecund fishes with extensive free-living egg and larval stages, most variation results from variations in mortality of these early stages (Figure 6.4). Since the mortality rates of these early stages are so high, the duration of the stages can be very influential in determining year-class strength, and ultimately population size and variation (Houde 1987). Pepin and Myers (1991) found that although neither egg size nor length at hatch was correlated with recruitment variability, amount of change in length (duration of the larval period) during the larval period was: the longer the larval period the more recruitment variation. Some fishes have a prolonged juvenile stage that may require more limited habitat than other life stages (e.g., estuaries, coral reefs, eelgrass beds, sandy beaches), and variations in recruitment might be caused by events that occur during this stage. Density-dependent effects can be expected in species whose juveniles require such a spatially limited habitat.

Very few freshwater fishes spawn pelagic eggs; most produce demersal eggs and provide them some protection (nests, etc.). Eggs of freshwater fishes are generally larger than those of marine fishes. Houde (1994) reviewed larval characteristics of freshwater and marine fishes and found that freshwater larvae are larger than marine larvae. Survivorship of freshwater larvae was greater than that of marine larvae. He concluded that larval dynamics of

190 Population Dynamics and Recruitment

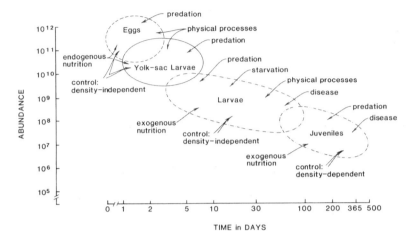

Figure 6.4. A conceptualization of the recruitment process in fishes including the sources of nutrition, probable sources of mortality, and hypothesized mechanisms of control for four early life-history stages. Log_{10} scales are used on both axes (from Houde 1987).

marine fishes would be more important in recruitment variability than in freshwater fishes where juvenile stage dynamics would be more important.

POPULATION FLUCTUATIONS

We begin with a general question that has been asked about marine fish populations almost since fisheries science began: Why is there so much variability in abundant fish stocks from one year to the next—that is, could year-class strength be predicted from one year to the next if adequate background knowledge were available? As recently as 40 years ago it was generally believed that year-class strength depended mainly on the abundance of the adult spawning population; however, it is now recognized as being much more complex than that. Most of the present hypotheses (see later discussion of present hypotheses) on variability in fish production have been based on the pioneering work of the great Norwegian fisheries scientist, Johan Hjort (1914), who studied fluctuations in the Norwegian Atlantic cod (*Gadus morhua*) and Atlantic herring (*Clupea harengus*) (Figure 6.1). Hjort was the first to recognize that a critical factor important to survival rates was whether or not there was starvation of larvae during a "critical period," or the so-called "Critical Period Concept." Hjort also recognized that another factor might be dispersal of larvae to unfavorable areas, or what is now often called the "Offshore Transport Hypothesis" (Box 6.1).

> BOX 6.1. Foundation of the Critical Period Concept (Hjort 1926)
>
> "As factors, or rather events which might be expected to determine the numerical value of a new year-class, I drew attention to the following two possibilities:
>
> (1) That those individuals which at the very moment of their being hatched did not succeed in finding the very special food they wanted would die from hunger. That in other words the origin of a rich year-class would require the contemporary hatching of the eggs and the development of the special sort of plants or nauplii which the newly hatched larva needed for its nourishment.
>
> (2) That the young larvae might be carried far away out over the great depths of the Norwegian Sea, where they would not be able to return and reach the bottom on the continental shelf before the plankton in the waters died out during the autumn months of their first year of life."

Recently it has been recognized that predation on larvae can be a major factor in survival rates, and finally there has been recognition of physical oceanography and meteorological events being much greater factors in population fluctuations.

Scales of Population Fluctuations

Most population structure in fishes has been associated with year-class phenomena; that is, interannual variation in recruitment has been the most important source of variations in population abundance. However, population abundance also varies on timescales longer than annual. Some populations seem to increase in abundance by having several successive good recruitment years. Likewise, if they have several successive poor recruitment years, the populations gradually decrease in abundance over several years.

In some cases marine ecosystems seem to fluctuate between more than one state. These states are characterized by different physical oceanographic conditions, levels of production, and relative abundances of several populations. Such ecosystem states have been called "Regimes," and changes from one state to another have been called "Regime Shifts" (Steele 1996; Beamish et al. 1999). Since different species thrive in different regimes, it seems that recruitment of some species is enhanced in one regime, but diminished in another. The mechanisms associated with regime shifts and the biological responses to them are poorly understood but are subjects of ongoing research.

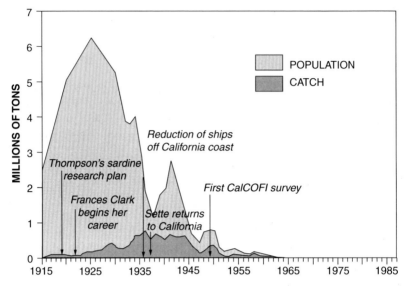

Figure 6.5. Time series of abundance and commercial catch of Pacific sardines (*Sardinops sagax*) off the coast of California since 1915 (from Hewitt 1988).

Ranges of Population Fluctuations

Variation in recruitment success is the norm among fishes, but recruitment varies more in some species than in others (see Rothschild 1986, Table 2.1). Tenfold variations in year-class strength are common in fishes. In the 1930s Pacific sardine (*Sardinops sagax*) supported a large fishery off the coast of California. By the 1960s the fishery was closed and sardines were rare (Figure 6.5). The population went from over 10 million metric tonnes to practically none (Hewitt 1988). Since the 1990s sardines have become abundant enough again for a limited fishery to open.

Species with more limited fluctuations in recruitment include Pacific halibut (*Hippoglossus stenolepis*) and sablefish. These species are long-lived, so the population is composed of many year-classes.

Concordant and Disconcordant Fluctuations

Populations of several species in an area sometimes show the same patterns of variations in abundance. For example, Cushing (1980) noted that several species of codfishes in the North Sea showed similar increases in abundance in the late 1960s and called it the "gadoid outburst." Koslow et al. (1987) found coherence in abundance trends of Atlantic cod and haddock

(*Melanogrammus aeglefinus*) stocks in the Northwest Atlantic. Hollowed et al. (1987) showed that among several common fishes in the Northeast Pacific, some showed concordant population trends, whereas others showed disconcordant trends. The anchovies and sardines of the eastern boundary currents in several parts of the world show disconcordant patterns of abundance (Lluch-Belda et al. 1992). Fish scales deposited in anaerobic basins off the coast of California show that these patterns of abundance in Pacific sardines and northern anchovies have occurred over the 2,000-year duration of the record (Baumgartner et al. 1992). Thus, although such fluctuations in abundance may be influenced by harvest, at least some occur in the absence of fishing.

Since population size is largely determined by year-class success, these concordant and disconcordant fluctuations in abundance indicate that conditions for survival of early life-history stages vary in the same ways for these species. For concordantly varying species, environmental conditions that are favorable or unfavorable for one of these species are likewise favorable for the other species. For disconcordantly varying species, conditions that are favorable for one group of species are unfavorable for the other species. For example, Turrell (1992) suggested that inverse trends in population size in Atlantic herring and sand lances (ammodytids) in the North Sea might be due to differences in currents that transport their larvae.

Causes of Population Fluctuations

The search for causes of population fluctuations has occupied much of fishery science for almost 100 years, since it became apparent that fluctuations were largely due to events occurring during early life history, which resulted in variations in recruitment. Rarely have single causes been found to explain variations in recruitment. The broad conclusions of most studies are that several physical and biological factors are involved, acting in nonlinear ways at various times to cause mortality in the egg, larval, and juvenile stages. These factors will be discussed in some detail after we look at the relationship between spawners and recruitment.

Besides the natural causes of fluctuations, the activities of humans can cause changes (usually decreases) in fish abundance. Fishing reduces the number of fish that produces eggs, and thus the total egg output of the population. Some fisheries catch disproportionate numbers of larger, more fecund fish, and thus have a large impact on population fecundity. Some fisheries are conducted on fishes that are aggregated for spawning, so in addition to removing fish from the spawning population, spawning migration and behavior might be compromised.

A number of pollutants can kill fish eggs and larvae directly, or cause changes in the productivity that reduce food available for the larvae.

Critical habitat for spawning and rearing of fishes can be reduced or destroyed by a variety of human activities. Dredging and filling in coastal and freshwater areas can limit the habitat available for fish rearing. Dams can prevent anadromous fishes from making spawning migrations, or young catadromous fishes from returning to fresh water.

SPAWNER-RECRUIT RELATIONSHIP

General Theory

A straightforward appraisal of the spawner-recruit relationship would seem to indicate that, in general, higher fecundity implies that there will be greater recruitment (i.e., reproductive success) and that more eggs and larvae will survive to the adult stage (Myers and Barrowman 1996). Many fisheries scientists believe that the greater the parental stock size, the better the recruitment, and in some cases it is true, but clearly for some species it is not true—that is, you can have a very high number of adults spawning, yet very low recruitment, or very low numbers of spawners and very high recruitment. (Of course, there is no argument that if there are few to no spawners present, there is little to no recruitment from those spawners.)

The relationship of fecundity to population densities is still unclear and probably is a very species-specific relationship or even a subpopulation relationship. As an example, Bagenal (1973) showed that based on 25 years of data there was tremendous variation in fecundity from year to year, citing data for plaice (*Pleuronectes platessa*) (48%), Norway pout (*Trisopterus esmarkii*) (250%), haddock (56%), and Atlantic herring (34%). He found that lower fecundities were associated with years of high population densities, and higher fecundities were associated with years of low population densities. With this data he suggested that fecundity variations form a density-dependent population regulating mechanism to prevent wide fluctuations in recruitment. Rice et al. (1987) found that although egg deposition in bloater (*Coregonus hoyi*) in 1983 was only 57% of that in 1982, larval recruitment from those fewer eggs in 1982 was 2.4 times greater than larval recruitment from the 1983 egg deposition.

Some density-dependent or compensatory mechanisms (i.e., may vary depending on population densities) are age at first maturity, intraspecific competition for food and spawning space, and cannibalism.

Density Dependence

If recruitment were proportional to the number of eggs produced at all population sizes, populations would continue to increase indefinitely. This is not the case, so some means must be in effect to keep populations in check. The mechanisms for changes in rates of recruitment at different population levels are called density-dependent mechanisms. Such mechanisms include cannibalism, changes in growth rate, changes in time of maturity, and changes in fecundity. By these mechanisms recruitment rate is increased at small population sizes and reduced at large population sizes. Some of these mechanisms change vital rates whereas others cause compensatory mortality. It has been difficult to demonstrate density-dependent or compensatory mortality of pelagic eggs and larvae of fishes. Eggs of fishes that spawn demersally on specific spawning beds may become so dense that some of the eggs may suffocate. Cushing (1983) demonstrated mathematically that larvae may be dense enough in some cases to affect their food supply of zooplankton. At the reduced population levels of most commercial fishes at this time such density-dependent effects would not be expected. Juveniles may be dense enough to cause density-dependent mortality by consuming more food than is produced, or by crowding each other in nursery areas. As juveniles settle, they go from living in a three-dimensional space (the water column) to living in a two-dimensional space (the bottom). At higher densities some settlement might occur in marginal areas, so those juveniles will be subject to increased predation and may not find sufficient food for optimal growth.

Environmental Influences

Density-Independent Effects. Environmental influences cause the same amount of mortality regardless of the population size; thus they are considered density-independent effects. Such effects include changes in temperature, which change the rate of metabolism of eggs and larvae and might affect production of their prey, and changes in currents that transport eggs and larvae to or away from suitable nursery grounds.

Thus, density-dependent and possibly density-independent factors contribute to population regulation during early stages in fishes. Anything that causes direct mortality or decreases growth rate, which prolongs the vulnerable egg and larval stages, will reduce the number that eventually recruit to the adult population. Interannual changes in the factors that affect mortality during these early stages lead to differences in recruitment, and thus year-class strength.

Empirical Relationships

Over the years several empirical and theoretical relationships have been proposed to describe the relationship between the biomass of spawners (size of the stock) and the number of recruits that is produced annually (size of the year-class). These tend to be curvilinear to reflect the idea of density dependence (Figure 6.6). The degree of curvilinearity varies among the formulations. With the Ricker curve, the level of recruitment peaks at intermediate

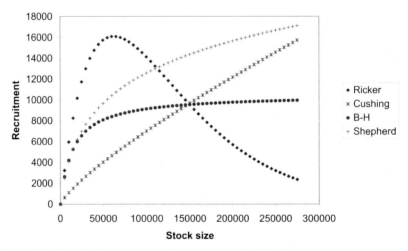

Figure 6.6. Empirical stock/recruitment curves (original).

levels of spawner biomass. With the Beverton-Holt curve, recruitment levels reach an asymptote at high level of spawner biomass. With the Cushing curve, recruitment increase slows at high spawner biomass levels. Thus the Ricker curve demonstrates overcompensation, whereas the Beverton-Holt and the Cushing curves demonstrate compensation. Shepherd developed a flexible empirical stock recruitment model that can be used to fit difficult data sets. Iles (1994), Cushing (1995a), Myers et al. (1995), and Myers (2001) provide good summaries of these relationships. Actual data of spawners plotted against recruits show considerable noise however (Myers et al. 1995), which is thought to reflect the effects of the environment on survival of the egg and larval stages, before recruitment (year-class strength) is set. Frequently, there are also significant errors associated with both the estimates of spawning stock size and numbers of recruits.

It would be ideal to plot the number of eggs produced by the population (population fecundity) against the number of recruits that results from the spawning. However, in many cases, the best estimate of spawning population size is the biomass of the population. This estimate deviates from the ideal in that the sex ratio, age at maturity, and size/fecundity relationships are not accounted for.

The number of recruits used to plot these relationships is also problematic. If a survey estimate of numbers of settled juveniles is used, it assumes that the year-class size is fixed before the juveniles are sampled. Differences in mortality of the juveniles after they are sampled could still change the year-class strength to varying amounts. In some cases, an index of juvenile abundance is used. Bias in this estimate can result from interannual differences in distribution of the juveniles relative to the survey area, or in the timing of the survey relative to the seasonality of the early life-history events.

Following are symbols for stock-recruitment relationships:

R = Recruitment
S = Spawners
α = Slope of function at $S = 0$
β = coefficient of density dependence (> 0)
K = Threshold biomass (> 0)
γ = Degree of compensation ($> 0, < 10$)

Shepherd (1982): Shepherd wrote the spawner-recruitment relationship in a general form from which earlier formulations can be derived:

$$R = \alpha S / 1 + (S/K)^\gamma$$

Ricker (1954, 1958): Ricker proposed that the relationship between spawner biomass (a surrogate for number of eggs produced) and recruitment could be expressed by a logistic function:

$$R = \alpha S^{-\beta S}$$

This curve has an asymmetric dome shape. Recruitment increases rapidly as spawners increase at low stock abundance. After peaking at relatively low stock abundance, recruitment slowly decreases at higher spawner abundances. This relationship was developed mostly for salmonids (salmon and trout) where there is limited spawning area. Thus, the density-dependent factor was amount of spawning habitat. This relationship has been applied to many other groups of fishes since its introduction.

Beverton and Holt (1957): Soon after Ricker proposed his formulation, Beverton and Holt proposed that the spawner-recruit relationship could be expressed as an asymptotic curve:

$$R = \alpha S / (1 + (S/K))$$

With this curve, recruitment increases with number of spawners at low spawner levels, but the increase slows and eventually levels off at larger stock sizes. Such a relationship would be realized if nursery grounds had a constant carrying capacity.

Cushing (1971): Cushing proposed that the spawner-recruitment relationship could be expressed as an allometric or power function:

$$R = \alpha S^\beta$$

With this curve, the rate of increase in recruitment slows at high spawner biomass levels.

Recruitment Theories

Before 1914, the general theory about why there was such variation in fish populations was that the populations of fish just moved elsewhere. This idea about recruitment variation changed with the critical period and offshore transport concepts of Hjort (1914), and then with subsequent expansions and refinements of these concepts.

Critical Period Theory. Johan Hjort is responsible for the famous critical period concept, which indicated population fluctuations were due to events during early stages. Hjort suggested that the year-class strength of a particular fish species would be determined at the earliest larval stages since it is the most critical time during the life history of a fish. In his classic 1914 paper, Hjort proposed two major factors that could increase larval mortality: (1) lack of prey items during a "critical period" within which the larvae must feed or die ("Critical Period Hypothesis"), and (2) larvae could be dispersed by advective currents to unfavorable areas where they may not grow or would not be recruited to the adult population and would perish ("Offshore Transport Hypothesis").

The Critical Period Concept became popular in the late 1960s and early 1970s, leading many researchers to do lab and field work on various aspects of larval ecology. May (1974) critically reviewed research addressing the Critical Period Concept by considering: (1) survival curves for natural populations of larval fishes, (2) evidence for the starvation of larvae at sea, and (3) the sensitivity of larval fishes to lack of food. The following discussion is based on May's considerations, summarizing his review and adding some of the more recent literature pertaining to this subject.

SURVIVAL CURVES FOR NATURAL POPULATIONS OF FISHES. Although one might expect survival for larvae to show a steep decrease after yolk sac absorption (Figure 6.7) in natural populations, there are few survival curves available

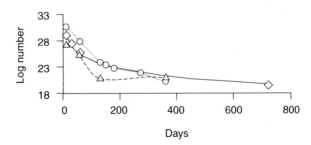

Figure 6.7. An example of the time course of mortality from eggs through juveniles of plaice (from Bailey et al. 2005).

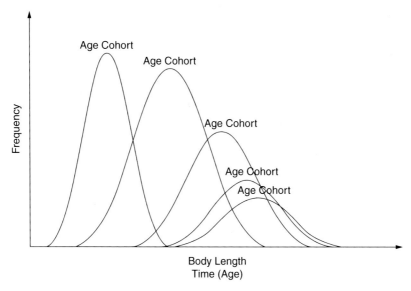

Figure 6.8. Hypothetical illustration of the problem of estimating growth rate of larval fish from length frequency diagrams because of overlapping curves. Daily growth ring analysis is now almost exclusively used to age larval fish.

and May (1974) could only find a few cases of this in the literature. In addition, there are sampling problems associated with ichthyoplankton surveys—for example, patchiness, filtration, larval vertical distribution, and net detection and avoidance by visual or neuromast detection. Finally, in order to generate survival curves for natural populations from these data, larval growth rates are needed, and these are often poor. Growth rates were usually estimated from length frequency modes, which are unreliable because of indistinct separation of growth from separate spawnings based on length alone (Figure 6.8). Larvae must be aged using otoliths and the daily growth rings present; this technique has been used extensively, but is not foolproof. McGurk (1984) has critiqued studies and shown that many first-feeding larvae in the lab that are starving or growing slowly do not meet two basic assumptions: (1) that the first ring is deposited at a fixed age in each species, and (2) that the rate of ring deposition is one ring per day. Based on survival curves available when May did his review, data were not convincing whether mortality is concentrated at the end of the yolk sac stage. Leggett and Deblois (1994) did a more recent review of the literature and also concluded that the evidence does not support the critcal period concept.

EVIDENCE FOR STARVATION OF LARVAE AT SEA. Is the incidence of feeding larvae low at sea? If larvae caught at sea have little or no food in their guts, it might indicate that larvae are starving. There are real sampling problems with doing larval feeding studies including that some larvae may have nocturnal feeding patterns whereas most sampling is done in the daytime; healthy, well-fed larvae may be able to avoid the net much easier than starving larvae; there may be a rapid rate of digestion of larval gut contents; and upon capture, defecation or regurgitation may immediately occur, especially if the specimens are put directly into formalin. But still, the question remains as to whether there is even qualitative data that there are emaciated or dead larvae associated with low quantities of phytoplankton and zooplankton. In general, where more known food items are located, more larvae are healthy—at least for some species (e.g., American plaice [*Hippoglossoides platessoides*]). Histological techniques were developed for northern anchovy (Theilacker 1987) and also applied to other species. If larvae do not initiate feeding shortly after yolk sac absorption, starvation results in degeneration of the intestinal, pancreatic, and liver cells and worsens with time. This has been observed in both the lab and field, and corroborated with RNA/DNA ratios. (The amount of DNA in cells is fixed, but the amount of RNA is related to the amount of protein synthesis taking place: the more RNA the more synthesis. RNA/DNA ratios have been found to indicate recent feeding conditions experienced by larvae [see Clemmesen 1996].)

EVIDENCE FOR THE SENSITIVITY OF LARVAL FISHES TO LACK OF FOOD. Is there a "point of no return" for fish larvae? If one delays larval feeding, is there a point in time when the larvae will not feed and will die even if they are offered the option to feed? Lasker et al. (1970) found that if larval northern anchovy were deprived of food for more than 1.5 days after yolk sac absorption, irreversible starvation would occur. Many species do have a point of no return. On the other hand, starved California grunion (*Leuresthes tenuis*) larvae (May 1974) will survive if fed as long as they are still alive to feed—that is, grunion larvae never passed a point of no return (even after 20 days), and larval bloaters in Lake Michigan appear to not have a point of no return (L.B. Crowder, personal communication). The amount of food present in the ocean is usually orders of magnitude below the optimal amounts shown for larval fish in lab studies. In the open sea there are 13–40 nauplii per liter and 1–7 copepodites per liter, whereas in estuaries there are up to 200 nauplii per liter (Table 6.1). Many scientists have attempted to correlate plankton biomass with larval abundance by looking at whether years with high production are also years with good, strong year-classes. A somewhat successful correlation is seen with northern anchovies (Figure 6.9) where a reasonably good relationship is found, but it should be noted that a correlation does not prove a causal relationship because there may be other factors at work. Searching ability and feeding efficiency have also been studied to determine the efficiency of larvae in locating and capturing food items. The research done indicates searching ability and feeding efficiency depend on such things as prey size, prey density, prey type, success

TABLE 6.1. Food Density Thresholds for Marine Fish Larvae (Hunter 1981)

Species and Common Name	Container Volume (Liters)	Duration (Days)	Food Type	Stock Density (No./L)	Density (No./L)	Percent Survival
Plaice[a]						
Pleuronectes platessa	5	14	Artemia nauplii	50 (larvae)	1,000	72[b]
					500	72
					200	54
					100	32
Northern Anchovy[c]						
Engraulis mordax	10.8	12	Wild zooplankton (nauplii)	10 (eggs)	4,000	51
					900	12
					90	0.5
					9	0
Bay Anchovy[d]						
Anchoa mitchilli	76	16	Wild zooplankton (nauplii-copepodites)[f]	0.5–2 (eggs)	4,700[e]	65
					1,800	50
					110	10
					60	5
					30	1
Sea Bream						
Archosargus rhomboidalis	76			0.5–2 (eggs)	2,600[e]	75
					890	50
					130	10
					0	5
					50	1
Lined Sole						
Achirus lineatus	38			0.5–2 (eggs)	610[e]	75
					220	50
					30	10
					20	5
					9	1
Haddock[g]						
Melanogrammus aeglefinus	37.8	42	Wild zooplankton (nauplii)	9[h] (larvae)	3,000	39
					1,000	22
					500	3
					100	0
					10	0

[a] Wyatt (1972).
[b] Survival was 100% at 50/L for first 7 days without a decrement in length; see also Riley (1966).
[c] O'Connell and Raymond (1970).
[d] Houde (1978).
[e] Estimated food density for indicated survival levels.
[f] Plankton blooms of *Chlorella* sp. and *Anacystis* sp. maintained in rearing tanks.
[g] Laurence (1974).
[h] Estimated by adjusting for hatching success.

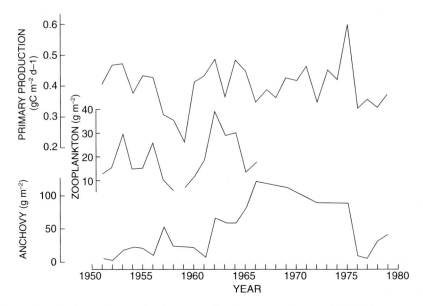

Figure 6.9. Estimated annual primary production, annual mean CalCOFI zooplankton concentrations, and northern anchovy (*Engraulis mordax*) total biomass in Southern California Bight (from Smith and Eppley 1992).

rate of capture, and others, but as is the case, there is disparity over what is found in the lab and what is seen in the field. One thought is that larvae may be able to actively seek prey of the proper quantity and quality (e.g., patches of food) and stick with it, thus increasing their chances for survival.

There are certainly other factors than just starvation at work in determining larval recruitment strength, the most obvious of these being the effect of predation, which is discussed later in this chapter. Some scientists also argue that starvation and predation are strongly influenced by density-dependent factors: the more fish larvae and eggs there are, the greater the competition for food resources among larvae, and also the easier it is for predators to feed on eggs and larvae.

To summarize Hjort's first hypothesis (Critical Period Concept), the presence of a critical period depends on a number of environmental and species-specific factors, and does occur in some fish populations; however, the work done since 1914 does not support a generalization of Hjort's first hypothesis for even a major portion of fish populations.

Cushing (1975) took the critical period concept a step further in his "Match/Mismatch Hypothesis" (Figure 6.10). The mechanism (like the critical period concept) is only applicable in temperate regions where discontinuous production occurs. Reproductive success of a fish species depends on the amount of time and abundance overlap between larval production

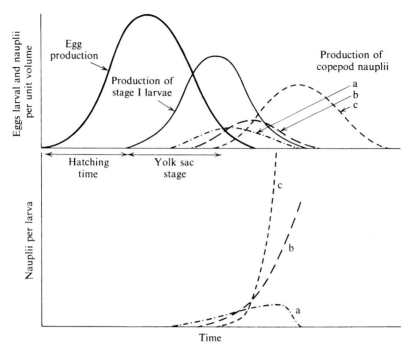

Figure 6.10. The match or mismatch of larval production to that of their larval food. The numbers of nauplii/larvae represent the degree of feeding success. The three curves represent three conditions of production and hence three conditions of feeding success (from Cushing 1975).

and larval food production. The length and time of the spawning season, which is a set period, is important for synchronization with food. Recruitment *variability* is associated with timing of phytoplankton production—that is, it is driven by physical factors, especially water temperature (but also others). The review by Leggett and Deblois (1994) found general support for the match/mismatch hypothesis, but the support was weak. Stronger support was given by Cushing (1990) and by specific studies of Atlantic cod (Ellertson et al. 1989; Mertz and Myers 1994a) and Atlantic herring (Fortier and Gagne 1990).

Lasker (1981) argued that not only is the temporal coexistence with an adequate food supply important, but also the spatial overlap of fish larvae and food so that there is the right kind and quality of food in both the horizontal and vertical profile of the water column ("Stable Ocean Hypothesis," Figure 6.11). Lasker hypothesized that upwelling and storms could disrupt and dilute

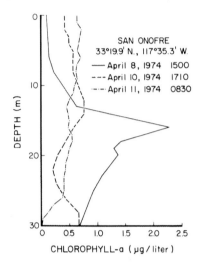

Figure 6.11. Chlorophyll maximum layers before (8 April 1974) and after (10 and 11 April 1974) a violent windstorm near San Onofre, California (from Lasker 1975).

high-quality food aggregations found in chlorophyll maximum layers. To test his hypothesis, Lasker looked at several year-classes of northern anchovy and compared their relative strengths with the prevailing physical conditions at the time of spawning with the asumption that good physical conditions would result in good phytoplankon and zooplankton production which would result in a good year-class, and poor conditions would result in a poor year-class. Lasker tested his hypothesis and it held for 1975 and 1976. With this information, Lasker reasoned that vertical stability of the water column, and suitability of food particles available to first-feeding larval anchovies, must be far more important than the magnitude of primary production by itself, and that in years with high primary production there is usually a predominance of diatoms, which are unsuitable for first-feeding anchovy larvae.

With his theory considering these spatial factors, Lasker predicted that the 1978 year-class would be weak since the 1977–1978 winter was one of the stormiest recorded and because there was no stratification of the water column until late in the spawning season. However, this was not the case as shown by Methot (1983), who tested this prediction by comparing the seasonal variation in the survival of anchovy larvae in the spring of 1978 and 1979 with the population size of juvenile anchovies caught in the autumn of 1978 and 1979. Methot's results showed that the relative survival of larval anchovies increased during 1978 and decreased during 1979. The increase in survival during the last two months of the 1978 spawning season was sufficient to cause a considerable increase in recruitment. No matches with short-term environmentally favorable conditions for either year were found to explain the difference, and Lasker's prediction of a poor 1978 year-class was incorrect. It turned out to be a very strong year-class in spite of the fact that water column stability, together with food aggregations, was not present early in the season.

An explanation for the variable recruitment of 1978 and 1979 was that coastal upwelling and offshore transport resulted in higher larval mortality. In 1979, upwelling was greater than in 1978 during the latter part of the spawning season, and this may have resulted in the displacement of many larvae to areas unfavorable for development or recruitment. Thus entrainment of larvae in spawning areas may be essential to their survival and ultimate recruitment. Methot concluded that the variability found in recruitment is probably the result of several factors, but the offshore transport hypothesis is a definite possibility of major importance—that is, Hjort's second hypothesis.

Offshore Transport-Retention Theory. Besides Hjort's Offshore Transport Hypothesis, an intriguing hypothesis about populations that emphasizes retention for survival, rather than transport for mortality, is the "Member-Vagrant Hypothesis" or more specifically, "The Herring Hypothesis" (Sinclair 1988, Sinclair and Trembley 1984). The hypothesis is based on empirical evidence that larvae of discrete herring populations are retained under the physical regime in which the adults spawn, and that the larvae are behaviorally and physiologically adapted to staying in the physical regime where they are spawned (Table 6.2 and Figure 6.12A–C). If it is a "poor" region (in terms of productivity) adults spawn in the summer and autumn, the larval phase is long, and the larvae metamorphose into juveniles within the productive season window (spring and summer). If it is a "good" region in terms of productivity, the adults spawn in winter and spring, the larval phase is short, and larvae metamorphose within the productive season window (spring and summer).

Sinclair concludes there is no apparent link between variation in year-class strength and the timing of the production cycle (in contrast to Cushing, Lasker, and others). This is a hypothesis to account for the number of herring populations, their specific location of spawning, and their mean absolute abundance. Atlantic herring deposit adhesive, demersal eggs in or at the edge of a water column of a physical oceanographic system that exists in the same geographic location from year to year. This system, coupled with the behavior of the larvae themselves, enhances aggregation in larval distribution during the first few months of the early life history—that is, the larvae are not adapted to drift with a residual current to a particular nursery area, but are doing the exact opposite: they are adapted to maintain an aggregated distribution in spite of the diffusive state of the physical environment.

TABLE 6.2. Peak Spawning Periods for Atlantic Herring Populations (Sinclair 1988)

Population	Peak Period
Clyde Sea	February 20–28
Norwegian	February 18–March 18 (27 years)
Minch	March[a]
Blackwater estuary	April[a]
Schlei Fjord–Kiel Bay	April[a]
Magdalen Island	May 9 (27 years)
Southwestern Gulf of St. Lawrence	May 14–18 (27 years)
Chedabucto Bay	May[a]
Southeastern Gulf of St. Lawrence	May 29–June 6 (26 years)
St. Lawrence estuary	June 15–July 7 (5 years)
Scots Bay	July 20–August 3 (6 years)
Northwestern North Sea	August
Southern Gulf of St. Lawrence	July 31–September 13 (27 years)
Southwestern Nova Scotia	August 25–September 10 (5 years)
Grand Manan (Historical)	July–September
Minch	September 5–25
Banks and Dogger Bank	September[a]
Manx	September
Mourne	Late September–October
Coastal Gulf of Marine	
Eastern section	September 15–October 17
Western section	October 1–21 (5 years)
Jeffery's Ledge	September 29–October 25 (2 years)
Donegal	October
Georges Bank	October 5–23 (8 years)
Nantucket Shoals	October 12–November 2 (8 years)
Dunmore	September–October, December–January
Downs	December
Plymouth	January

[a] Time derived by taking midpoint of spawning period.

Thus the mean abundance of spawning populations is a function of the size of the spawning ground and of the larval (oceanographic) retention area (the main reason depressed herring populations cannot be "mitigated" by just dumping herring larvae into the sea), no biotic interactions are needed, nor probably are they very important, in regulating population size—the main thing is the physical oceanography (spatial process) at the spawning site ("Physics over Food Chains"). The variability in abundance is from losses of individuals from the retention/spawning area where physical oceanographic processes dominate in generating variability in losses from the retention area.

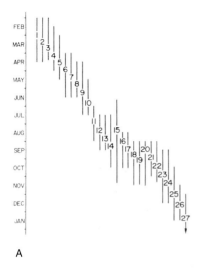

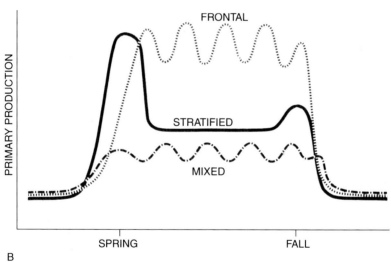

Figure 6.12 A-B. Evidence for member-vagrant hypothesis with Atlantic herring (*Clupea harengus*) (from Sinclair 1988; Sinclair and Trembley 1984). A. Spawning periods of Atlantic herring populations. (Populations correspond to numbers given in Table 6.2.) B. Schematic representation of seasonal cycles of primary production in frontal, mixed, and stratified areas.

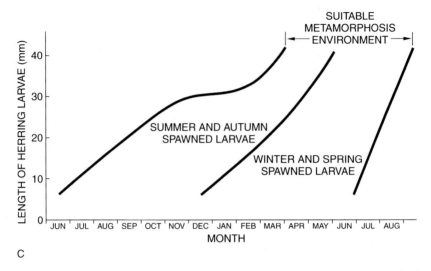

Figure 6.12C. Generalized depiction of growth rates and timing of metamorphosis for herring larvae spawned in different seasons.

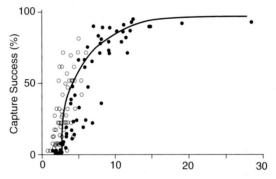

Figure 6.13. Capture success as a function of predator to prey length ratio. Points are original data from regressions. Open circles represent invertebrate predators and solid circles represent fish predators. The equation for the fitted line is capture = $100 - ((\text{ratio} + 3.37)/4.76)^{-2.28}$ (from Crowder, in Miller et al. 1988).

Growth-Mortality Theory. It is now clearly recognized that predation is very important in the survivability of larvae. Crowder and colleagues (in Miller et al. 1988) basically said that in order to decrease predation it is theoretically an advantage to be as big as possible ("Bigger Is Better Hypothesis," Figure 6.13), whereas Houde (1987) modified this idea to say that it was not only important to grow big but also to grow fast and to develop through the early developmental stages ("Stage Duration Hypothesis"). Experiments have shown that lack of food prolonged larval growth and made larvae

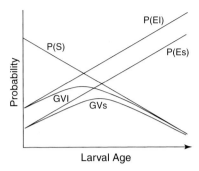

Figure 6.14. Conceptual model of the influence of age-related changes in encounter rates on the gross vulnerability of larval fish of different sizes to predation, where P(Es) = probability of encounter for small larvae; P(El) = probability of encounter for large larvae; P(S) = susceptibility of both large and small larvae to predation (combined probability of attack and capture); GVs = gross vulnerability of small larvae to predation; and GVl = gross vulnerability of large larvae to predation (Litvak and Leggett 1992).

vulnerable to predation longer. However, Litvak and Leggett (1992) stated that bigger is not better because the larvae become more conspicuous to predators that they want to evade; they have some lab evidence to support this theory (Figure 6.14). Hare and Cowen (1997) tested these growth-mortality hypotheses, using the otolith record of larval and pelagic juvenile stages of the bluefish (*Pomatomus saltatrix*), and found that larger and faster-growing individuals had a higher probability of survival, whereas the evidence was equivocal with regard to the stage duration hypothesis. Campana (1996) also found that body size at age at the pelagic juvenile stage was highly correlated with year-class strength, the explanation perhaps being that low growth leads to higher predation and greater variation in year-class strength.

Perhaps at this time the most likely outcome would seem to be that with regard to larval size, growth, starvation, and predation, it is those larvae that ". . . consistently see and hear better, swim faster, and respond more quickly and more vigorously when exposed to a threat . . . " which are more likely to avoid predation (or starvation); reportedly there is now laboratory evidence that "larval athletes" are less likely to be eaten by predators (Cowan and Shaw 2002).

FACTORS AFFECTING SURVIVAL OF EGGS AND LARVAE

As Hjort suggested nearly 100 years ago, both biological and physical factors probably influence the survival of fish eggs and larvae. After nearly a century of research we can improve little on Hjort's suggestions. The general conclusion is that mortality can be the result of a complex combination of factors acting on all early life-history stages (Anderson 1988; Leggett and Deblois 1994). For example, at low temperatures production of food might be delayed, development is delayed, and growth is slowed, so the stages most vulnerable to predation last longer: lack of food slows growth and increases predation as an indirect result of lower temperature.

Physical Factors

Many environmental factors probably affect survival of early life-history stages, and thus recruitment success. Temperature has an obvious influence on physiological rates including developmental rates and stage duration. Eggs and larvae may encounter lethal temperatures, but more likely temperature will affect their metabolism in more subtle ways. For example, development is slowed at lower temperatures, increasing the duration of the egg stage. The eggs will then be subject to predation for longer periods at a lower temperature. Currents can transport eggs and larvae to or away from areas suitable for further development. Salinity may play a role in egg and larval survival but it is probably minor in all cases except in some estuarine situations. Storminess and turbulence can affect food production and the ability of larvae to feed. Currents and features associated with them (e.g., eddies, gyres, fronts) can concentrate larvae and their food and aid in transport of eggs and larvae from spawning areas to nursery grounds.

Temperature. Francis (1993) found a strong positive relationship between year-class strength of New Zealand snapper (*Pagurus auratus*) and temperature during their first year. Although the mechanisms for this relationship are unknown, he speculated that faster growth at a higher temperature during the larval stage would reduce the time that larvae were exposed to predation, and that faster growth at a higher temperature would increase overwinter survival of the juveniles. Because of the small size of larvae, temperature can even influence the viscosity of the water and have an impact on larval swimming speeds (Hunt von Herbing 2002).

Salinity. Laboratory experiments have investigated the influence of salinity on egg and larval development (Alderdice and Forrester 1971). However, salinity is probably not the direct cause of mortality of most fish eggs and larvae, although eggs deposited intertidally in estuarine areas could be subjected to wide fluctuations in salinity.

Salinity was indirectly linked to recruitment failures in a number of fish stocks in the Northeast Atlantic. Cushing (1995a) showed that in the late

1960s and early 1970s, salinity and temperature were reduced and surmised that production of prey was delayed for larvae of the fishes that experienced recruitment failures. However, Mertz and Myers (1994b) found no evidence of such a decrease in temperature, and found little connection between factors that lead to prey production and fish recruitment in this area.

Storminess and Turbulence. Considerable research has focused on the impacts of turbulence and storminess on larval survival through effects on feeding rates (see Dower et al. 1997). Results of field (Ellertsen et al. 1984), laboratory (Landry et al. 1995), retrospective (Cushing 1995b), and modeling (Rothschild and Osborn 1988; MacKenzie et al. 1994) studies seem to indicate that moderate levels of turbulence enhance encounter rates between larvae and their food, without disrupting the feeding process (Cury and Roy 1989). However, larvae of some species seem to survive better in calm conditions that allow dense layers of their food to accumulate (Lasker 1975). Checkley et al. (1988) found that Atlantic menhaden (*Brevoortia tyrannus*) spawned in a region and time so their larvae would drift shoreward during frequent storm events off the coast of North Carolina, USA. Eckmann et al. (1988) found that early stratification of the water column (implying early onset of primary production) was positively correlated with year-class strength in the common whitefish (*Coregonus lavaretus*). Bailey and Macklin (1994) found that walleye pollock (*Theragra chalcogramma*) larvae reaching the first-feeding stage during stormy conditions had low survival.

Currents and Flow Features. Currents and flow features also impact larval feeding and survival. Fronts and gyres have been important in the early life history of several species. In some cases larvae and their prey accumulate along fronts (Govoni and Grimes 1992). Larvae near the fronts are in better condition than those away from the fronts. Larvae in gyres are also in better feeding regimes and are not swept away from suitable habitats by prevailing currents (Canino et al. 1991). The life histories of many species depend on the eggs and larvae being transported from the spawning area to nursery areas (e.g., Pacific hake [*Merluccius productus*], Bailey 1981; Atlantic menhaden [*Brevoortia tyrannus*], Checkley et al. 1988; bluefish, Hare and Cowen 1996). Changes in the flow of these currents can impact larval survival and ultimately year-class strength. For example, Polacheck et al. (1992) found that the exceptionally large 1987 year-class of haddock on Georges Bank resulted from larvae that were transported in unusually strong along-shelf surface flow to the Middle Atlantic Bight where they settled as juveniles, which then recruited back to Georges Bank. Also, Caputi et al. (1996) found that the strength of the Leeuwin Current had a major influence on recruitment of several invertebrates and fishes.

Biological Factors

Biological factors affecting egg and larval survival include feeding and predation (Pepin 1990; Leggett and Deblois 1994). Most research has dealt with feeding, but recently studies on predation have increased.

Feeding. Food production in the sea and larval feeding behavior and nutrition have been the objects of a large amount of research to understand larval survival. Larvae rely on production of particular kinds of food for survival. Production of this food is usually dependent on physical conditions and is often associated with conditions fostering the spring bloom (Cushing 1990). These conditions include warming of surface waters and calm seas to create a stratified surface layer where phytoplankton can thrive. Copepods feeding on the phytoplankton reproduce, and their larvae (nauplii and copepodites) are the food that most fish larvae require. The timing and duration of spawning in relation to the occurrence of the spring bloom are critical factors in larval feeding (Mertz and Myers 1994a).

Inadequate food has frequently been found to limit larval growth and directly lead to starvation or slow growth and thereby increase predation which leads to mortality. Interannual differences in larval growth rates and survival of lake whitefish (*Coregonus clupeaformis*) were directly related to abundance of prey (Freeberg et al. 1990).

Predators. The other main biological cause of population fluctuation is predation on early life stages. Feeding is only important for larvae after their yolk sac has been exhausted, but predation acts on all life-history stages. In recent years there has been much research on predation of fish eggs and larvae (Bailey and Houde 1989). Eggs and larvae are subject to predation from a wide variety of organisms. Demersal eggs are preyed upon by a host of invertebrate predators, as well as other fishes. Major predators of planktonic fish eggs and larvae include chaetognaths (arrowworms); crustaceans such as amphipods, euphausiids, and copepods; and siphonophores; jellyfishes; and ctenophores.

Planktonic fish larvae also feed on other larvae. Cannibalism has been found in walleye pollock (Walline 1985) and Atlantic mackerel (*Scomber scombrus*) where Grave (1981) estimated that the larvae constituted 83% of the diet of larger mackerel larvae. As the abundance of these predators varies temporally and spatially, their impact on survival of young fish will vary. It may also vary depending on what other food is available for the predators. Juvenile fishes are vulnerable to an additional suite of predators including birds and other fishes. In fact, predation may be the most important source of juvenile mortality in many species.

TABLE 6.3. Sources of Mortality of Larvae by Stage
(from Hewitt et al. 1985)

Stage	Major Source of Mortality
Hatching to yolk sac larvae	Predation
First feeding to 15 days	Starvation
> 15 days	Predation

Although larvae succumb to a wide variety of predators, lack of sufficient quantities of suitable food is more frequently implicated in their mortality. In some cases, insufficient food slows growth and prolongs the time of maximum vulnerability to predators. In the laboratory, Pepin and Shears (1995) found that size of larvae had a significant impact on mortality rates from predation.

Hewitt et al. (1985) conducted field and laboratory experiments on jack mackerel (*Trachurus symmetricus*) in an attempt to separate the portion of total mortality that was due to predation and the part that was due to starvation. In the field, ichthyoplankton collections were used to determine larval condition, growth, net retention, and production. In the lab, experiments determined growth and body shrinkage due to preservation treatment, and starvation (histology) was determined for various aged (by otoliths) larvae. Total mortality rates were estimated and predation was inferred as the difference between the two (Table 6.3). General estimates were that losses from fertilization to yolk sac absorption ranged from 99.5 to 99.9%, and that predation was the dominant source of mortality but was reduced once larvae had successfully fed and developed mobility.

IMPORTANCE OF JUVENILE STAGE

Although most of the focus of recruitment studies has been on the egg and larval stages, it has become apparent that in many fishes mortality during the young-of-the-year juvenile stage (0 age group) can have a significant effect on year-class strength. Following the planktonic larval stage, a dramatic metamorphosis occurs in many fishes that involves not only a morphological transformation, but also behavioral and ecological changes. For example, in flatfishes one eye migrates across the skull of the planktonic larvae, so the juveniles have both eyes on one side of the head. This is accompanied by development of superficial pigment on the eyed side, and settling to the bottom, moving from a three-dimensional world to a two-dimensional one. Juveniles of some flatfishes require specific sandy bottom habitats, whereas others move from offshore spawning areas into estuaries.

Juveniles are more capable of directed swimming than larvae, and thus are no longer considered planktonic. They can actively migrate to nursery areas, sometimes aided by prevailing currents.

The habitat of juveniles can be different from that of the adults and is quite varied. Juveniles of some species are neustonic and develop countershading and other morphological characteristic of epipelagic fishes. Other species associate with other organisms: eelgrass, macro-algae, molluscs, or corals and associated fauna. As the juvenile stage is reached, schooling begins in many species that school as adults.

In some species the juvenile stage is so different from both the larval stage and the adult stage that some consider there to be two metamorphoses: one between the larval and juvenile stage, and another between the juvenile and adult stage. With such remarkable changes in structure and habitat, survival becomes problematic for reasons that are not identical to those associated with the larval stage. Mortality factors for juveniles can reduce the size of year-classes that had good survival as eggs and larvae.

When Is Recruitment Set?

The meaning of the term *recruitment* varies among researchers working on different groups of fishes. In fishes that have distinct year-classes and are subjects of fisheries, recruitment refers to the event when the year-class enters the fished portion of the population. This may occur several years after the fish settle out of the plankton and become juveniles. In these fishes, mortality during the juvenile stage can impact the level of recruitment.

In tropical fishes with nearly continuous reproduction, recruitment means the event when larvae settle out of the plankton and become juveniles. In these fishes, the question of whether recruitment is determined by larval supply or mortality after the larvae settle has generated considerable research (Holm 1990; Kaufman et al. 1992; Tupper and Hunte 1994; Schmitt and Holbrook 1996).

For high-latitude fishes, a major problem in population dynamics is to determine when year-classes become established. Surveys of abundance of newly recruited fish could enable managers to adjust regulations to account for the size of incoming year-classes. It is generally thought the year-class size is determined during the planktonic egg and larval stages, but evidence from many species indicates that year-classes can be reduced by mortality during the juvenile stage (Myers and Cadigan 1993; van der Veer et al. 1990). In an exploratory critical factor analysis of mortality of various early life-history stages of walleye pollock from the Gulf of Alaska, Bailey et al. (1996) found that recruitment levels could be established at any stage, given sufficient supply from prior stages.

Factors Affecting Survival of Juveniles

As with eggs and larvae, both biological and physical factors cause mortality of juvenile fishes. With their larger mass, juveniles can withstand poor feeding conditions longer than larvae can; however, as with larvae, reduced growth rates increase the time that juveniles are subject to predation. Habitat can be limited and create the possibility of density-dependent mortality when larval supply provides more larvae to the juvenile settling area than it can support (Nordeide et al. 1994). Larvae settling in a less-than-ideal habitat can be subjected to increased predation, inadequate food supplies, or inhospitable physical conditions.

Many temperate fishes that spawn in spring become juveniles in late summer, and then enter a period of much reduced growth during winter. The condition of juveniles entering their first winter can affect their overwintering survival, associated with starvation, predation, or intolerance of extreme physical conditions by smaller members of the cohort (Sogard 1997; Hurst and Conover 1998; Hales and Able 2001). Generally, the larger the juveniles are at the beginning of the overwintering period, the better their chances for survival.

Food. With the transformation from the larval stage, a change in diet often occurs. This might be considered a second critical interval in the life history of fishes (Thorisson 1994). Juveniles that settle to the bottom often start eating demersal crustaceans and infauna. Bluefish change from a diet of zooplankton to one of fishes, which is accompanied by a large increase in growth rate (Juanes and Conover 1995). Although larvae are generally too dilute to affect their food supply, juveniles that settle in restricted habitats can become abundant enough to deplete their food supply and cause density-dependent reduction in growth. During the gadoid outburst in the 1960s and 1970s in the Northeast Atlantic, there was evidence that the food supply of juveniles of some species was in short supply (Hislop 1996). With the larger body size of juveniles compared to larvae, the risk of starvation is much reduced. However, food deprivation can reduce growth rates and lead to smaller sizes at the end of the first growing season, which can result in increased overwintering mortality.

Predators. Predation is generally thought to be the most important source of mortality in juveniles (van der Veer et al. 1990). Juveniles are preyed upon by a wider range of predators than are larvae. Fishes, including conspecifics, are major sources of predation as are birds, marine mammals, and a host of invertebrates (e.g., shrimp, crabs). Koslow et al. (1987) found that haddock recruitment in the Northwest Atlantic was negatively correlated with abundance of young Atlantic mackerel, which may be due to predation by the mackerel in winter. Since schooling commences in some fishes as juveniles, mobile predators can be very effective once schools are located.

Currents. The juveniles of many species require transport of many kilometers, whereas others need minimal transport to recruit successfully. For those that depend on currents to transport them from spawning areas to nursery grounds, any alteration in these currents can have deleterious effects on their survival. Increasingly it is becoming apparent that behavior of the juveniles is critical for them to utilize currents for directed transport, or to remain at a given location.

Larval bluefish off the U.S. East Coast are moved northward at the edge of the Gulf Stream from their South Atlantic Bight spawning area. They reach the Middle Atlantic Bight as pelagic juveniles and are transported to the shelf edge by Gulf Stream warm-core ring streamers, or more developed juveniles might swim to the shelf edge. This occurs at a time when the surface waters are warming and the slope/shelf temperature front is dissipating. They then actively swim from the shelf edge across the Middle Atlantic Bight to reach their estuarine nursery areas (Hare and Cowen 1996).

On the other hand, there is evidence of limited dispersal in some fishes, with juveniles recruiting to areas very near where they were spawned. This lack of dispersal has been reported in rocky intertidal fishes (Marliave 1986), and in reef fishes of isolated oceanic islands (Schultz and Cowen 1994).

Juvenile Habitat Requirements (Nursery Grounds). Many species have specific habitat requirements as juveniles. The question of whether larval supply or processes at and shortly after settlement control recruitment levels has received considerable attention (Richards and Lindeman 1987; Milicich and Doherty 1994; Levin 1996). In many such species it seems that the amount of high-quality juvenile habitat is quite important in determining population sizes (Tupper and Boutilier 1997).

A number of species off the U.S. East Coast, as well as elsewhere, occupy estuaries as juveniles (Allen and Barker 1990; Able and Fahay 1998). Complex behavioral interactions of larvae with tidal currents and other physical processes allow them to enter the estuaries where they quickly transform into juveniles (Figure 6.15). The general phenomenon is called "tidal stream transport" and involves the larvae changing their vertical distribution so that they move into the estuary and upstream in the estuary during periods of rising tides.

The juveniles of many flatfishes have specific settling areas that are often shallow areas with sandy substrate (van der Veer et al. 2000). The amount of such habitat can control the size of populations of such species.

Larvae of many fishes that associate with hard substrate as adults settle onto similar habitats as juveniles. These can be coral reefs in the tropics, or encrusted rocky outcrops at higher latitudes.

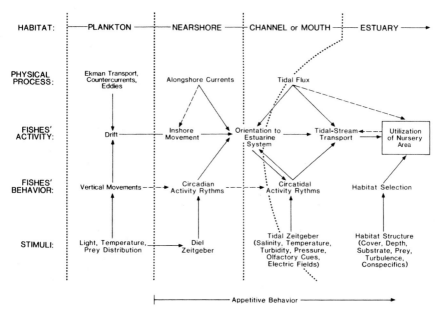

Figure 6.15. A conceptual model for the role of stimuli and behavior in fish movement to estuarine nursery areas (from Boehlert and Mundy 1988).

RECRUITMENT STUDIES

History of Recruitment Studies in the United States
(From Kendall and Duker 1998).

The history of recruitment research in the United States can be traced back to Hjort through Henry Bigelow at Harvard University. Bigelow had two graduate students in the 1930s who studied the causes of fluctuations in fish abundance: L.A. Walford worked on haddock, and O.E. Sette worked on Atlantic mackerel. It is evident that the ideas of Hjort on the importance of early life history were foundational in their work.

Walford (1938) reported on an investigation of haddock early life history on Georges Bank. He justified this study by stating:

"Judging from present studies of the Bureau of Fisheries, natural fluctuations in abundance of the American haddock over a wide area are due not to migrations of the adult population away from the fishing grounds but to actual changes in the numbers of fish. Furthermore, these changes do not usually affect the population as a whole but rather individual year broods, which, during the first year of their life, are subject to varying fortunes that determine their success or failure . . . the critical time when the success or failure of a year brood is determined occurs during the period of the embryonic, larval and post-larval stages . . . causes in fluctuations in abundance can be found by intensively studying the biology of these early stages and at the same time by observing changes in the environmental elements."

Although Walford did not cite Hjort's papers on causes of fluctuations in fish abundance, it is evident from the above that he considered the same factors important. The basis of his study was collections of planktonic eggs and larvae made during several cruises aboard the *Albatross II* in the Georges Bank area during 1931–1932. He also used data collected by Sette in the same years, but farther south to Chesapeake Bay. Based on these collections, and associated physical data, he postulated that interannual changes in drift patterns could carry variable amounts of eggs and larvae away from their nursery areas on Georges Bank. More recent work by the Federal Fisheries Agency and others on haddock early life history and fisheries oceanography has followed this pioneering work of Walford (e.g., Chase 1955; Colton and Temple 1961; Koslow et al. 1985).

Sette (1943a) reported on a study of the life history of Atlantic mackerel off the Atlantic coast. The objective of his study was to understand the fluctuations in abundance of this population. After a preliminary cruise in 1926, he collected eggs and larvae from 1927 through 1932 aboard the *Albatross II*. His field work ended abruptly in June 1932 when the *Albatross II* was taken out of service as an economy measure. (Mackerel would be safe from having their eggs and larvae collected in any large-scale manner by Federal fisheries biologists until more than 30 years later, when in 1966 the *Dolphin* conducted plankton surveys off the Northeast Atlantic coast [Berrien 1978].) Sette concluded, by measuring interannual changes in drift and mortality of cohorts of larvae, that increased mortality during the transition from yolk to exogenous food sources was not a major contributor to variations in mortality, but that variations in drift, caused by winds, seemed to be correlated with year-class strength. Thus Sette rejected Hjort's first hypothesis (the critical period) but concurred with his second hypothesis (drift). This was a pioneering study in that population estimates of larval growth and mortality rates were computed.

The California Cooperative Oceanic Fisheries Investigations (CalCOFI) have been the largest and most prolonged of the fisheries oceanography studies that the Federal Fisheries Agency has participated in (Scheiber 1990). Concerns in the early 1930s over declining catches of Pacific sardines off the coast of California (Figure 6.5) caused the California Department of Fisheries to seek regulations limiting the fishery. The industry was influential in getting the Federal government involved in studies to determine the causes of these declines: was it overfishing, or was it natural fluctuations in abundance? "The ship operators . . . resorted to a plan (used before and since) by which . . . legislation could be postponed by asking for a special study of the abundance of sardines . . ." (Scofield 1957, quoted in Radovich 1982). Sette was detailed to California in 1937 to work with other fisheries scientists on this problem (Powell 1972, 1982). Before long Sette presented a plan for the study of all aspects of the life history of the sardine, in relation to the fishery (Sette 1943b) (Figure 6.16). Although each participating agency had its own agenda, a working relationship was established, and in 1947 the California legislature established the Marine Research Committee (MRC) with representatives of the industry, and several scientific agencies, with Sette as scientific advisor (Baxter 1982; Radovich 1982). The MRC developed the California Cooperative Sardine Research Program, which in 1953 was renamed the California Cooperative Oceanic Fisheries Investigations (CalCOFI). From the beginning, the program had a broad emphasis: to study the sardine and its environment, and the effects of fishing on the species: "1. Physical-chemical conditions in the sea. 2. Organic productivity of the sea and its utilization. 3. Spawning, survival, and recruitment of sardines. 4. Availability of the stock to the fishermen (behavior of the fish as it affects the catch)—abundance, distribution, migration, behavior. 5. Fishing methods in relation to availability. 6. Dynamics of the sardine population and fishery" (MRC 1948, quoted in Scheiber 1990). This program involved collecting large amounts of physical and biological data over a huge section of ocean off the U.S. West Coast (Hewitt 1988). As the sardines continued to decline and the fishery ceased, CalCOFI broadened its objectives to include more species and larger-scale oceanographic processes. "Ultimately we hope to be able to predict what the effect of the environment is on spawning success" (Lasker 1965). Northern anchovies seemed to replace sardines as the dominant, small, coastal pelagic fish of the California Current. The impacts of large-scale circulation variations on the California Current became appreciated, starting with the 1957–1958 El Niño. Understanding the causes of fluctuations in abundance of sardines, including the role of man, continued to be an elusive goal (Marr 1960; Radovich 1960; Murphy 1961). The issue was complicated by finding that there were several stocks of sardines with separate centers of distribution and seasonal migratory patterns. The relation between the anchovy and sardine (Figure 6.17) was open to much speculation: were they competitors at some stage in their life history? as adults? as larvae? Did one prefer slightly different oceanographic conditions than the other? Did cooler temperatures favor anchovies? Were sardine numbers reduced by fishing to a point that anchovies could "take over"? The answers are still not apparent. Sardines are more abundant now than in the 1970s–1980s, so if today's tools were applied to the study of their early life history it may be possible to gain an understanding of their population dynamics.

Late in its first century, the Federal Fisheries Agency also undertook some smaller, less ambitious fisheries oceanographic studies. "In 1953, the

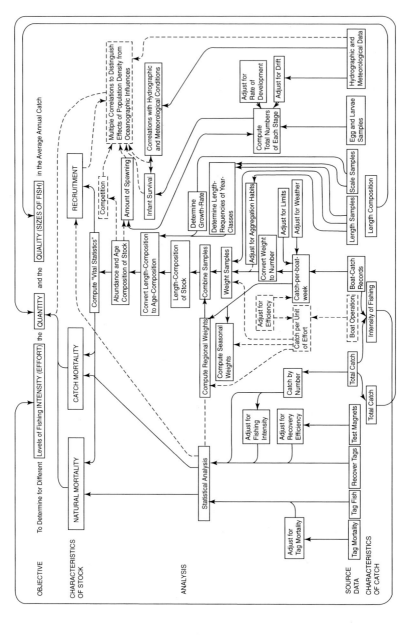

Figure 6.16. Conceptual model of Pacific sardine (*Sardinops sagax*) dynamics (from Hewitt 1988, based on Sette 1943b).

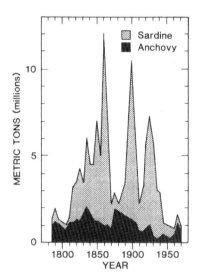

Figure 6.17. Estimate of the spawning biomass of northern anchovy (*Engraulis mordax*) between 1790 and 1970 (180 years at 5-year intervals) derived from the scale-deposition rate in two anoxic basins (Soutar and Isaacs 1974) and adjusted to the spawning biomass determined between 1980 and 1985 for northern anchovy (Methot and Lo 1987), and estimated between 1951 and 1969 (Lo 1985). The Pacific sardine biomass from Smith 1978 (from Smith and Moser 1988).

Fish and Wildlife Service inaugurated a program to study the early life history of haddock in the Gulf of Maine in an attempt to relate spawning location and the pattern of drift to the success of year class. The *Albatross III* made cruises during the spring of 1953, 1955, 1956, and 1957" (Colton 1964). On the basis of these cruises Colton and Temple (1961) "concluded that under average conditions most fish eggs and larvae were carried away from Georges Bank and that only under unusual hydrographic conditions were eggs and larvae retained in the area" (Colton 1964).

From January 1953 through December 1954, nine cruises were conducted off the southeast coast of the United States aboard the M/V *Theodore N. Gill* to "ascertain the potential productivity of those waters adjacent to our coast from Cape Hatteras on the north to the Florida Straits on the south" (Anderson et al. 1956). In 1965 through 1968 a survey of eggs and larvae of fishes off the U.S. East Coast was conducted during 12 cruises aboard the R/V *Dolphin* in order to study offshore distribution patterns of eggs and larvae of fishes that are estuarine dependent as juveniles (Clark et al. 1969, 1970).

Discussions in 1968 among several Federal scientists from around the country involved in early life-history studies resulted in a nationwide program: MARMAP (Marine Resources Monitoring, Assessment, and Prediction). Among its several goals, this program intended to standardize collection of fish egg and larval data as well as environmental data "to integrate assessments of living marine resources and environmental information for the purpose of predicting future production and yields of those resources" (Sibunka and Silverman 1984).

Recent Recruitment Studies

Fisheries-Oceanography Coordinated Investigations (FOCI). FOCI is a NOAA-funded research program focusing primarily on recruitment dynamics of walleye pollock, the economically important gadid of the North Pacific Ocean and the Bering Sea. The most complete studies by FOCI involved the pollock spawning that occurs in Shelikof Strait, Gulf of Alaska (Kendall et al. 1996). Here a combination of field, laboratory, and modeling studies on factors that could influence year-class strength have led to annual forecasts of recruitment that are used in the management of the population. The studies, ongoing since 1985, are conducted mainly by scientists from NOAA's Pacific Marine Environmental Laboratory and Alaska Fisheries Science Center.

The pollock spawning in Shelikof Strait was discovered in 1980 and became the source of a vigorous fishery in the 1980s. It was later found that the large population was sustained by several exceptionally strong year-classes that were produced in the late 1970s. The mass spawning of planktonic eggs is consistent in time (early April) and place (lower Shelikof Strait), and the resultant larvae drift mainly to the southwest in the Alaska Coastal Current during a larval period that lasts several weeks. These population characteristics provided an optimal situation for an intensive recruitment study. Numerous cruises to the area over the years have investigated larval distribution, growth, and mortality. Coordinated studies of the physical environment investigated the influence of variations in currents, storms, and eddies on the larvae and their food source (mainly copepod nauplii). Concurrent laboratory studies validated otolith aging of the larvae, and calibrated larval condition indices. Laboratory behavior studies on the larvae investigated ontogenetic changes in the capabilities of larvae and how they respond to various environmental factors (such as light, temperature, and prey concentrations). A significant product of these studies was a conceptual model of factors that influence survival of the early life stages of pollock in this population (Figure 6.18). This model guided subsequent research and served as a basis for the annual projections of year-class strength generated by FOCI.

Some of the conclusions of FOCI studies in Shelikof Strait are that little variation in mortality occurs during the egg stage. Following the yolk sac stage, larvae may be subject to inadequate food supplies for the next two

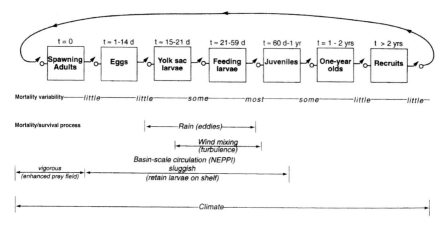

Figure 6.18. Conceptual model of Gulf of Alaska walleye pollock (*Theragra chalcogramma*) survival at different life stages. Relative mortality, important environmental processes, and the life stages that they affect are indicated (from Megrey et al. 1996).

weeks. Larvae in mesoscale eddies, which frequently form in lower Shelikof Strait, are in better condition than those outside of eddies, at least on some occasions. Storms that occur during the first-feeding stage of the larvae are detrimental to their survival. In some years, year-class strength is established after the larval stage, during the early juvenile stage. Thus, the stage at which recruitment is set varies among years (Bailey et al. 1996). There may be a relationship between El Niño years and year-class strength (poor year-classes are produced in El Niño years, but larger year-classes are produced in the two or three subsequent years).

South Atlantic Bight Recruitment Experiment (SABRE). SABRE was a multidisciplinary research program funded by NOAA from 1991–1997 (Crowder and Werner 1999). The focus of SABRE studies was factors that led to recruitment variation in the fishes of the Atlantic coast of the United States that characteristically spawn offshore, but whose juveniles spend some time in the adjacent, expansive estuaries of the region (estuarine-dependent fishes). The studies dealt primarily with Atlantic menhaden and to a lesser extent with spot (*Leiostomus xanthurus*). Field, laboratory, and modeling studies were performed and coordinated through competitive selection of projects and oversight by a project management committee. Academic as well as Federal investigators participated in the program. Some emphasis was placed on developing and using new technology in these studies.

Four life stages with critical junctures between them were studied: eggs and larvae on the continental shelf, late-larvae entering estuaries, juveniles in the estuaries, and juveniles leaving the estuaries. The approach followed was an analysis of survivors from each stage to subsequent stages (e.g., within the spawning period, what dates produced eggs that survived to become juveniles entering estuaries? What were the environmental characteristics of that period that may have promoted survival?). Most offshore field sampling for eggs and larvae took place in Onslow Bay (the continental shelf region between Capes Lookout and Fear, North Carolina). A better understanding of the ocean physics of the area was sought through field and modeling studies. A newly developed underway egg sampler aided in locating and characterizing spawning areas. Other studies focused on the oceanic distribution of larvae, and mechanisms they used to find and enter the relatively small inlets of the estuaries. Studies on late larvae and early juveniles investigated ingress into and growth and mortality in estuaries. The site of most of these studies was Beaufort Inlet, North Carolina, and the estuarine system shoreward of the inlet.

Before SABRE studies, menhaden spawning habitat was poorly known, and their eggs had been collected only sporadically. Using the newly developed Continuous Underway Fish Egg Sampler (CUFES), menhaden spawning was mapped, and eggs were found near the surface in patches in water with depths between 20–40 m on the continental shelf. Spawning takes place primarily in fall and winter in water with surface temperatures between 12 and 25°C. Larval behavior included daily vertical excursions to the surface to fill the swimbladder at sunset. Larvae responded to changes in temperature and salinity in a way that would keep them in the upper part of the water column. Modeling studies and empirical data showed the importance of wind direction on larvae being transported shoreward and alongshore to reach the estuaries. Larval immigration through the inlets was related to tides and winds, as well as large-scale water circulation on the shelf. Once in the estuaries, late larvae use selective tidal stream transport (they are abundant in the water column on rising tides at night) to move farther into the estuaries. Year-class strength seems to be most strongly related to survival of late larvae and early juveniles, indicating the importance of the estuarine phase in the life cycle of menhaden.

Global Ocean Ecosystem Dynamics (GLOBEC). GLOBEC is an international program investigating the global ocean ecosystem and how it responds to physical forcing, to forecast biological responses to global climate changes. The main focus of GLOBEC is zooplankton, which is critical prey for larval and juvenile fishes. Because of this role, zooplankton production influences fish population dynamics by affecting survival of larvae and thus recruitment levels. GLOBEC undertakes retrospective, observational, and modeling projects to achieve its objectives. GLOBEC studies focus on several kinds of marine ecosystems around the world such as upwelling areas. Within this international program, several national programs have developed, and ICES (International Council for the Exploration of the Seas), the long-standing group interested in fisheries of the North Atlantic, has a regional program.

ICES: COD AND CLIMATE CHANGE. The GLOBEC project undertaken by ICES is called Cod and Climate Change (CCC). With its focus on dynamics of recruitment and growth of Atlantic cod stocks, it seeks to understand the effects of climate variability. The program has included many field, laboratory, modeling, and retrospective studies on early life history of Atlantic cod. Comparative studies on the many cod stocks around the North Atlantic are also included. Scales of investigation range from effects of small-scale turbulence on encounter rates and feeding of individual cod larvae to effects of large-scale interdecadal changes in wind on circulation, transport of heat, and young cod. It is hoped that these studies will lead to prediction of major changes in cod distribution and productivity under different physical regimes. The program started in 1990 and continues to focus on research on Atlantic cod and climate by member countries, to conduct workshops, and to produce reports on the subject (visit their website at: http://www.ices.dk/globec/).

U.S. GLOBEC: GEORGES BANK. A U.S. GLOBEC program that includes a focus on recruitment dynamics of fishes is being conducted in the Georges Bank region of the Northwest Atlantic Ocean. The fishes of concern are Atlantic cod and haddock. The goal is to be able to predict how these species might respond to climate change. The program includes field studies, both large-scale surveys and process studies, environmental monitoring, and modeling efforts. During the process studies larval fish feeding rates are measured aboard ship. Altogether, 30 academic and governmental organizations are participating in this program. The program began in 1993, and through 1999 over 100 cruises had been conducted.

Future Directions of Recruitment Studies

Timescales of Variability and Biological Responses. Although most recruitment research to date has focused on interannual variations in year-class strength of individual species or stocks, it is becoming increasingly apparent that interspecies and interyear patterns of recruitment exist. For example, consider sardine and anchovy recruitment varies over a cycle of about 60 years in eastern boundary currents around the world (Baumgartner et al. 1992; Lluch-Belda et al. 1992; Klyashtorin and Smirnov 1995), the "gadoid outburst" in the North Sea (Cushing 1980), shifts in dominance between Atlantic herring and Atlantic mackerel in the Northwest Atlantic (Skud 1982), a correlation between recruitment of

Pacific cod (*Gadus macrocephalus*) and walleye pollock in the eastern Bering Sea, and the inverse relationship between production of several Pacific salmon stocks in Alaska waters and those off the West Coast of the United States (Mantua et al. 1997). Besides interannual variability and local conditions, recruitment is likely affected by environmental processes that vary on large spatial and low-frequency temporal scales. Future recruitment studies should not ignore these environmental processes and how they affect recruitment.

Significant advances have been made in the last two decades in understanding climate variation and its impact on the ocean. The atmosphere varies on several timescales from diel, to event (storm), to seasonal, to interannual, to decadal, and longer (global climate change). The importance of decadal variability on marine communities and productivity is just now being recognized (Francis et al. 1998). Decadal variability is due to a number of processes that can act together or oppose each other to produce conditions observed at any one time. Extreme conditions occur when all of these processes push the climate in the same direction. In the North Pacific, the El Niño-Southern Oscillation is the best-understood large-scale process that affects the climate and ocean. El Niños are now being forecast and their influence on the weather and physical ocean conditions are predicted (Chen et al. 1995). They impact ocean productivity through disturbance of coastal upwelling. Distributions of marine organisms are changed due to warming of the water. El Niños seldom last more than a few months and occur once or twice a decade. Although El Niños originate in the tropics, the effects of some of them on the ocean can be measured at high latitudes. These effects reach high latitudes through direct transfer of energy and through teleconnections between atmospheric conditions in the tropics and at higher latitudes.

The next longer scale of climatic variation in the North Pacific is characterized by the Pacific Decadal Oscillation (PDO) (Mantua et al. 1997). The phase of the PDO changes at frequencies of about 20–30 years. Negative PDO values are associated with relatively warm sea-surface temperatures in the Northeast Pacific. PDO was negative through most of the 1960s and 1970s and has been mainly positive since 1977. The abrupt change in magnitude and sign of the PDO between 1976 and 1977 (Miller et al. 1994) coincided with dramatic changes in population levels of numerous species in the Northeast Pacific and has been considered a regime shift (Beamish et al. 1999). The entire ecosystem of the region may have shifted from one steady state condition to another at that time. Enhanced primary productivity caused by higher temperatures and shallow, mixed layer depths following the 1977 regime shift may have worked its way up the food chain through increased zooplankton abundance (Brodeur and Ware 1992) to produce the high abundances of Alaska salmon that were present in subsequent years.

At an even longer timescale, current increasing temperatures may signal global climate change. Schumacher et al. (2003) have reported recent increases in temperature and decreases in sea ice in the Bering Sea, and Wiles et al. (1998) have reported recent warming in the Gulf of Alaska. Several groups of North Pacific and Bering Sea marine animals are showing long-term monotonic changes in abundance (e.g., Roemmich and McGowan 1995; Hill and Demaster 1998; Brodeur et al. 1999). Whether these are responses to conditions associated with gradual large-scale increases in temperature, and whether these changes are partially or wholly man-induced are subjects of considerable ongoing research.

A New Paradigm for Recruitment Research. Given that lower-frequency (longer than interannual) variations in recruitment are commonplace, and may be the most important mode in determining abundances of many stocks, and that recruitment patterns of co-occurring species are often interrelated, how should we study recruitment? Statistical procedures examining patterns of recruitment in relation to environmental variables may point to areas of importance, but they are unlikely to produce an understanding of biological responses to the environment. They will be no more effective for understanding or predicting recruitment than stock-recruitment-based studies have been in the past. An understanding will involve examining mortality/survival of eggs, larvae, and juveniles: what are the age-specific rates, how do they vary, and what are the causes of variation? "We should try to pry open the black box of recruitment since this holds the key to longer term trends" (Steele 1996).

Thinking that one factor acting on one life stage is responsible for recruitment variation will continue to be unsuccessful. Even for one species, the factors and stages that are critical may vary with time. Therefore, to understand recruitment well enough to produce forecasts that managers need, research must consider year-class variation in light of longer timescales and ecosystem processes.

TIMESCALES. Single-season or short-term projects will not provide the answers. Recruitment in many species is autocorrelated: year-class strength varies on timescales longer than interannual. Studies of even five years' duration may only observe one general level of recruitment, or at best the transition from one level to another. Although there are interannual variations in the environment, these occur in the context of longer scales of variability. Climatic events such as regime shifts may create perturbed ecosystems that take several years to regain a state of maturity, where biological processes are more important than physical processes in determining population levels. There are variable lags between changes in abundance of long-lived predators and their prey in their response to low-frequency environmental conditions. Recruitment studies must take into account the climatic state of each year under investigation. Clearly, long timescales (20–40 years) must be considered. Retrospective, laboratory and modeling studies can partially augment field studies to extend the time horizon.

MULTIPLE TROPHIC LEVELS. The importance of both bottom-up (food) and top-down (predation) processes probably varies among species and stages within a species, and may change under different environmental conditions. The basic productivity (carrying capacity) of ocean systems seems to change under different climatic regimes (e.g., Roemmich and McGowan 1995; Polovina et al. 1994). Interannual and longer timescale temperature differences may impact the timing of the spring bloom and rate of reproduction and development of copepods, which provide food for larvae and cause a mismatch between food availability and occurrence of larvae. The abundance and distribution of larval and juvenile predators may likewise be impacted by large-scale atmospheric and oceanic conditions, and thus influence survival. Also the effects of parasites and diseases cannot be ruled out as sources of mortality. Therefore, recruitment studies must examine factors affecting production and distribution of trophic levels both above and below the early stages of fishes.

MULTISPECIES FOCUS. Since recruitment of several species in an area often seems to be related either positively or negatively, more understanding should be gained by studying them together (Koslow 1984). In fact, the recruitment process of one such species may never be understood without taking into consideration co-occurring species. In some cases, different temperature regimes seem to favor one species over another, but what is the actual causal mechanism? For example, is it development rate of the eggs and larvae, or does temperature affect the production of larval food: timing and preferred species? Once a species becomes dominant, can it prevent another species from assuming that role through predation or competition?

CHAPTER 7

Habitat, Water Quality, and Conservation Biology

HABITATS
Variety of Habitats Occupied by Fish Eggs and Larvae
  Open-Ocean Pelagic
  Coastal Ocean Pelagic and Bottom
  Estuaries
Habitats May Be Stage Specific
  Demersal Eggs and Pelagic Larvae
  Some Stages May Be Neustonic
  Anadromous and Catadromus
  Estuarine Dependent

HUMAN IMPACTS AND WATER QUALITY
Anthropogenic Physical Changes in Habitat
  Modified Runoff
  Coastal Landscape Changes
Exotic Introductions
Pollutants and Their Effects
  Eutrophication and Oxygen Depletion
  Oil and Oil Dispersants
  Effluents and Organochemicals
  Metals

CONSERVATION BIOLOGY AND MARINE PROTECTED AREAS (MPAs)
Definitions
  Conservation Biology
  Marine Protected Areas
Marine Protected Areas (MPAs) or Marine Reserves
  Fisheries Mismanagement
  Marine Reserves, a Partial Solution
  A Model of Fisheries Management Objectives of Marine Reserves
  The Track Record of Marine Reserves
  Marine Reserves and Recruitment
  Conclusions

HABITATS

Variety of Habitats Occupied by Fish Eggs and Larvae

Fishes occur in almost all natural waters on Earth: from caves and hot springs, to the deepest parts of the ocean, and under the sea ice of the Antarctic Ocean. Fish eggs and larvae also occur in an incredible array of habitats. The habitat of the early stages is not necessarily the same as that of the adult. A number of spawning migrations take place so that eggs may be laid in areas far from those inhabited by the adults. For example, most salmon (*Oncorhynchus* spp.) spend their adult life in the open ocean, but migrate to the upper reaches of freshwater streams to deposit eggs in the gravel. The young emerge from the gravel sometime after hatching and spend various amounts of time moving downstream and eventually out to sea to complete the cycle. The freshwater eels (anguillids) of Japan and the North Atlantic have a reverse life cycle: the adults are found in fresh water, and they migrate to the open ocean to spawn. The eggs and larvae develop at sea and move slowly over a several-year period to the coast and then upstream to complete the cycle.

In general, marine fish eggs seem well adapted to the various marine environments. Lindquist (1970, Figure 7.1) found in the Skagerak (which is characterized by an inflow of high-salinity North Sea water and an outflow of low-salinity Baltic Sea water) that in the shallow coastal waters the majority of species produced demersal eggs. In the nearshore pelagic zone, pelagic eggs without oil droplets predominate, whereas the central "oceanic" waters are occupied by pelagic eggs with oil droplets. In other marine areas this same zonation may be spread over a much wider horizontal distance.

Open-Ocean Pelagic. Fishes of the open ocean (e.g., lanternfishes [myctophids] and lightfishes [gonostomatids]) usually have pelagic eggs and larvae, regardless of the depth of occurrence of the adults. However, even in the open-ocean pelagic researchers find that distinct larval assemblages are seen in specific areas, often related to the physical characteristics of the water masses they are in, but also to the horizontal and vertical behavior of the larvae themselves (Hare et al. 2001). Thermal habitat off Oahu, Hawaii, was particularly important to the temporal and spatial distribution of several species of tuna (Thunnini) larvae (Boehlert and Mundy 1994).

Some groups that include species that occur in the deep ocean, such as the eelpouts (zoarcids), lay demersal eggs that hatch at an advanced stage and thus forgo a pelagic dispersal stage. Some fishes of the deep ocean are among the rarest animals on earth, and finding mates is problematic. Some deepsea anglerfishes (ceratioids) have solved this problem in an unusual way in that the males are parasitic on the females. The larvae of these fishes are sexually dimorphic: the males have enlarged nasal organs and modified teeth to find and attach themselves to females.

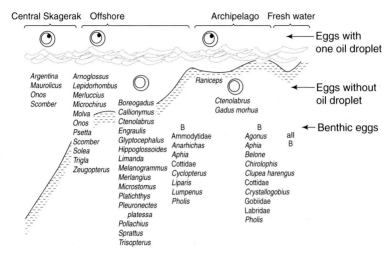

Figure 7.1. The distribution of different types of eggs in the various zones of the Skagerak (from Lindquist 1970). B = benthic (demersal) eggs.

Coastal Ocean Pelagic and Bottom. Fishes that spawn pelagic and demersal eggs are found in coastal waters. Although most species have pelagic larvae, some of the species that are most abundant (e.g., Atlantic and Pacific herring [*Clupea* spp.]) hatch from demersal eggs. Often pelagic eggs are released by coastal spawners in the early evening or at night. Commonly, the earliest stages of coastal pelagic larvae are most abundant near the surface, whereas later larvae (postflexion) are lower in the water column, often near the thermocline and/or chlorophyll maximum layer. Some larvae migrate vertically on a diel basis, and some are neustonic. Coastal pelagic larvae also show distinct assemblages, often related to temperature, wind, and salinity regimes (e.g., Laprise and Pepin 1995), but turbidity may also be an especially important characteristic (e.g., Harris et al. 2001). Larvae of many demersal (e.g., see Olney and Boehlert 1988) and live-bearing species, such as nearshore Pacific rockfishes (*Sebastes* spp.), settle as early juveniles to distinct habitats (e.g., seagrass and eelgrass beds, kelp, shell banks).

Estuaries. Tidal flushing and drying at low tide are problems associated with spawning in estuaries, so many fishes that spawn there have demersal eggs, which are often attached to plants or to gravel. Larvae occurring in estuaries often undergo vertical migrations associated with the tidal cycle currents so that they can remain in the estuary, or move horizontally in the nursery with the tide. The larvae or early juveniles of some species that spawn offshore migrate into estuaries to spend their juvenile period. Demersal eggs predominate in freshwater portions of estuaries, as they do in all fresh waters. Some anadromous fishes (e.g., white perch [*Morone americanus*] and striped bass [*Morone saxatilis*]) spawn in the freshwater reaches of estuaries. A larval mixture of estuarine, saltwater, and freshwater fishes can be found in estuaries, although the number of species (but not individuals) in estuaries is usually much less than in the nearshore environment (e.g., see Harris et al. 2001).

Habitats May Be Stage Specific

The area of occurrence of each stage (adults, eggs, larvae, juveniles) in the life history of fishes may differ considerably.

Demersal Eggs and Pelagic Larvae. Most species with demersal eggs have pelagic larvae.

Some Stages May Be Neustonic. Several species are neustonic (i.e., associated with the very surface of the water) as eggs, larvae, or juveniles.

Anadromous and Catadromus. Some species (e.g., salmon, lampreys [petromyzonids], sturgeon [*Acipenser* spp.]) migrate from the ocean to fresh water to spawn (anadromous), whereas others (e.g., American shad [*Alosa sapidissima*] and freshwater eels) migrate from fresh water to the ocean to spawn (catadromus).

Estuarine Dependent. Quite a few coastal species (e.g., U.S. West Coast English sole (*Parophrys vetulus*) [Gunderson et al. 1990], U.S. Puget Sound yellowtail rockfish (*Sebastes flavidus*) [Moulton 1977], and U.S. East Coast menhaden (*Brevoortia tyrannus*)), spot (*Leiostomus xanthurus*), and bluefish (*Pomatomus saltatrix*) [Able and Fahay 1998], often or always spawn in the ocean, but their larvae or early juveniles move into estuaries to spend their juvenile period.

HUMAN IMPACTS AND WATER QUALITY
Anthropogenic Physical Changes in Habitat

The coastal habitat in many areas of the world has been subjected to man-induced changes.

Modified Runoff. Damming of rivers and otherwise changing the amount and cycle of runoff can have negative impacts on fishes that use these areas, including wetlands that are very important spawning and egg and larval rearing areas for many marine species. An obvious example is that eggs laid in areas that become dry when flows are reduced are killed, but less obvious are the changes to substrates for egg disposition and to water current patterns responsible for dispersal and retention of pelagic larvae. Abnormally high flows can flush larvae out of river systems and wetlands. Increased sediment loads resulting from erosion of disturbed lands can suffocate eggs that are laid on or in the bottom.

Coastal Landscape Changes. Dredging as well as filling and building breakwaters, groins, and dikes ("shoreline armoring") has occurred in many marine coastal areas. This has had serious negative impacts on fishes that use these areas in their unmodified state as habitat for spawning and the early life stages. These nearshore shoreline marine areas generally have a particularly high diversity of fishes and are a very important spawning habitat, especially for demersal spawners. Demersal spawners have evolved so that the spawning site location along the shore is not only due to the type of substrate needed for the depositing and rearing of their eggs but also to be in the correct proximity for transport to the larval rearing areas with the needed physical and biological attributes. "Mitigation" efforts, such as transplanted eelgrass beds for Pacific herring (*Clupea pallasi*) spawning, have yet to demonstrate that they are successful in the real sense of establishing abundant returning adults (not just showing that eggs will be spawned on such substrate). Horn and Allen (1981) showed in a study in Southern California that their data did not support the "mitigation" proposition that harbors can adequately replace estuaries as inshore fish habitats. In contrast, Stephens et al. (1994) make a convincing case that an artificial breakwater that was constructed in Southern California has continued long term to support a significant fish assemblage, in spite of major mortality from entrapment of adults and entrainment of larvae at a power plant.

Exotic Introductions

Non-native fishes have been introduced intentionally and unintentionally to freshwater systems throughout the world, and even some introductions to coastal systems have become established (e.g., American shad and striped bass from the U.S. East Coast to the U.S. West Coast). In some cases these introductions have been of early life-history stages. For example, the ruffe (*Gymnocephalus cernuus*) was apparently introduced to the Great Lakes system of the United States and Canada through eggs or larvae in ballast water of ships carrying cargo from Europe. Although there is a lack of research documenting marine fish eggs and larvae being found in ballast water of ships (as opposed to invertebrate eggs and larvae), there is little doubt that they are present, as recently seen in the ballast water of ships traveling cross-Pacific and entering Puget Sound, and from San Francisco

Bay to Valdez, Alaska (Jeff Cordell, University of Washington, pers. comm. 2003).

The marine fauna of Hawaii includes non-native species that have established breeding populations (Randall 2007). Problems for native fishes posed by exotic species include introduction of diseases, destruction of habitat, competition for food, additional predation pressure, disruption of breeding (competition for nest sites), and hybridization.

Pollutants and Their Effects

Coastal and inshore waters are quite susceptible to pollution. Although introduction of pollutants is now generally closely regulated or prohibited in many areas, illegal activities and accidents still result in various unwanted chemicals entering natural waters. Point source pollution (e.g., discharge from a single factory or sewage plant) is generally easier to identify and control than chronic pollution from dispersed sources. The breakup of large ships carrying crude oil or other bulk chemicals is the most noteworthy form of pollution, but low-level, continuous introduction of chemicals from runoff is often more damaging. In regard to the effects of pollution on early life-history stages of fishes, although pollutants may be detrimental or lethal, their impact depends on their concentration and proximity to the eggs and larvae. Since in temperate areas most fishes reproduce seasonally, the impact of pollution events depends on their timing. A tanker grounding in fall may have little impact on fishes that spawn in spring. The *Exxon Valdez* was the largest tanker accident in U.S. history, which unfortunately occurred in the inshore waters of Prince William Sound, Alaska, in March 1989, at a time at or just before spawning for the majority of marine species (including Pacific herring), and just before juvenile pink salmon (*Oncorhynchus gorbuscha*) were emerging from the gravel—a species that utilizes the beach shoreline (a meter or two of water depth) as its nursery area. Needless to say, the results were significantly negative to early life stages of many of the marine fishes in Prince William Sound.

It is critical to establish carefully the importance of the nearshore areas to eggs and larvae, and map them. Although this sounds straightforward, it has seldom been done, at least partially because it is difficult to do and results are not necessarily repeatable. For example, in contrast to reports of an even distribution of larvae off the shores of the Hawaiian Islands, Miller (1974) found much higher densities of larvae nearshore and said that poor water quality would have a considerably greater effect than would be predicted from inshore density surveys of adults.

Finally, larvae lend themselves well to behavioral methods for assessing sublethal impacts such as temperature on schooling behavior of larvae (Williams and Coutant 2003) or of contaminants eliciting responses such as changes in swimming activity, respiratory rate, feeding ability, or predator avoidance, although the relationship between the laboratory findings and field observations is not always obvious. The best studies combine laboratory and field work to assess the effects of contaminants on larval behavior changes.

Eutrophication and Oxygen Depletion. Eutrophication is particularly well-known in freshwater situations where nutrient overenrichment results in low-oxygen conditions and poor or deadly habitats for the early life stages of fishes. However, eutrophication is also, unfortunately, well-known for marine systems, especially where there is poor circulation and flushing and sewage discharge. The brackish Baltic Sea is a classic example: the pelagic larvae of Atlantic cod (*Gadus morhua*) require the higher salinity waters below the halocline. However, because of eutrophication as a result of sewage, there is not enough oxygen below the halocline for normal larval development, with a resulting decline of the cod in the Baltic. However, since 1986 sewage has been diverted to the open sea and both the cod and other fishes have shown recovery from the decreasing effects of eutrophication and pollution (Urho 1989). Breitburg et al. (1999), using data from the Chesapeake Bay, used computer simulations to show that even in the absence of direct effects on larval mortality of low-oxygen values due to eutrophication, the indirect effect of predation (jellyfishes) mortality is considerably increased due to low-oxygen values.

Oil and Oil Dispersants. It is very useful to have baseline surveys done just before oil contaminates an area, but unfortunately this is usually not done until after the spill when it is less useful (Norcross and Frandsen 1996). Carls (1987) found that Pacific herring larvae exposed directly to the water-soluble fraction of crude oil were affected rapidly with high mortality, reduced swimming ability, and rapid reduction in feeding rates. However, larvae that were exposed indirectly, through oil-contaminated prey, did not show similar reductions and it was not felt to be an important source of larval contamination in the marine environment. Mangor-Jensen (1986) did not find that the osmoregulatory ability of Atlantic cod eggs and larvae was affected when the eggs and larvae were exposed to the water-soluble fraction of North Sea crude oil (50–280 ppb).

Several researchers have found that when dispersants are used to disperse oil, the oil-dispersant combination is much more toxic to larvae than the water-soluble fraction (e.g., Gulec and Holdway 2000), which brings to question the advisability of using dispersants after oil spills.

Effluents and Organochemicals. The effects on larvae of sewage, pulp mill, and other effluents are often a problem, but frequently it is not possible to determine exactly what pollutants are causing the problems. A common finding is that there are deformities of various kinds to the larvae found in the sewage plume (e.g., Kingsford et al. 1997). But if effluents are well diluted, or if the larvae spend little time in the plume, there may be no observable effects (e.g., Leslie and Kelso 1977), although in at least one case negative impacts on spawning fish have been measured for pulp mill effluents (Lehtinen et al. 1999). Chlorine is often a component of effluents and increasing concentrations have been shown to inhibit larval development and produce abnormal larvae (Morgan and Prince 1978).

Laboratory bioassay studies (Misitano et al. 1994) on larval surf smelt (*Hypomesus pretiosus*), beach spawning fishes with nearshore larvae, showed that when the larvae were exposed to contaminated Puget Sound sediment, mortality increased if the sediments were acutely toxic. Sublethal effects in less toxic sediments were seen in the form of decreased DNA content. Also, the ratio of normal to abnormal larvae was specifically related to increased concentrations of polynuclear aromatic hydrocarbons (PAHs) and polychlorinated biphenyls (PCBs). Fay et al. (1985) also found that lesions in the kidneys and gill tissues of gizzard shad (*Dorosoma cepedianus*) larvae were correlated with PCB body burdens.

Metals. Heavy metals are known to cause reduced survival in eggs and larvae, but the effects of particular metals on particular species of fish will vary. Gong et al. (1987) found that the toxicity of six heavy metals to the survival of larvae of two marine flatfishes was in the order of greatest to the least toxicity, namely: Hg >Cu >Cd >>Zn >Pb >Cr; eggs tolerated higher concentrations than larvae, the supposition being that the envelope (chorion) of the egg protects it against heavy metal penetration.

However, results of testing the effects of an antifouling paint containing copper, triazine, and tributyltin oxide (TBTO) on the early development of embryos in Atlantic cod indicated that in known field concentrations of the antifouling agent there were no detectable effects on fish embryonic and larval development (Granmo et al. 2002).

CONSERVATION BIOLOGY AND MARINE PROTECTED AREAS (MPAs)

Definitions

Conservation Biology. The discipline that treats the content of biodiversity, the natural processes that produce it, and the techniques that sustain it in the face of human-caused environmental disturbance.

Marine Protected Areas. An area set aside to protect marine resources and biodiversity, and which has ecological, historical, educational, or aesthetic significance.

Marine Protected Areas (MPAs) or Marine Reserves

Fisheries Mismanagement. The 20th century was marked by increased exploitation of living marine resources and parallel increases in attempts to manage these resources for long-term sustainability. We have gone from considering the ocean's resources as unlimited and available for unbridled exploitation, to trying to manage fisheries. In spite of these management efforts, widespread overfishing occurred. With the widespread depletion of fish stocks, a call has gone out for conservation and restoration of the resources of the ocean, with some even calling for preservation measures, as in the case with marine mammals.

Recently we have become increasingly aware of the impact of overharvest on living marine resources. Whether it is tropical reefs and their community of fishes being decimated for the aquarium trade, or the collapse of large-scale industrial fisheries such as the Atlantic cod fishery in the Northwest Atlantic, even the general public has become aware of serious problems throughout the marine ecosystems (Agardy 1999).

Besides limiting harvest to stabilize fished populations, various attempts have been made to enhance them, through hatcheries at first, and now though setting aside marine reserves. Now we are looking for new ways to prevent further damage to marine ecosystems and to restore populations to their former states. There is a basic change in philosophy toward a precautionary approach in fisheries management (Lauck et al. 1998) and protecting part of the habitat to enable stocks to rebuild, which is gaining considerable support. Some are even suggesting a basic reversal in thinking—protected areas should be the rule and fishing areas the exception (Walters 2000).

Marine Reserves, a Partial Solution. With the shortcomings of other fisheries management schemes, creating no-take marine reserves is increasingly gaining support, and such have already been established in several places around the world (Allison et al. 1998). Marine reserves have been established in both tropical and temperate areas and are generally in areas where reef-associated populations have been severely overharvested. Many are located in tropical coral reefs around the world to protect this type of habitat as well as to rebuild the communities associated with it. In temperate areas marine reserves have also been established offshore over hard and soft bottom. The recent closure of large areas of Georges Bank had dramatic effects on the populations of scallops as well as several species of harvested fishes (Murawski et al. 2000). The Channel Islands off the coast of California have been a marine sanctuary for some time, but as rockfish populations have declined, areas there are being proposed as marine reserves. As an example of a temperate, nearshore marine reserve, the Edmonds Underwater Park in Puget Sound, Washington, was established in 1970 as an artificial reef that was closed to fishing so scuba divers could enjoy observing the community of large fishes that took up residence there (Palsson 1998).

Thus, excluding certain marine areas from harvest has a long history, and has recently become more widely accepted as a means to reverse the downward trend in abundance of fishes and other marine life worldwide. Also, with the destruction of habitat by fishing practices (e.g., Koenig et al. 2000), there are calls for setting aside areas for reestablishment of unperturbed marine ecosystems, for conservation and enhancement of marine resources, and to allow study and casual observation (Agardy 1999).

Several objectives have been cited for establishing marine reserves (Dugan and Davis 1993a; Jones 1994). The potential effects of marine reserves on target populations include increased abundance, increased individual size and age (Roberts 1995), increased reproductive output, enhanced nearby populations (Auster and Shackell 2000), enhanced recruitment inside and outside the reserve (Dugan and Davis 1993b), increased resilience to overharvest (Guenette et al. 1998), and insurance against recruitment failures. Potential community benefits include increased species diversity (Western 1995), increased habitat complexity, increased community stability (Dugan and Davis 1993b), restored ecosystems to preharvest structure (Babcock et al. 1999), and unspoiled areas provided for viewing fishes and associated life (ecotourism).

The justification for marine reserves that relates most directly to early life history is that fish in the reserves grow larger, become more fecund, and thereby locally increase larval supply. It is reasoned that through larval transport, some of these larvae will settle elsewhere and thus will enhance juvenile recruitment over an area much larger than the reserve itself. Theoretical studies have demonstrated this effect of marine reserves, but field work is generally lacking.

Enhancing nearby populations is the most compelling aspect of marine reserves for fishers and fisheries managers, although the effectiveness of even this function is still under debate (Conover et al. 2000). Enhancement can occur by two means: spillover of adults from the reserve to nearby areas, and transport of larvae produced in the reserve to fished areas. Spillover of adults has been documented in some cases (e.g., Russ and Alcala 1996; Roberts et al. 2001), but the recruitment effect is largely conceptual and has mainly been the subject of modeling studies (e.g., Dight 1995; Guenette et al. 1998; Stockhausen et al. 2000; Botsford et al. 2001). The enhancement value results from recruits that originated from eggs and larvae produced by the larger, more fecund adults in the reserve dispersing to fished areas and settling and recruiting there (Man et al. 1995).

Of course not all fishes are suitable candidates for protection by marine reserves. Most reserves are designed for rather sedentary fishes that live in association with bottom structure (e.g., coral reefs or temperate rocky-reefs). Fishes with restricted movements and localized home ranges during at least part of their life cycle will benefit most from reserves (Kramer and Chapman 1999). Long-lived fishes will benefit from protection by continuing to grow and increase in fecundity (Davis 1989). A reproductive pattern that includes a relatively long dispersal period, as eggs or larvae or both, is required to increase recruitment in adjacent areas. Other life-history characteristics, such as sequential hermaphroditism and spawning migrations and aggregations, should be included in plans for establishing marine reserves. Clearly, much basic life-history information about the species that are to be protected is needed in deciding if reserves will be effective, where they should be located, how large they should be, and how many there should be (cf. Lindeman et al. 2000). For example, Martell et al. (2000) found that life-history characteristics of lingcod (*Ophiodon elongatus*), which include seasonal spawning migrations, necessitate reserves being large and permanent.

Marine reserves are presently one of the most active topics in marine science. There is a rapidly growing body of literature on them, several websites exist (e.g., http://www.panda.org/resources/publications/water/mpreserves/; http://mpa.gov/; http://depts.washington.edu/mpanews/issues.html), and frequently they have been the subject of conferences (e.g., Dugan and Davis 1993a; Yoklavich 1998; Conover et al. 2000; NRC 2001).

A Model of Fisheries Management Objectives of Marine Reserves. The conceptual model of the effects of marine reserves is easily understood (Figure 7.2). Marine reserves are supposed to conserve the marine life within their boundaries and enhance it in nearby unprotected areas. The conservation role of reserves in recruitment is included in this diagram. Eggs produced by the larger, more fecund adults in the reserve result in more recruits that settle in the reserve. In fact, the population in the reserve could also be increased by settlement of larvae produced elsewhere. This benefit of the reserves assumes that the target populations are recruitment limited, and in fact that recruitment is limited by the number of offspring produced by the adult population (larval supply). If other factors are responsible for the reduced adult abundance in the area, the reserves may have little benefit.

The Track Record of Marine Reserves. In a review of marine protected areas in Canada, Jameson and Levings (2001) found that the term has not been rigorously defined, and in some cases little or no protection of fishes from harvest actually occurs in these areas. Thus the objectives and regulations associated with marine reserves need to be clearly stated for them to be effective.

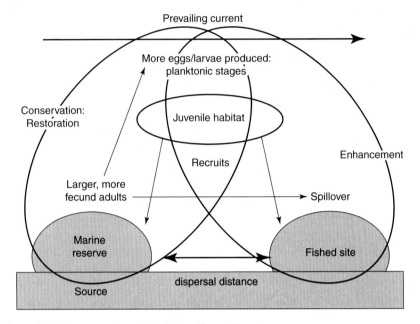

Figure 7.2. Conceptual model of the effects of marine reserves.

Halpern and Warner (2002) reviewed empirical studies on 80 marine reserves and found rapid and lasting effects of protection from harvest demonstrated by increases in abundance, size, and diversity of fishes within the reserves; effects on populations outside the reserves were not investigated. Cote et al. (2001) examined the effects of marine reserves on diversity and abundance of fishes and found that species' richness was consistently greater in the reserves, but abundance was greater only for those species that were subject to nearby fishing pressure. Jennings (2000) modeled the recovery of fish populations that were protected from harvest in marine reserves and found that with small areas, it might not be possible to distinguish between redistribution and population growth. He did not address the possible benefits that might accrue due to increased recruitment.

Marine Reserves and Recruitment. One of the primary objectives of marine reserves is to increase recruitment of target species both within the reserves and in adjacent areas (Figure 7.3). More fish will then be available for harvest in these adjacent areas that are open to fishing. The idea is that adults in marine reserves which are free from harvest will live longer and grow larger. Since fecundity is directly related to fish size, roughly to length cubed, the larger fish will produce many more eggs. In most marine fishes

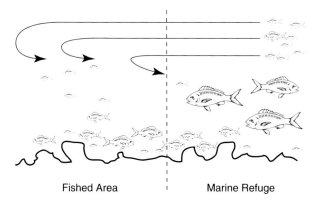

Figure 7.3. Conceptual model of the relationship of marine reserves to recruitment of targeted populations (Wayne Palsson, pers. comm.).

the eggs planktonic and, and along with the planktonic larvae, are the primary dispersal phases in sedentary fishes. Thus the eggs and larvae produced in the marine reserve will settle in the reserve and in adjacent areas to enhance recruitment both within the reserve and elsewhere (Carr and Reed 1993). However, in a review of 31 empirical studies on the effects of marine reserves on target populations (both finfishes and invertebrates), Dugan and Davis (1993b) found only three that considered recruitment effects: one of these showed positive effects and two did not demonstrate effects. More recently, Planes et al. (2000) found an exceptionally low number of studies specifically addressing recruitment processes in MPAs. Stoner and Ray (1996) conducted one of the few studies to date examining the effects of a marine reserve on larval production. Abundance of queen conch (*Strombus gigas*) larvae appeared to be directly associated with increased abundance of adults in the marine reserve. Enhanced recruitment in the reserve was due to arrival of settling-competent larvae from upstream areas outside the reserve. Contributions of larvae produced in the reserve to nearby fished areas were not demonstrated.

In order for recruitment enhancement to occur, the fished area must be within the dispersal distance of the eggs and larvae produced in the marine reserve (Guenette et al. 1998; Botsford et al. 2001; Gunderson and Vetter 2006; Figure 7.4). However, little is known about dispersal of larvae, and although some probably travel long distances others seem to be retained near the area where they were produced (Swearer et al. 1999; Warner et al. 2000). If larvae disperse from where they were spawned, prevailing currents must be toward the fished area for a reserve to act as a source for recruits to a fished area (cf. Dahlgren et al. 2001). If currents run from the fished area to

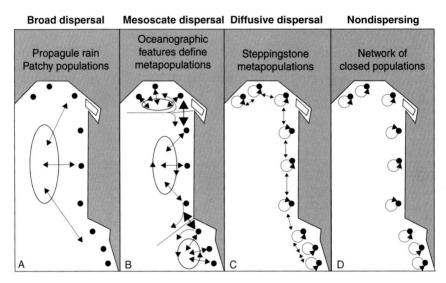

Figure 7.4. Propagule dispersal models and population structure. A. broad advective dispersal typical of species with early planktonic stages. B. mesoscale dispersal typical of species whose early life stages develop in the plankton but are retained in an oceanographic feature such as the Southern California Eddy. C. diffusive dispersal describes nearshore species whose eggs and larvae remain in nearshore boundary layers subject to diffusive rather than advective flows. D. nondispersing describes species that produce large, precocious young capable of swimming and not subject to passive dispersal in currents (from Gunderson and Vetter 2006).

the reserve, it could be considered a sink rather than a source of recruits and would not enhance recruitment in the fished area (Roberts 1998; Crowder et al. 2000). Valles et al. (2001) found that larval supply was greater over a fished area than over a nearby marine reserve, pointing out the need to carefully site reserves so they have an adequate larval supply for enhanced settlement and eventual recruitment to occur. Since little is known about larval dispersal, networks of reserves which will act as sources of larvae are recommended (Roberts et al. 2001).

Another consideration regarding the enhanced recruitment value of marine reserves is that the juveniles may not settle directly onto the habitat occupied by the adults. For example, if they settle onto soft bottom or in vegetation and move to the reefs later, these juvenile habitats may need to be protected also to allow conservation and enhancement of the target species (cf. Hill and Creswell 1998). Lindholm et al. (2001) found that habitat change caused by trawling and dredging resulted in increased predation on juvenile Atlantic cod, and suggested that excluding some juvenile habitats from fishing could ameliorate these effects. However, reserves in reef areas might have a negative effect on settled juveniles, if they are subjected to increased predation from the greater abundance of piscivorous adults in the reserves (Tupper and Juanes 1999; Garcia Rubies 1997 [cited in Planes et al. 2000]).

Conclusions. The concept of marine reserves is very attractive as a fisheries management tool and for other ecological and social reasons. Marine reserves have been shown to increase the number and size of fishes within them, and increase the diversity of the community associated with them. The spillover of adults from reserves to nearby fished areas has been documented in several places. Theoretical benefits of reserves relative to recruitment have been elucidated, however few field studies have been conducted to measure these effects. Increased egg and larval production resulting from establishing reserves should be detectable by plankton sampling once the reserves are functioning. The recruitment benefits of the reserves, both within the reserves and in adjacent areas, might be more difficult to demonstrate, since many factors besides egg and larval production are involved in the recruitment process. For marine reserves to be effective in enhancing recruitment, the population must be recruitment-limited in the first place, and protection of juvenile as well as adult habitat might be required. Recruitment considerations and studies should accompany plans for establishing and monitoring marine reserves.

CHAPTER 8

Rearing and Culture of Marine Fishes

REARING MARINE FISHES
TO ENHANCE OR REPLENISH
WILD STOCKS
  Mass Releases of Yolk Sac Larvae
    The Late 1800s and Early 1900s
    Few Advances Until the 1960s
  Mesocosm Studies
  *Large-Scale Culturing and
    Rearing*

EXPERIMENTAL CULTURE
  METHODS
  Obtaining Gametes
  Artificial Fertilization
  Rearing Containers
  Water
  Temperature Control
  Light

Food
  Culturing Rotifers
  Obtaining Wild Food

TYPES OF EXPERIMENTAL STUDIES
  *Embryonic Development
    and Endogenous Nutrition*
  *Food Requirements and Feeding
    Behavior*
  Environmental Responses
  Predation
  Condition Index Validation
    Histology
    RNA/DNA
    Cytometry
  Otolith Increment Validation
  Pollution Effects

REARING MARINE FISHES TO ENHANCE OR REPLENISH WILD STOCKS

Mass Releases of Yolk Sac Larvae

The Late 1800s and Early 1900s. In the 1860s, G. O. Sars artificially fertilized eggs of Atlantic cod (*Gadus morhua*) and followed their development through hatching, leading to culturing and releasing large numbers of yolk sac larvae into the sea off Norway, which he felt might ameliorate the occurrence of unfavorable year-classes (see Kendall and Duker 1998). Johan Hjort fought hard, but largely in vain, to have sampling conducted to determine the value of these releases. In the United States, propagating fishes soon became the major activity of the U.S. Fish and Fisheries Commission after it was established in 1870, although it was not in the original work plan. The first commissioner, Spencer Fullerton Baird, originally rejected this idea, although he soon actively promoted it, and in June 1872 funds were appropriated for hatcheries. Baird viewed hatcheries as a way to dampen fluctuations in abundance. Culture of freshwater fishes was emphasized during the first few years. Common carp (*Cyprinus carpio*) from Germany were introduced throughout the country, and shad (*Alosa sapidissima*) were introduced to West Coast rivers from the East Coast. Fish propagation was very popular and appropriations were soon three times the amount for all other activities of the Commission, justified by the idea: ". . . it is better to expend a small amount of public money in making fish so abundant that they can be caught without restriction . . . than to expend a much larger amount in preventing the people from catching the few that still remain" (Goode 1883). Following successful breeding of Atlantic cod and haddock (*Melanogrammus aeglefinus*) in 1878, propagation of marine species became an increasingly large part of the culture efforts. The first laboratory of the Commission, at Woods Hole, MA, was built as a hatchery in 1885, and it was followed quickly by laboratories at Boothbay Harbor, ME, and Gloucester, MA. The first two ships of the Commission, the *Grampus* and the *Fish Hawk,* were floating fish hatcheries. Marshall MacDonald, who had been head of the propagation efforts, was appointed commissioner in 1888, and fish culture became even more prominent. From 1872 through 1940, over 65% of the budget went to hatcheries, during which time more than 200 billion fish eggs and fry were released. By 1940, 98% of the eggs and 75% of the larvae released were of marine species. Atlantic cod, haddock, winter flounder (*Pseudopleuronectes americanus*), and American lobster (*Homarus americanus*) were the primary marine species cultured. In 1929, at the height of hatchery operations, over 2.5 billion cod eggs alone were released. The effectiveness of culturing marine fishes was never objectively proven, and the last U.S. marine hatchery activities ceased in 1952.

Although stocking of young freshwater fishes obviously led to self-sustaining populations in some cases, and continual stocking provides fish that would not have been there otherwise, the situation in marine waters is very different. With the enormous fecundity of most marine fishes, extreme mortality must occur during these egg and early larval stages. Rather than rearing marine fish to a size larger than when most natural mortality occurs, their eggs and early larvae were released into the sea. The premise for this practice was that there is a strong correlation between the number of eggs produced and the number of young recruited to the population. This idea was seriously questioned by Hjort in Norway, and most British fisheries scientists of the day. In Norway, experiments failed to prove beneficial effects of releasing eggs and early larvae of cod, but such experiments were not conducted in the United States (see Solemdal et al. 1984). Rather, the Fish Commission relied on hearsay accounts of increased numbers of fish after eggs had been released. In fact, the large numbers of young cod caught in Gloucester Harbor, MA, five years after culture efforts had started were referred to locally as "Commission cod."

Few Advances Until the 1960s. Although culture of freshwater fishes and salmon (*Oncorhynchus* spp.) became routine, there were few technical advances in rearing of marine fishes during the first half of the 20th century. Although larvae of a variety of fishes were reared on occasion, sometimes by brute strength methods, no consistent techniques were established. Finding a ready source of suitable food remained a problem until the use of brine shrimp (*Artemia* spp.) larvae was initiated in the 1940s. Some rearing was done in the 1950s with an aim toward understanding the ecology and physiology of developing fish larvae, rather than toward augmentation of natural populations. In the early 1960s a concentrated effort was made to develop rearing techniques for plaice (*Pleuronectes platessa*), the success of which paved the way for similar studies on many other fishes (Shelbourne 1964, 1965). Although brine shrimp nauplii were an effective food for some marine fish larvae, they were too large to be used with many other species. The development of culture techniques for the rotifer *Brachionus plicatilis* provided a dependable source of food for smaller larvae, and their culture became feasible.

Mesocosm Studies

Once rearing of marine fish larvae in the laboratory was well established, it became clear that more realistic ecological experiments could be performed using larger rearing containers. Laboratory studies had indicated that wall effects and water quality of small rearing containers were problems that could be lessened by increasing the container size. Experiments in large containers (mesocosms) could be used to bridge the gap between field and laboratory work (Solemdal 1981; de Lafontaine and Leggett 1987; Fuiman and Gamble 1988). Enclosures such as concrete rearing ponds on land and large plastic bags floated in protected waters have been used for these experiments. Growth and mortality of larvae are investigated by stocking the mesocosm with known numbers of fish eggs or early larvae of one or more species, and adjusting conditions such as prey and predator types and densities (Øiestad and Moksness 1981). Effects of pollutants can also be investigated by introducing them into the mesocosms at known concentrations. Conditions in the mesocosms are monitored, and samples of larvae may be collected during the experimental period, which may last a few days or weeks. At the end of the experiment, the enclosures are typically emptied and all the larvae collected to measure such factors as size, condition, and gut contents. Such large rearing containers are also used to grow larvae to the juvenile stage when they are released into the wild to supplement natural populations (van der Meeren and Naas 1997).

Large-Scale Culturing and Rearing

Large-scale culturing and rearing of marine fishes is now being accomplished for several species around the world (Furuyu 1995; Battaglene 1996; Travis et al. 1998; Howell et al. [eds.] 1999; Seikai 2002). Besides providing juveniles for fish farming operations, the major objective of these programs is to replenish depleted local fished populations. The premise is that releasing reared juveniles along with other remedial efforts such as pollution abatement, habitat restoration, and catch restrictions will enable depleted stocks to recover to former levels of abundance. Typically, eggs are stripped and fertilized from adult brood stock, larvae are reared beyond the transformation stage, and young juveniles are released into the wild. A few examples of such programs include production of young Atlantic cod in Norway (Svåsand 1998), red drum (*Sciaenops ocellatus*) in the United States (McEachron et al. 1998), and Japanese flounder (*Paralichthys olivaceus*) in Asia (Masuda and Tsukamoto 1998). As in the earlier programs, their success can only be determined by recognizing the cultured fish among the wild stocks. In order to accomplish this now, cultured fish of some species are marked by producing chemical (Nordeide et al. 1992, 1994; Blom et al. 1994; Secor and Houde 1998) or stress (such as short-term temperature shock) checks on their otoliths. The blind side of cultured Japanese flounder is more heavily pigmented than the wild fish, which provides an easy way to spot reared fish, although this abnormal pigmentation reduces their market value (Iwata and Kikuchi 1998). Genetic markers have also been used as a means of recognizing cultured Atlantic cod (Jørstad et al. 1987).

With the development of the technical capability to enhance marine stocks by releasing cultured juveniles, comes other issues regarding the feasibility and advisability of such activities. Was the decline in the stock due to poor recruitment which resulted from a lack of reproducing adults, or was the carrying capacity of the area decreased for the species under consideration? Will the released juveniles actually enhance the wild population, or will they displace wild juveniles? Will the released juveniles alter the genetic makeup of the wild population if they breed with wild fish? Is it economically feasible to rear the species: what is the cost-benefit ratio? Are management options other than releasing reared juveniles more cost effective in rebuilding an overfished stock?

Many technical questions must also be addressed as a rearing/release program is developed. What is the best source for brood stock? The brood stock generally should be taken from the area where the releases will occur, in order to produce fish adapted to local conditions, and fish that are genetically close to the wild population. What size (age) should the juveniles be when released? This may become an economic question: What is the return from maintaining the fish longer in terms of reduced mortality when they are released? Where, when, and how should the juveniles be released? The releases should be designed to minimize mortality after release and have minimal impact on the wild stock. Detailed information on habitat and food requirements of juveniles as well as on potential predators is required for releases to be effective. Predation is a major source of mortality of juvenile fishes and it has been found that reared juveniles introduced to predators before release survive in the wild better than naive juveniles (Olla et al. 1998). In general, the ability of hatchery-reared fish to survive in the sea is less than that of wild fish (Howell 1994).

EXPERIMENTAL CULTURE METHODS
Obtaining Gametes

Experimental and production culture methods have evolved over the years and are generally established on a species-by-species basis (see Hunter 1984; Tucker 1998 and literature referred to therein). Gametes for rearing can be obtained by collecting ripe adults on the spawning grounds. Although this is very effective in some cases, it also can limit the time of year for experimental work and contribute a high degree of risk, since finding spawning adults is not always predictable. Males are usually ripe for a longer period than females and the condition of the sperm does not seem to be as critical as the condition of eggs. Thus, the availability of ripe females is more problematic than finding ripe males. Also, testes can be kept cool and away from seawater for up to about a day, and still be used to fertilize eggs if necessary. Sperm can even be frozen using cryoprotectants if necessary. Eggs should be stripped from females as soon after capture as possible. Although in some cases both sexes can be held alive for some time after capture and the gametes will still ripen and be viable, in other cases egg quality decreases rapidly after capture.

Methods have been developed to induce ripening of gonads in several species, to allow production of gametes over a longer period of time and in a more predictable manner. Providing adequate diet and gradually changing temperature and photoperiod can aid gonad maturation and induce spawning in some species. Hormone injections have also proven effective in other species.

Artificial Fertilization

Artificial fertilization of pelagic fish eggs is usually accomplished by extruding the eggs into a clean glass or plastic bowl at ambient rearing temperature. The bowl can be moist with ambient water, but does not need to contain much water. The adult fish should be rinsed before extruding the eggs, to clean off any foreign substances that may contaminate the bowl. Also, care needs to be taken so no feces is extruded into the bowl while squeezing the eggs out. The eggs are extruded from the female by gently rubbing the gonad toward the vent. The first few eggs are allowed to fall away from the rearing container, and then the eggs are directed into the rearing container. Fully ripe eggs flow out of the fish under little pressure, and are clear and not cloudy. Sperm from one or more males is added to the eggs immediately by extruding it in the same way that eggs were extruded from the females, using the same cautions for cleanliness. A little clean ambient water can be added to the bowl and the eggs and sperm mixed by gently swirling the bowl. After a few minutes more water can be added. The eggs and water can be poured into glass jars for rearing within an hour. Most fertilized eggs float high in the container after being fertilized, so the underlying water with unfertilized eggs and debris can be siphoned off. Clean water is added and this process is repeated until the water in the rearing container is no longer cloudy from the sperm. Artificially spawning demersal eggs is similar, except the eggs are usually adhesive so they are extruded onto a substrate on which they will develop. For laboratory work, glass plates can be effective substrates for demersal eggs.

Rearing Containers

Containers for rearing pelagic eggs can be 1- to 4-liter glass jars. They can be placed in a water bath to stabilize temperature. Eggs can be stocked at densities of 500–1,000 per liter. As larvae grow, their densities need to be decreased. Larvae survive and grow better in larger tanks, about 100–400 liter, and cylindrical black fiberglass containers are often used.

Water

Water for rearing should simulate in salinity the water where the eggs and larvae normally occur in the wild. Salinity of rearing water can be adjusted by adding distilled water to lower salinity or by adding aquarium salts to raise salinity. Water can be filtered (1–5 μm) to remove debris and organisms. It can also be sterilized by passing it under an ultraviolet light. Antibiotics can be used to reduce bacterial growth on the eggs, but they may add an additional source of confusion to experimental results. Filtered, sterilized seawater can be held for some time and used as needed in the laboratory, allowing marine fish rearing to occur at some distance from the source of seawater.

The rearing water for pelagic eggs should be static (not aerated or circulating; with demersal eggs, water can be aerated or circulated to maintain high oxygen levels), but it should be renewed daily with ⅓ to ½ the volume of clean water. Debris should be siphoned off the bottom of the rearing container as the water is renewed. Water quality decreases (low oxygen, high ammonia) in time due to buildup of dead prey and metabolites. Water quality deteriorates faster at higher temperatures, so water exchanges should be more frequent when higher rearing temperatures are used. Recently, techniques have been developed to circulate rearing water through biological filters that rely on a community of bacteria to remove metabolites. To prevent damage to the eggs and larvae, very low flow rates are used as the filtered water is continuously exchanged with water in the rearing containers.

Temperature Control

Temperature has a great effect on developmental rate and hatching success of fish eggs. In culture, temperature can be maintained by placing the rearing containers in a water bath that is thermostatically controlled. Laboratory water heaters can be used to increase temperature, and electric water coolers can be used to cool the water. Experiments can be performed to determine the effect of temperature on several ecological factors involved in fish egg and larval survival and development.

Light

Light for rearing is usually provided by fluorescent electric lights suspended above the rearing containers. Since ultraviolet rays of sunlight are detrimental to fish eggs, unfiltered sunlight should not be used in rearing, except in experiments testing its effects. Light levels can be adjusted to simulate levels that the eggs and larvae would experience at their depth of occurrence in the wild by placing screens or other filters over the rearing containers. Light levels of 3.0 to 3.5 μ molphoton/m^2/s^1 at the surface of the rearing container are typically used. Experiments can be performed to examine the effects of various light levels on egg and larval survival and development. Larvae reared under higher-than-normal light levels are often more heavily pigmented than wild-caught larvae. Photoperiod can be set to simulate that in the wild at the time of the year the eggs and larvae normally occur, or can be used as an experimental factor.

Food

Providing food for marine fish larvae in culture has proven to be the greatest impediment to success in rearing. The size of the gape of the mouth of larvae determines the maximum size of the prey that can be ingested.

The width of the mouths of first-feeding marine fish larvae are typically 20–150 μm. Brine shrimp nauplii (~ 400–700 μm at hatching) can be cultured easily, and they are a suitable size for larger first-feeding marine fish larvae. Newly hatched brine shrimp contain a large amount of yolk to nourish the larvae, but the brine shrimp must be fed enriched yeast or cultured phytoplankton to remain nutritious for the larvae. Brine shrimp larvae are a good food source for later larvae, for those species too small to take them at first feeding.

For successful rearing, food densities must be considerably higher than expected in the field. Rotifers are generally stocked at densities of about 5–10 organisms per milliliter, wild plankton at about 1–3 organisms per milliliter, and brine shrimp nauplii at about 1–2 organisms per milliliter. Food density, type, and quality can be used as experimental variables.

Culturing Rotifers. For larvae too small to take brine shrimp nauplii at first feeding, culture of the rotifer *Brachionus plicatilis* (~ 60 μm) may provide suitable food. These rotifers can be cultured on a diet of yeast and phytoplankton, which in turn can be cultured using nutrient-enriched seawater (Fontaine and Revera 1980; Yoshimura et al. 1996). Adding cultured phytoplankton directly to the larval fish rearing containers can improve water quality and provide food to keep the rotifers or brine shrimp nutritious.

Obtaining Wild Food. Wild food in the form of microzooplankton can be used to culture marine fish larvae. Wild food is usually collected by towing fine-meshed (50–100 μm) plankton nets through areas with high concentrations of reproducing zooplankton. The collected zooplankton is filtered through a screen (~ 200 μm) to remove larger, potentially predatory organisms. Since a wide range of taxa are usually collected, there is less control over exactly what prey are available and consumed by the larvae than when using cultured food. Additionally, finding a place where sufficient quantities of appropriate zooplankton are consistently available is not always easy. Although wild zooplankton can be kept and fed cultured phytoplankton for a few days, it gradually loses its vitality, and fresh material must be collected frequently to provide adequate food for the larvae.

TYPES OF EXPERIMENTAL STUDIES

With the development of culture techniques for marine fish larvae, they have become the objects of a wide range of experimental studies. Most studies have been designed to learn more about the ontogeny of the larvae, to investigate the responses of larvae to manipulation of various environmental factors to help understand how the larvae interact with their environment, and to develop large-scale culture techniques.

Embryonic Development and Endogenous Nutrition

Studies on egg development have been conducted since early in the 20th century. Fertilizing planktonic eggs of marine fishes and rearing them to hatching is quite easy, and so it has been done for many species. Early studies focused on describing embryonic development and providing characters so the eggs could be identified in plankton samples. The rate of development of fish eggs in relation to temperature was then investigated for many species (e.g., Ryland and Nichols 1975). These studies usually include describing developmental stages of eggs and the time to reach each stage in relation to temperature, so the age of wild-collected eggs could be estimated, if the temperature of the water where they occurred is known. This information is helpful in estimating egg mortality rates and back-calculating spawning time and place (knowing the currents in the area where the eggs were collected). The effects of other environmental variables (such as salinity) on rate of development and survival have also been studied for many species (e.g., Alderdice and Forrester 1971; Laurence and Howell 1981; Liu et al. 1994). Yolk utilization and the effects of egg size and quality on larval survival have also been investigated (Knutsen and Tilseth 1985; Quantz 1985; Fukuhara 1990; McEvoy and McEvoy 1991). Effects of pollutants on egg development and mortality have also been studied.

Food Requirements and Feeding

As larvae exhaust their yolk supply, their feeding becomes an object of experimental studies. Many of these studies explore the effects of various feeding regimes on larvae to relate to differences in type and quantity of food that they might encounter in the wild (e.g., MacKenzie et al. 1990). Effects of starvation and suboptimal food supply have frequently been considered (Margulies 1989). The relationship between feeding rate and prey density has been studied (Figure 8.1). Feeding behavior (Munk and Kiørboe 1985), growth, and survival on various foods and under various conditions have been investigated (Houde and Schekter 1980). Food consumption and gut evacuation rate have been determined to relate to abundance of food in the wild and larval condition (Canino and Bailey 1995).

Development of the structure and function of the digestive system has been documented in a number of studies (e.g., Govoni et al. 1986; Oozeki and Bailey 1995).

Behavior

Environmental Responses. Development of sensory organs in relation to feeding behavior and predator avoidance has been the subject of several studies (e.g., Houde and Schekter 1980; Blaxter 1986; Batty 1989;

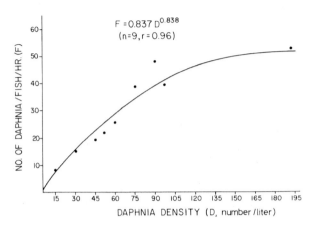

Figure 8.1. The relationship between larval walleye (*Stizostedion vitreum*) feeding rate and prey density (from Mathias and Li 1982).

Pryor-Connaughton et al. 1994; Stearns et al. 1994). The effects of starvation on escape speed by larval fish have also been investigated. Recent advances in video techniques, including using infrared light, have greatly facilitated such studies (Blaxter and Fuiman 1990).

Predation

Larval fishes have been subjected to a variety of predators in the laboratory, including their own species (e.g., Brownell 1985). The responses of larvae to fishes as well as invertebrate predators have been investigated (e.g., Bailey 1984; Purcell et al. 1987; Monteleone and Duguay 1988; Cowan and Houde 1992; MacKenzie and Kiørboe 1995).

Condition Index Validation

Since inadequate food is thought to be a major cause of mortality in larval fishes, considerable research has been devoted to finding ways to measure the condition of field-caught larvae. Probably larvae that have actually starved are quickly consumed, but those that have experienced poor feeding conditions may show signs of nutritional deficits, which may indicate that starvation is imminent.

A variety of biochemical and histological condition indices have been developed. Many of these indices are size-dependent. The sensitivities of the indices vary in terms of the amount of time between when the larva

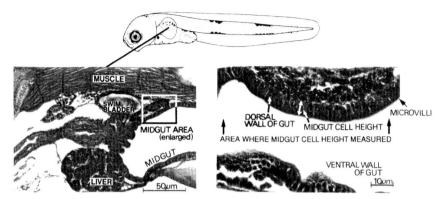

Figure 8.2. The location where the midgut cell height measurement was taken in larval walleye pollock (*Theragra chalcogramma*) (from Theilacker and Porter 1995). The left photomicrograph shows a sagittal section of the midgut and surrounding organs (larva 8 d after hatch, 4.88 mm SL in Bouin's fixative); the right photomicrograph shows the area where midgut cell height was measured (larva 12 d after hatch, 5.68 mm SL in Bouin's fixative).

experiences food shortages, and when the index reflects the poor conditions. To calibrate condition indices, larvae are reared in the laboratory under various feeding regimes, usually from no food to being fed ad libitum, and then their condition is measured by the technique under development. The amount of time under various feeding conditions is also varied. Once the index has been calibrated, it is applied to field-caught larvae to see what proportion are in jeopardy of starvation (Robinson and Ware 1988).

Histology. Laboratory experiments have demonstrated that changes in various tissues are associated with suboptimal feeding conditions (e.g., Theilacker and Porter 1995). Liver, brain, and digestive tract tissues provide evidence of prior feeding conditions and recovery from poor feeding conditions. The degree of deterioration varies among the tissues for a given amount of time without food. The height of the cells lining the midgut has proven to be an accurate index of larval condition (Figure 8.2).

RNA/DNA. Techniques have been developed to measure the amount of RNA and DNA in individual larvae (Buckley 1980; Clemmesen 1996). The amount of DNA in cells is constant, but the amount of RNA is a function of the amount of protein synthesis (Figure 8.3). Low amounts of RNA indicate

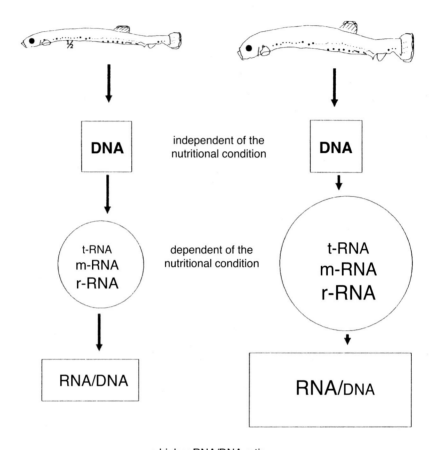

Figure 8.3. Theoretical reaction of RNA and DNA content in starving (left) and well-fed (right) fish larvae. The size of the letters reflects the amount of nucleic acids. Bigger letters mean higher concentrations. t-RNA = transfer RNA, m-RNA = messenger RNA, r-RNA = ribosomal RNA (from Clemmesen 1996).

that a larva is experiencing poor feeding conditions since it is not actively synthesizing protein. The amount of DNA in a larva is dependent on its mass, so the size of the larva must be taken into consideration when applying the RNA/DNA ratio to measure larval condition. Laboratory studies have documented the relationship of DNA to larval size, and the effects of temperature and various feeding conditions on RNA/DNA of larvae of several species (Buckley et al. 1990; Canino 1994; Johnson et al. 2002).

Cytometry. An advanced technique to measure larval condition involves using flow cytometry to measure the proportion of cells that are actively dividing (Theilacker and Shen 1993). The more cells that are dividing, the better the condition of the larva. This technique has been applied to determine the proportion of brain cells that are dividing in larvae under various feeding conditions in the laboratory.

Otolith Increment Validation

Considerable experimental work with larval fishes has involved investigating the microstructure of their otoliths (see Campana and Moksness 1991). It has been determined that, in general, daily increments are laid down in the otoliths. However, this must be validated for each species under study. Also, there is variation in the timing of the first increment: in some fishes it is laid down before hatching, whereas in others it is at a later time; this must be established for each species through laboratory experiments. In some studies, experimental marks have been introduced into the otoliths by exposure to chemicals (e.g., alizarin, oxytetracycline) or temperature to help validate the nature of increment formation. Such otolith tags have also been used to identify larvae that have been released into the wild. The effects of various environmental conditions (such as low food) on otolith increment formation must also be investigated in the laboratory (Bailey and Stehr 1988). The width of increments seems to be related to food supply in at least some species (see Clemmesen and Doan 1996). Distinctive marks on the otolith are sometimes associated with life-history transitions, such as settling in flatfishes (pleuronectiforms) and reef fishes; laboratory experiments are needed to interpret these marks.

Pollution Effects

The effects of pollutants on larval development and survival have been the object of many experimental studies to the point that larval development can be used as a sensitive bioassay tool in some cases (e.g., Goksoyr et al. 1991). Malformed larvae can result from exposure to low levels of some pollutants, such as those associated with crude oil and oil products.

LABORATORY EXERCISES

Exercise 1. Ova measurement to determine number of spawnings and introduction to writing a scientific paper
Exercise 2. Use of histology to determine if multiple spawnings are occurring
Exercise 3. Fecundity determination
Exercise 4. Artificially spawning, fertilizing, and rearing planktonic fish eggs
Exercise 5. Culturing marine fish larvae
Exercise 6. Conducting ichthyoplankton tows
Exercise 7. Sorting and identifying fish eggs and larvae in ichthyoplankton samples
Exercise 8. Diversity/identification of planktonic fish eggs and larvae
Exercise 9. Preparing and examining larval fish otoliths for daily growth
Exercise 10. Larval fish gut contents
Exercise 11. Histological condition index: midgut cell height
Exercise 12. Clearing and staining larval fish

LABORATORY EXERCISE ONE

Ova Measurement to Determine Number of Spawnings and Introduction to Writing a Scientific Paper

Purpose To use ova diameter to determine the frequency of spawning. (Ova diameter frequency modes may indicate separate spawning events.)

Reference If possible, use a reference that pertains to the species you are working on, such as Hinckley, S. 1987. The reproductive biology of walleye pollock, *Theragra chalcogramma*, in the Bering Sea, with reference to spawning stock structure. Fish. Bull. 85: 481–498.

Materials
Dissecting microscope (preferably a "zoom" model) with an ocular micrometer in one eyepiece.
Stage micrometer.
Preserved ovary.
Scalpel.
Needle probes.
Petri dish.
50% glycerin.

Procedures
1. Adjust the "zoom" knob on the scope to a setting that enlarges the largest ovum to be measured to a size just within the ocular micrometer range.
2. Use a stage micrometer (calibrated slide) to calibrate the ocular micrometer: Obtain a conversion factor which (when multiplied by any "raw" ocular micrometer reading) yields a measurement in mm.
3. Take a small subsample of the ovary by cutting a cross section (or take a subsample of about 100 eggs from a beaker of swirling eggs—see Lab Ex. 3). Subdivide the cross section into wedges if necessary. Always cut through the center of the cross section.
4. Add 50% glycerin to the subsample in the petri dish. Tease the eggs apart so that one layer evenly covers the bottom of the petri dish.
5. Measure one egg at a time (and only one time per egg) with the ocular micrometer and record the value in "raw" units. It is best to have a grid underneath the petri dish to keep track of the measured eggs. Measure 50–100 eggs. Measure all eggs that are clearly visible within one grid before moving to the next grid, so all egg sizes will be represented proportionally in your sample. Once you have measured the desired number of eggs you may stop and your sample is completed.
6. Repeat the procedure for another 50–100 eggs from the same ovary (for a replicate sample).

7. Write the exercise as a brief (not more than five text pages) scientific paper on the spawning frequency (per year, not during spawning period) of this species, based on ova diameter measurements and the literature. Be sure to include the following:

>An appendix with the raw data from today's exercise.
>
>A description (in Methods and Materials) of the micrometer calibration and the formula used to calculate standard deviation of the ova diameter data.
>
>Convert your raw micrometer data to mm with your conversion factor.
>
>Make a table of simple statistics—number (N), Mean, Range, standard deviation (SD)—for each size group of eggs.
>
>Plot a histogram of the ova diameter.
>
>Be sure to include at least three references.

LABORATORY EXERCISE TWO
Use of Histology to Determine If Multiple Spawnings Are Occurring

Purpose Multiple spawning events can occur in fishes with determinate or indeterminate fecundity. In both cases, knowledge of multiple spawnings is required for fecundity estimates used in biomass estimates, population dynamics models, and life-history studies. Histological analysis of ovarian slides provides the most information of ovarian and oocyte development. In this exercise, the presence of multiple spawnings during a single spawning season will be determined by examining ovary cross sections using a light microscope.

The following abbreviations are used in Figures Ex. 2.1 through Ex. 2.7.
UO   unyolked oocytes
YG   yolk globule
ZR   zona radiata
N   nucleus
FOL   follicle
POF   post-ovulatory follicle

References

Hunter, J. R. and B. J. Macewicz. 1985. Measurement of spawning frequency in multiple spawning fishes. In: An egg production method for estimating spawning biomass of pelagic fishes: application to the northern anchovy, *Engraulis mordax* (R. Lasker ed.), pp. 67–78. U.S. Dep. Commer., NOAA Tech Rep. NMFS 36.

McDermott, S. F. and S. A. Lowe. 1997. The reproductive cycle and sexual maturity of Atka mackerel, *Pleurogrammus monopterygius,* in Alaskan Waters. Fish. Bull. 95: 321–333.

Nichol, D. G. and E. I. Acuna. 2001. Annual and batch fecundities of yellowfin sole, *Limanda aspera,* in the eastern Bering Sea. Fish. Bull. 99: 108–122.

Pearson K. E. and D. R. Gunderson. 2003. Reproductive biology and ecology of shortspine thornyhead rockfish, *Sebastolobus alascanus,* and longspine thornyhead rockfish, *S. altivelis,* from the northeastern Pacific Ocean. Environ. Biol. Fish. 67: 117–136.

Materials

Slides of ovary cross sections from multiple mature fish of the same species during spawning season. Cross sections need to be stained with Harris' hematoxylin stain followed by eosin counterstain (H&E stain).

Light microscopes, each with an objective capable of visualizing structures in an individual oocyte, and an objective capable of visualizing several oocytes at once.

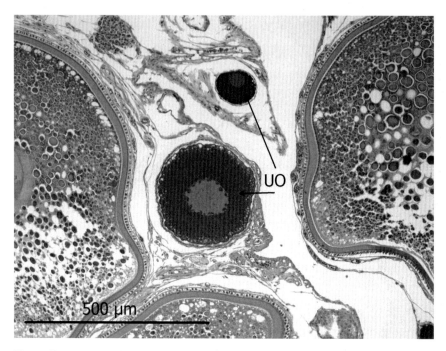

Figure Ex. 2.1. Unyolked oocytes from a Greenland turbot (*Reinhardtius hippoglossoides*). Note absence of eosin-stained yolk globules.

Procedure Systematically examine each slide and record the oocyte development stages present (stage information follows), and the presence or absence of postovulatory follicles. The following stages are based on stages described for northern anchovy (Hunter and Macewicz 1985). Some differences for other species are noted.

UNYOLKED: These are relatively small oocytes. The cytoplasm stains purple from hematoxylin. Eosin-stained (pink) yolk droplets are absent (Figure Ex. 2.1).

PARTIALLY YOLKED: Overall oocyte size increases. This is the early stage of vitellogenesis. This stage begins with development of eosin-stained (pink) yolk droplets within the cytoplasm. The droplets become larger as this stage continues. The zona radiata develops and widens (Figure Ex. 2.2). Authors have selected different arbitrary measurements of yolk development to differentiate this stage from the advanced yolked stage. Some mark the end of this stage when the yolk globules make up more than 50% of the area of the oocyte cross section (Pearson and Gunderson 2003). For northern anchovy, the yolk globules first appear in the periphery of the cytoplasm, and later extend toward the nucleus. The partially yolked stage is completed when yolk globules extend farther than 3/4 of the distance from cell periphery to the perinuclear zone. Unless you know the pattern of yolk deposition, use 50% of yolk volume as the end of this stage.

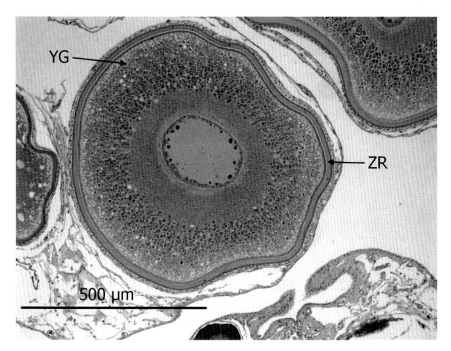

Figure Ex. 2.2. Partially yolked oocyte from a Greenland turbot showing eosin-stained yolk globules.

ADVANCED YOLKED: Vitellogenesis continues in this stage. Yolk droplets compose greater than 50% of cell volume. Oocyte size continues to increase. The nucleus is roughly in the center of the oocyte (Figure Ex. 2.3).

MIGRATORY-NUCLEUS OOCYTES: The nuclear membrane dissolves and the nucleus migrates from the center of the oocyte to the micropyle (Figure Ex. 2.4). In some species, the yolk droplets fuse to form yolk plates prior to nuclear migration (McDermott and Lowe 1997). In these species, this stage may be combined with the hydrated oocytes stage.

HYDRATED OOCYTES AND OVA: This stage begins when the yolk droplets fuse or coalesce to form yolk plates. During hydration, the oocyte undergoes rapid growth due to fluid uptake. In the early stages of hydration the oocyte remains inside the follicle (Figure Ex. 2.5), but after ovulation, the ova are free from the follicle (Figure Ex. 2.6). Dehydration during histology will make ova appear severely misshapen.

POST-OVULATORY FOLLICLE (POF): These are remnants of the follicles after ovulation. The granulosa and theca cells in POFs remain visible relatively soon after ovulation but can deteriorate and be resorbed quickly (Figure Ex. 2.7).

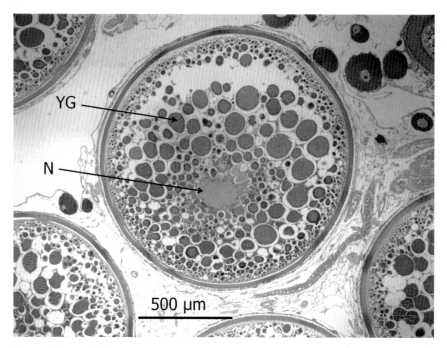

Figure Ex. 2.3. Advanced yolked oocyte from a Greenland turbot with nucleus in center of oocyte and nucleus held in oval shape by nuclear membrane.

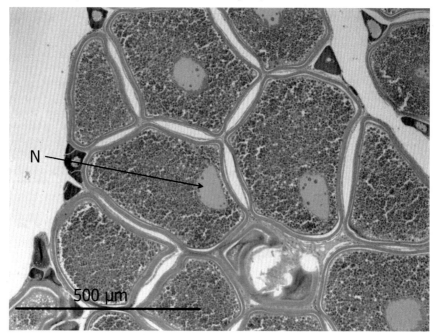

Figure Ex. 2.4. Migratory nucleus from a yellowfin sole (*Limanda aspera*). Nucleus has migrated from center of oocyte to one side. Slide courtesy of Dan Nichol, NMFS.

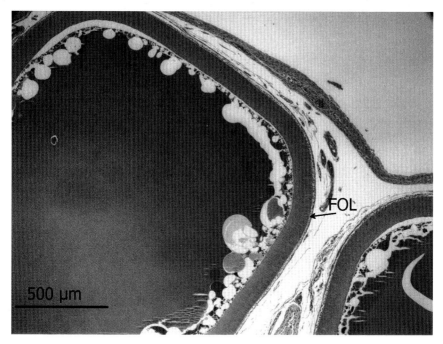

Figure Ex. 2.5. Early hydrated oocyte from an Atka mackerel, *Pleurogrammus monopterygius*. Yolk globules are coalescing, however oocyte remains in the follicle.

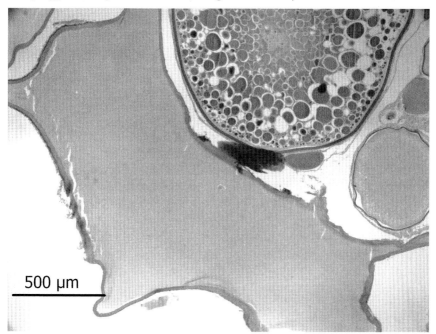

Figure Ex. 2.6. Hydrated and ovulated ovum from a Greenland turbot. Note effect of dehydration during histology on the shape of the ovum.

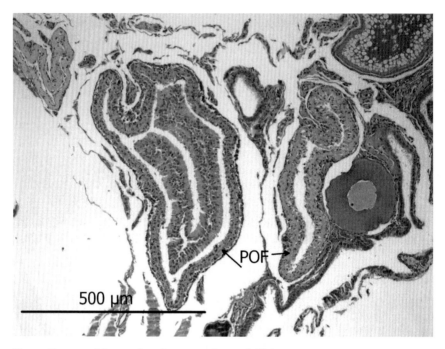

Figure Ex. 2.7. Atka mackeral, post-ovulatory follicles.

Make a table with the following information:

Number of fish with vitellogenic oocytes.
Number of fish with POFs and vitellogenic oocytes.
Number of fish with hydrated oocytes and less advanced vitellogenic stages.
Number of fish with migratory-nucleus stage oocytes and less advanced vitellogenic stages.

For fishes with multiple spawnings, successive batches of oocytes will hydrate, ovulate, and be spawned. The presence of migratory-nucleus oocytes, hydrated oocytes, or post-ovulatory follicles combined with earlier stages of vitellogenic oocytes is indicative of multiple spawnings. Migratory-nucleus oocytes, hydrated oocytes, and POFs may be very short duration and not found in a large percentage of fish.

Fish that spawn once per spawning season develop a fairly synchronous group of vitellogenic oocytes. Hydrated oocytes, migratory-nucleus oocytes, and large numbers of post-ovulatory follicles would not be found in the same ovary with developing vitellogenic oocytes.

Given the evidence you have, tell whether the fish species you examined is a batch or multiple spawner.

LABORATORY EXERCISE THREE

Fecundity Determination

Purpose To determine the number of eggs in the ovaries.

Reference If possible, use a reference that pertains to the species you are working on, such as Hinckley, S. 1987. The reproductive biology of walleye pollock, *Theragra chalcogramma,* in the Bering Sea, with reference to spawning stock structure. Fish. Bull. 85: 481–498.

Materials
Measuring board.
Scalpel.
Beaker.
Magnetic stirrer.
Pipette.
Needle probes.
Paper towels.
Top-loading balance.
Analytical balance.

Procedures
  A. Volumetric Method
   1. Obtain a fish and measure total length (TL) in millimeters, carefully open the fish, and dissect the ovaries. If the ovaries have already been removed and are preserved in formalin, take the ovaries out of formalin and wash in fresh water.
   2. Free all eggs from ovarian tissue.
   3. Place eggs in large beaker with water so that eggs + water = 2,500 ml.
   4. Stir magnetically—begin slowly.
   5. Withdraw a 5-ml sample using a pipette.
   6. Count all eggs in sample.
   7. Replace eggs in 2,500-ml beaker.
   8. Replicate sampling and counting.
   9. Calculate fecundity using the following formula:

$$F = \frac{\text{Mean count of subsamples} \times 2{,}500}{5 \text{ ml}}$$

   (or, in general, F = mean count per ml × total ml in beaker)

B. Wet Weight Method
 1. Obtain a fish and measure TL (in millimeters), carefully open the fish, and dissect the ovaries. If the ovaries have already been removed and are preserved in formalin, take the ovaries out of formalin and wash in fresh water.
 2. Drain and/or blot dry (paper towel) the ovaries, and then weigh the ovaries on the top-loading Mettler balance to the nearest 0.1 gram and record. (Be sure the arrest is all the way off the Mettler balance.)
 3. From each ovary carefully obtain about 500 eggs (a proportionally representative ovarian issue) and combine the eggs for a total subsample of about 1,000 eggs. Take a subsample by cutting a cross section through the ovary. If the sample is too big you may cut wedges if necessary. Always cut through the center of the cross section. This ensures that eggs are proportional to ovarian tissue in the subsample. (A very important factor for precision is to figure out a way to hold the subsample until the time of weighing at the same relative moisture content as the total ovaries were when they were weighed.)
 4. Weigh the subsample on a Mettler analytical balance to the nearest 0.0001 gram (probably means transferring the subsample to a preweighed weighing dish).
 5. Count the exact number of eggs in the subsample by teasing them apart from the tissue as much as possible.
 6. Calculate fecundity using the following formula:

$$F = \frac{(\text{Total subsample count})(\text{Wt. of the paired ovaries})}{\text{Wt. of subsample}}$$

LABORATORY EXERCISE FOUR
Artificially Spawning, Fertilizing, and Rearing Planktonic Fish Eggs

Purpose To obtain and maintain live fish eggs through artificial fertilization of eggs from adults.

Note Incubation period of fish eggs varies with species from a few hours to several weeks. Within species, incubation period may experience a twofold variation. Thus the timing of these procedures should be appropriate to the species and temperatures under study.

Reference
Hunter, J. R. 1984. Synopsis of culture methods for marine fish larvae, pp. 24–27. In: H. G. Moser et al. (eds.), Ontogeny and systematics of fishes. Am. Soc. Ichthyol. Herp., Spec. Publ. No. 1.

Materials
2–3 round, white plastic bowls (about 8 inches diameter).
Plastic soup ladle.
Flat aquarium net on frame (use aquarium net frame, attach netting so it provides a flat surface).
Wash bottle.
Sieve made of 4-inch section of 6-inch PVC pipe with nitex mesh, of a size small enough to hold eggs (about 0.5 mm), glued to one end.
Quart or gallon jars and netted plastic lids (to create the netted lid, cut a large hole in lid and place a piece of pantyhose under lid as it is screwed onto jar).
Long plastic pipettes (8–10 inches).
Clean water from area where adults were collected.

Procedures
1. Collecting adults: Collect live spawning adults by trawling or other appropriate means. Maintain in running water until they are stripped. Select females that are running ripe or will extrude eggs with gentle pressure. Select males that are leaking milt or will release milt with gentle pressure. Use 1–2 males per female, depending upon availability. Preserve adults after spawning for future reference.
2. Stripping and fertilizing the eggs: Put about 1-liter ambient temperature seawater into a plastic bowl. Express a small amount (about 10 ml) of milt from one male into the bowl and mix gently by hand briefly. Express eggs from a female by applying gentle pressure along the ventral midline posteriorly toward the vent

with the base of your thumb. Position the female so that the eggs flow into the bowl as they are expressed. Good eggs are completely translucent and should flow freely from the vent. If you have to apply a lot of pressure, the female is probably not ready and you may also contaminate the rearing water with fecal material. Put eggs from only one female into the bowl at a time. Express milt from a different male into the bowl and mix briefly.

3. Rinsing the eggs: Let the eggs and milt sit about 5–10 minutes after mixing. Put the PVC sieve into another bowl that is filled with seawater and gently ladle surface eggs into the sieve. The best eggs will be in the top 1/2 inch of water in the bowl, so ladle eggs only from the surface. Rinse eggs by gently pouring seawater into the sieve until the bowl is full. Quickly lift sieve out of the bowl; dump rinse water; fill bowl with clean, filtered seawater; put sieve back into bowl; and pour more water over the eggs. Repeat this procedure until the rinse water is clear. Do not leave eggs exposed to air during this process.

    Fill another bowl with about 1-liter clean seawater. Transfer the eggs into the bowl by backwashing the sieve with a wash bottle filled with clean seawater. Ladle surface eggs onto the aquarium net in order to estimate number. About 1,500 eggs should then be transferred with a wash bottle into a gallon jar (300 into a quart jar), which has been filled 3/4 full with ambient seawater. Screw on plastic lid with netting. Maintain jars in a water bath (or refrigerator) held at an appropriate temperature for the species being reared. Label jars with female identity, spawning date, and time.

4. Maintaining the eggs: Change 50% of the water in the rearing jars every day with clean seawater. Pour old water out through the netting in the lid and add new water through the netting; there is no need to remove the lids except to remove dead eggs (a dead egg is white and sinks to the bottom of the jar). Daily, remove dead eggs with the pipette. By avoiding crowding the eggs, renewing the water daily, and removing dead eggs, you will provide sufficient oxygen for the eggs and prevent buildup of wastes and excessive bacterial growth.

LABORATORY EXERCISE FIVE

Culturing Marine Fish Larvae

Purpose To rear marine fish larvae in a laboratory setting.

Note Rearing of larvae of most marine fish is time consuming and difficult. This exercise, based primarily on procedures developed for the marine cold-water gadid, walleye pollock (*Theragra chalcogramma*), might not be appropriate for most classes but is offered here to acquaint students with the materials and procedures needed for such work.

References

Hunter, J. R. 1984. Synopsis of culture methods for marine fish larvae, pp. 24–27. In: H. G. Moser et al. (eds.), Ontogeny and systematics of fishes. Am. Soc. Ichthyol. Herp., Spec. Publ. No. 1.

Porter, S. M. and G. H. Theilacker. 1996. Larval walleye pollock, *Theragra chalcogramma*, rearing techniques used at the Alaska Fisheries Science Center, Washington. AFSC Proc. Rep. 96-06.

Tucker, J. W. 1998. Marine fish culture. Kluwer Academic Publishers, Boston.

Materials

ALGAL CULTURE

1. Rearing containers: 250-ml Erlenmeyer flasks and larger glass rearing containers; 32-gallon plastic garbage cans for mass culture of algae.
2. Balance and graduated cylinders to measure chemicals for rearing solutions.
3. Algae growth medium: (Organic-based commercial liquid plant fertilizers such as fish fertilizers can be substituted for the following solutions at a rate of 0.1 ml/l for fertilizers with 8% nitrogen.)
   A. Mix solutions 1, 2, and 3 below in proportions 2:1:2. Add 0.5 ml/l of this solution to water for rearing algae.
      1. Trace metals solution (995-ml distilled water plus 1 ml of each of the following solutions:
         Sodium molybdate: 12.6 g/l
         Zinc sulfate: 44.0 g/l
         Cupric sulfate: 19.6 g/l
         Cobalt chloride: 20.0 g/l
         Manganese chloride: 360.0 g/l
      2. Phosphate stock solution:
         Sodium phosphate (monobasic): 40.0 g
         Thiamin: 0.4 g

Vitamin B_{12}: 2.0 mg
Distilled water: 1 liter
3. Nitrate stock solution:
Sodium nitrate: 150 g
Distilled water: 1 liter
B. Add the following solution at 0.5 ml/l to the solution mixed in 1 above:
Algae iron solution: add to 1-l distilled water:
Ferric chloride: 1.8 g
EDTA: 6 g
Sodium hydroxide to pH of 7
4. Algal cultures: (A number of algal species can be reared to provide food for rotifers. The following have proved effective for rotifers used to feed walleye pollock larvae.)

SPECIES	SOURCE
Isochrysis galbana	University of Washington, Oceanography
Pavlova lutheri	University of Washington, Oceanography
Katodinium rotundatum	University of British Columbia

5. Microwave to sterilize culture medium in small rearing containers.
6. Household bleach to sterilize larger culture containers.
7. 1% sodium thiosulfate solution to neutralize bleach.
8. Fluorescent lights suspended above rearing containers.

ROTIFER CULTURE
1. Rearing containers: 4-liter jars, up to 32-gallon plastic garbage cans.
2. Appropriate equipment to aerate rearing containers.
3. Fluorescent lights suspended above rearing containers.
4. Glass or plastic tube siphon to clean bottom of rearing containers.
5. Stock culture of *Brachionus plicatilis*.

LARVAL FISH CULTURE
1. Filtered (5 μm), UV-treated water of appropriate salinity and temperature (salinity can be adjusted using distilled water or commercially available sea salts).
2. 120-l circular black fiberglass rearing tanks.
3. Temperature control units.
4. Fluorescent lights to provide 3.0–3.5 μmolphotons $m^{-2} s^{-1}$ at the surface of the water in the rearing tanks (may need to be adjusted to be appropriate for the species being reared).

Procedures

ALGAL CULTURE: Stock algal cultures can be maintained in 250-ml Erlenmeyer flasks. Fill flasks with 150-ml of Algal Culture medium and boil in microwave for 1–2 minutes. Cool to rearing temperature and add 1 ml of culture to flask. Use abut 50 ml of stock culture to inoculate 2.5-liter rearing containers (containing 1.5 liter of microwave-sterilized Algal Culture medium. After 1–2 weeks, inoculate larger containers with 500 ml of algal culture from the 2.5-liter rearing containers. Before inoculating the larger containers, water is added to the container and sterilized by adding 0.2-ml household bleach/l water, and letting it aerate overnight. The bleach is then neutralized by adding 1 ml/l thiosulfate solution and aerating. If bleach odor persists after 4 hr, add more thiosulfate. The bleach is considered neutralized when no odor can be detected. Following this sterilization, the nutrient and iron solutions are added, and the containers inoculated with algae.

ROTIFER CULTURE: Rotifers (*Brachionus plicatilis*) are reared at room temperature in aerated containers of various sizes (4-liter, 20-liter, 32-gal). Rotifers are added to rearing containers that have dense concentrations of several species of cultured algae. When the water in the rearing containers becomes clear (because the rotifers have eaten the algae), more algal cultures are added. Water is removed from the rearing containers, so more algal cultures can be added, using a plastic tube siphon with a 44-μm sieve over the end of it. A siphon tube is used to clean debris from the bottom of the rearing containers every other day. Fluorescent lights are suspended over the rearing containers.

LARVAL FISH CULTURE: Larvae are reared in closed (noncirculating) containers. Little if any aeration is used since larvae are very fragile and poor swimmers; they are injured by hitting the rearing containers in the turbulence caused by aeration. Larval rearing containers are filled with filtered water, placed in a temperature-controlled water bath and stocked with up to 5–10 eggs or early larvae per liter. Fluorescent lights on a 24-hour light/dark cycle to simulate natural conditions are placed above the rearing containers.

As the larvae approach the first-feeding stage, rotifers are offered as food. To minimize the negative effects of high ammonia concentrations, rotifers are washed before being introduced to the larval rearing containers. Rotifers should be stocked in the larval rearing containers at concentrations of about 10/ml to provide sufficient food for the larvae. Rotifers are concentrated by siphoning them from their rearing containers into a 44-μm sieve suspended in water. Move the sieve up and down in a bath of clean water to remove ammonia that has accumulated in the rotifer rearing container. Repeat this rinsing process 2–3 times. Add the washed rotifers to the larval rearing containers to achieve appropriate concentrations. The larval rearing containers are also stocked with algae for the rotifers to provide additional food for the larvae, and to help "clean" the water. More algae are added every few days.

Larval rearing container maintenance includes replacing 10–20% of the water each day by siphoning water from near the bottom of the containers (removing dead material in the process) and replacing it with clean water, which is introduced through a small hose. Care must be exercised to prevent larvae from being damaged or lost during these procedures.

LABORATORY EXERCISE SIX

Conducting Ichthyoplankton Tows

Purpose To collect plankton samples.

Note The methods used for ichthyoplankton tows will depend on the location of the collections, the boat available, and the collecting gear available. This exercise is written with coastal waters, a small research vessel equipped with a hydrographic winch, and a 60-cm bongo net in mind. The materials and procedures might have to be modified to accommodate other circumstances. For example, if a boat is not available, but a bridge over a tidal river is available, samples might be collected by suspending the net from the bridge into the river during tidal exchanges. If a boat is available, but it does not have an adequate winch, neuston (surface) tows could be made. In temperate waters, collections in spring usually yield the greatest number and variety of fish larvae.

References
Heath, M. W. 1992. Field investigations of the early life stages of marine fish. Adv. Mar. Biol. 28: 1–174.

Smith, P. E. and S. L. Richardson. 1977. Standard techniques for pelagic fish egg and larval survey. FAO Fisheries Tech. Pap. 175.

Materials
Boat with hydrographic winch with meter block to monitor amount of wire paid out. Boat should be capable of steaming at slow speed (~ 2 knots). Fathometer. Hose supplied with ambient water to wash down the plankton net. Laboratory space to process sample and fill out log sheets.

Plankton net; 60-cm bongo nets are preferred: they are capable of collecting mobile larvae because they are large and lack bridles. Mesh sizes of 0.3 to 0.5 mm are sufficient for most fish eggs and larvae. Smaller meshes clog easier and retain more nonfish material, which increases sorting time. Cod-end mesh should be equal to or smaller than the mesh in the net.

A 20–40-kg lead weight to suspend just below net.

Angle indicator to determine the angle of the towing wire so the depth of the net can be estimated (cosine of the wire angle times the amount of wire out).

Log sheet to record tow data (see Table Ex. 6.1), label paper for samples, pencils.

Sieve (about 10 inches in diameter) with mesh equal to or smaller than in the plankton net.

Containers for samples (quart glass jars with plastic lids work well).
Formalin: 50-ml concentrated formalin (37% formaldehyde) added to ambient water and the plankton sample (up to 1 liter). When collecting in salt water, buffer the sample with 20-ml saturated sodium borate/l. (Have adequate ventilation and use extreme care [protective gloves] with formalin.)

Procedures

OBLIQUE TOW: (Record appropriate data on the log sheet as the events occur.) Secure the net to the towing wire, and secure the weight just below the net. Have the vessel moving at about 2 knots, and slowly lower the net into the water. Adjust boat speed to achieve a 45° wire angle. Pay the wire out at up to 30 m/sec until the desired net depth is reached (estimate the depth of the net by multiplying the wire out by the cosine of the wire angle). Allow net to "settle" for a few seconds and then start to retrieve it at about 20 m/sec. When the net reaches the surface, wash the contents into the codend. Remove the codend and place the contents in the sieve to concentrate the plankton further. Once most of the liquid has been removed, pour the plankton into the sample container, add 50-ml concentrated formalin, buffer, and appropriate label.

HORIZONTAL TOW: Follow above procedures, except leave the net at the desired sampling depth (including surface for a neuston tow) for a specified period of time before retrieving it. Retrieve it using a fast winch speed.

TABLE EX. 6.1. Ichthyoplankton Collecting Log Sheet

Date	_____	Vessel	_____	Personnel	_____
Location	_____	Project	_____		_____
Weather	_____				_____
Gear	_____				_____
Tow number	_____				
		Start Time	_____	Position	_____
		Finish Time	_____	Position	_____
		Maximum tow depth	_____		
Operational comments		_____			
		_____			
		_____			
		_____			
		_____			
		_____			
		_____			

LABORATORY EXERCISE SEVEN
Sorting and Identifying Fish Eggs and Larvae in Ichthyoplankton Samples

Purpose   To remove fish eggs and larvae from the rest of the material in plankton samples, and determine the taxonomic identity of the eggs and larvae.

Reference
Smith, P. E. and S. L. Richardson. 1977. Standard techniques for pelagic ficheria egg and larval survey. FAO Fish Tech. Pap. 175.

Materials
SAMPLES
Plankton samples preserved in 5% buffered formalin; usually contained in quart glass jars.

CHEMICALS
Tap water.
5% buffered formalin.
70% Ethanol.
Concentrated formalin.

EQUIPMENT
Dissecting microscope (7–70×) with direct and reflected light, and ocular micrometer.
6–8 glass watch glasses or petri plates (50-mm diameter).
1-liter beaker.
1.5-liter beaker.
2,500-ml graduated cylinders.
Mesh cone of .333-mm nitex netting to fit on top of graduated cylinder.
Wash bottle.
Tablespoon.
Fine probes.
Fine forceps.
Small pipettes (one with inside diameter > 1 mm is useful for picking up eggs).
18-ml vials with polyseal caps.
Waterproof paper for labels.
"Dot" labels for the vial caps.
Rapidograph-style pen for use with India ink.
India ink.

Procedures Processing is carried out in several steps: measuring the volume of the plankton in each sample (optional); sorting and enumerating eggs and larvae; identifying the fish eggs and larvae; measuring fish larvae and curating the fish eggs and larvae. Much of this work is conducted in the presence of formaldehyde fumes, which are noxious and unhealthy. Therefore adequate ventilation must be provided, and care taken to avoid spills and contact with the skin. Laboratories routinely used for processing plankton samples should be equipped with exhaust hoses to remove fumes from each workstation.

VOLUME MEASUREMENT (OPTIONAL): After removing large and "nonplanktonic" material from the sample, remove excess preservative and pour the sample into a 500-ml graduated cylinder. Add or subtract enough preservative to bring the level of the liquid in the cylinder to an even milliliter. Place a mesh cone in another clean 500-ml graduated cylinder. Pour the plankton and formalin into the draining cone. The plankton stays in the cone while the liquid drains into the cylinder. The plankton is considered drained when the interval between drops from the bottom of the cone increases to 15 seconds. Draining times vary with the size and composition of the sample. Subtract the volume of the drained liquid in the cylinder from the initial volume of plankton plus liquid. The difference is the volume of the plankton. Return the plankton to the jar with preservative in preparation for sorting.

SORTING: Separate the plankton from its preservative by straining it through a draining cone. Gently rinse the cone containing the inside jar label and the plankton with tap water and then suspend it in a 1-liter beaker filled over ½ the way up the cone with fresh water containing a few drops of concentrated formalin. Spoon small amounts (about 30 ml) of plankton into 50-mm watch glasses for sorting.

Examine each dish under a dissecting microscope at about 7–10× magnification. Pick out all fish eggs and larvae with pipettes and/or forceps and transfer them to watch glasses labeled "fish eggs" or "fish larvae." When the fish eggs and larvae have been "sorted" and checked, pour the remaining contents of the dish into a "sorted" 1.5-liter beaker containing fresh water with a few drops of concentrated formalin. Repeat this procedure until the entire sample has been examined. When sorting is finished, place the fish eggs and larvae in 18-ml vials. Place labels in and on the cap of each vial, and cap the vials. Maintain a log of the sorting process (e.g., for each sample: who did the sorting, when sorting was started and finished, and how many eggs and larvae were found).

IDENTIFYING AND COUNTING FISH EGGS AND LARVAE;
MEASURING LARVAE AND STAGING EGGS

Fish Larvae All larvae in the sample should be identified to the lowest taxonomic category possible. Identifications are made by comparing morphological characters of the larvae (e.g., body shape, meristic features, pigmentation, special larval characters) with published descriptions (see Laboratory Exercise 8 on identifying fish eggs and larvae). Count the larvae of each taxon and store them in 70% ethanol in 3-dram vials. For each sample, record these identifications and counts on log sheets.

The preferred length measurement is standard length (SL): from the tip of the snout to the base of the hypural plate. In preflexion larvae notochord length (NL: tip of the snout to the tip of the notochord) is used. To perform the measurements, individual specimens of a species are arranged in rows in a plain petri dish. The dish is moved under the microscope and each specimen is aligned, and then measured with a scale, or ocular micrometer, to the nearest 0.1 mm. Usually it is sufficient to measure a subsample of 50 larvae, if more are present in a sample. However, if there is doubt that these data are representative of the length frequency distribution of the specimens in the entire sample, then more measurements must be made.

Fish Eggs Fish eggs are generally more difficult to identify than larvae. Thus, it may be neccessary to leave a large fraction of the eggs in a sample as "unidentified fish eggs." Manipulating eggs and the microscope techniques are similar to those used for larvae. Fish eggs are identified by comparing their morphological characters (e.g., egg diameter, oil globule presence and size, chorion characters, pigment, embryonic characters) with published descriptions (see Laboratory Exercise 8 on identification of fish eggs and larvae). Eggs are stored in 3% formalin in 1.5-dram vials.

In some studies identifying fish eggs is not a priority, and will not be attempted. Sometimes, only the eggs of a particular "target" species will be identified. Depending on the study, all eggs or eggs of each identified taxon should be counted. In samples with more than 5,000 eggs, as estimated by eye, a volumetric procedure can be used to estimate the number of eggs as follows. First, 100 eggs are picked out randomly and placed in a separate vial for staging. Then 900 eggs are counted and their settled volume is determined in a 5- or 10-ml graduated cylinder. The volume of the remainder of the eggs is determined. The total volume equals the sum of the volume of the 900 eggs plus the volume of the remaining eggs:

Total Eggs in Sample = ((Total Volume/Vol. of 900 eggs) 900) + 100.

Studies using fish eggs from plankton surveys for biomass estimates often require that eggs of the target species be staged, so that their age can be determined. Staging criteria are set up for each species of interest, and range from three (early, middle, late) for species with short incubation periods to over 20 stages (e.g., walleye pollock [*Theragra chalcogramma*]) in species that incubate for up to a few weeks. In samples with large numbers of eggs of the target species, a random subsample of about 100 eggs may be staged. Eggs are staged by comparing the developmental state of eggs in the sample with figures and descriptions of the stages of development of the species of interest.

LABORATORY EXERCISE EIGHT
Diversity/Identification of Planktonic Fish Eggs and Larvae

Purpose To become familiar with the diversity of forms of fish eggs and larvae and procedures and characters to identify them.

References

Fahay, M. P. 1983. Guide to the early stages of marine fishes occurring in the western North Atlantic Ocean, Cape Hatteras to the southern Scotian shelf. J. Northwest Atl. Fish. Sci. 4.

Fahay, M. P. 2007a. Early stages of fishes in the western North Atlantic Ocean (Davis Strait, southern Greenland and Flemish Cap to Cape Hatteras). Volume One: Acipenseriformes through Syngnathiformes. Monograph No. 1, North Atlantic Fisheries Organization. Dartmouth.

Fahay, M. P. 2007b. Early stages of fishes in the western North Atlantic Ocean (Davis Strait, southern Greenland and Flemish Cap to Cape Hatteras). Volume Two: Scorpaeniformes through Tetraodontiformes. Monograph No. 1, North Atlantic Fisheries Organization. Dartmouth.

Kendall, A. W., E. H. Ahlstrom, and H. G. Moser. 1984. Early life history stages of fishes and their characters, pp. 11–22. In: Moser, H. G. et al. (eds.), Ontogeny and systematics of fishes. Spec. Publ. 1, Am. Soc. of Ichthyol. and Herp.

Matarese, A. C., A. W. Kendall, Jr., D. M. Blood, and B. M. Vinter. 1981. Laboratory guide to early life history stages of Northeast Pacific fishes. NOAA Tech. Rep. NMFS 80.

Matarese, A. C. and E. M. Sandknop. 1984. Identification of fish eggs, pp. 27–30. In: Moser, H. G. et al. (eds.), Ontogeny and systematics of fishes. Spec. Publ. 1, Am. Soc. Ichthyol. and Herp. Allen Press, Lawrence, KS.

Powles, H. and D. F. Markle. 1984. Identification of larvae, pp. 31–33. In: Moser, H.G. et al. (eds.), Ontogeny and systematics of fishes. Spec. Publ. 1, Am. Soc. Ichthyol. and Herp. Allen Press, Lawrence, KS.

Materials

SPECIMENS

Reference collection of a variety of planktonic fish eggs and larvae.
Unidentified specimens of fish eggs and larvae.

EQUIPMENT

Dissecting microscope with direct and reflected light, and ocular micrometer.
Dissecting tools: fine forceps, probes, small pipettes.
50-mm watch glasses or petri dishes.
Literature on descriptions of fish eggs and larvae of the collection area.

Procedures

REFERENCE MATERIAL

1. Examine specimens in watch glasses using microscope at various magnifications.
2. Establish stage of development and measure length of specimens.
3. With eggs, note chorion features, perivitelline space, embryo features.
4. Count meristic features: myomeres and fin rays.
5. Note pigment patterns.
6. Note body shape and special developmental features (e.g., head spines, elongate or ornamented fin rays).
7. Compare illustrations and descriptions with specimens at hand.

UNIDENTIFIED SPECIMENS

1. Follow steps 1–6 above.
2. Note geographic area and season of collection.
3. Compare characters of specimen with those of various higher groups of fish (e.g., orders) found in the collection area.
4. Compare meristics and appearance of specimen with those found in collection area, with particular emphasis on those in the suspected higher group.
5. Once a species is suggested, rule out other potential species on the basis of meristic or other distinguishing characteristics.

LABORATORY EXERCISE NINE
Preparing and Examining Larval Fish Otoliths for Daily Growth

Purpose To dissect, mount, and examine larval otoliths for daily growth increments.

References

Campana, S. E. and J. D. Neilson. 1985. Microstructure of fish otoliths. Can. J. Fish. Aquat. Sci. 42: 1014–1032.

Jones, C. M. 1986. Determining age of larval fish with the otolith increment technique. Fish. Bull. 84: 91–103.

Secor, D. H., J. M. Dean, and E. H. Laban. 1991. Manual for otolith removal and preparation for microstructural examination. EPRI and Belle W. Baruch Inst. Mar. Biol. and Coast. Res. 85 p.

Dissecting and Mounting Otoliths

Materials

SPECIMENS

Larval fish fixed and preserved in 95–100% ethanol. Contact with formalin will destroy otolith structure.

CHEMICALS

Quantities approximate for 5–10 10-mm specimens:
Distilled water (100 ml).
95% ethanol (100 ml).
Clear fingernail polish (one bottle, much less needed).

EQUIPMENT

Dissecting microscope (6–50×) with direct light and polarizer.
20 glass microscope slides (frosted end).
4–5 clear plastic or glass 50-mm petri dishes.
Larval forceps.
Glass Pasteur pipette and bulb.
2 dissection probes.
2 fine-tip pins for probes.
Hog's eyelash probe.
Extra fine point permanent marking pen.
Slide box.

Procedures
1. Using larval forceps, remove larva from preservative and measure on microscope slide marked E (ethanol).
2. Place larva in 50-mm petri dish with distilled water until motion stops and larva sinks (1–2 min).
3. Place larva in a pool of distilled water on a microscope slide marked W (water).
4. Focus microscope on lateral view of head of larva and change polarizer until image is very dark.
5. Using dissecting probes, tear head just dorsal and posterior to the eyes. Otoliths (two pairs in small larvae: the lapilli and sagittae) should appear as bright spots in the dark polarized background. Move otoliths away from larva, but keep them in water.
6. After all otoliths have been located use a hog's eyelash probe to slowly move otoliths to edge of water, cleaning tissue from them in the process. Eventually otoliths should be together (but not touching) in their own pool, where they are allowed to dry (1–2 min).
7. Label (specimen number, etc.) a clean slide for mounted otoliths, and brush a patch of fingernail polish onto it.
8. Run probe through fingernail polish and pick up otoliths with probe (the dry otoliths should adhere to the polish). Place otoliths in the patch of fingernail polish.
9. Manipulate otoliths so they are concave side up. Use more fingernail polish if needed to move otoliths.
10. Draw a circle with fine-tipped marking pen around otoliths on undersurface of slide for future reference and set aside to dry (several hours).
11. When initial coat of fingernail polish has dried, apply another coat and let it dry for 24 hr before examining otoliths.

Examining Otoliths and Counting Growth Increments

Materials
SPECIMENS
Larval otoliths mounted on microscope slides.

CHEMICALS
Bottle of low fluorescence microscope immersion oil.

EQUIPMENT
Compound microscope equipped with polarizer, C-mount for video camera, and 10×, 40×, 100× objectives (final magnifications with 10× eyepieces of 100×, 400×, 1000×).
Video camera and monitor: black and white suggested due to more lines of resolution.

Procedures
1. Place a drop of immersion oil over the circle marked on a slide with otoliths.
2. With polarizer in place and the microscope at 100×, locate an otolith on the slide, and center it in the field of view.

3. Change to 1000×. Identify which otolith (sagitta or lapillus) is in view, based on size of nucleus (the area inside the hatch mark). In walleye pollock (*Theregra chalcogramma*) the nucleus of the lapillus is 19–23 μm, whereas the nucleus of the sagitta is 14–18 μm. The sagitta is larger than the lapillus in walleye pollock larger than 7 mm. Adjust lighting and focus up and down and count all visible rings starting with the hatch mark as 1.
4. Locate another otolith on the slide and repeat procedure to count rings, and confirm count obtained from first otolith.

LABORATORY EXERCISE TEN

Larval Fish Gut Contents

Purpose To dissect larval fish guts and examine ingested prey organisms.

References

Arthur, D. K. 1978. Food and feeding of larvae of three fishes occurring in the California Current, *Sardinops sagax, Engraulis mordax,* and *Trachurus symmetricus.* Fish. Bull. 74: 517–530.

Canino, M. F. and K. M. Bailey. 1995. Gut evacuation of walleye pollock larvae in response to feeding condition. J. Fish Biol. 46: 389–403.

Kamba, M. 1977. Feeding habits and vertical distribution of walleye pollock, *Theragra chalcogramma* (Pallas), in early life stage in Uchiura Bay, Hokkaido. Res. Inst. N. Pacific Fish., Hokkaido Univ., Spec. Vol., 175–197.

Last, J. M. 1978a. The food of four species of pleuronectiform larvae from the English Channel and southern North Sea. Mar. Biol. 45: 359–368.

Last, J. M. 1978b. The food of three species of gadoid larvae in the English Channel and southern North Sea. Mar. Biol. 48: 377–386.

Dissecting Larval Fish Guts

Materials

SPECIMENS

Larval fish fixed and preserved in 5–10% formalin. Ethanol-fixed larvae may be used but tissues tend to be dehydrated and more difficult to dissect.

CHEMICALS

Distilled water (100 ml).

EQUIPMENT

Dissecting microscope (6–50×) with lighted stage and direct light (optional).
Glass microscope culture slide (1 or 2 depressions 18–25 mm).
Clear plastic or glass 50-mm petri dish.
Pair of larval forceps.
Wash bottle for distilled water.
Glass Pasteur pipette and bulb.
2 dissection probes.
Hog's eyelash probe.

Procedures
1. Place larvae in petri dish, remove formalin, and cover with distilled water.
2. Place 3–4 drops of distilled water in depression on culture slide and transfer a larva to this pool with forceps.
3. Measure fish total length using ocular micrometer (optional).
4. Pin the fish on its side against the slide using one dissecting probe. Gently insert the other probe through the exterior muscle and skin where the dorsal side of the stomach meets the trunk of the fish. Use the probe to tear away the skin, working from the anterior to posterior end of the gut, to free it from the body. Be careful not to puncture or rip open the gut during this procedure and try to minimize shredding the muscle and skin into small fragments.
5. The gut should now be attached to the body only by the esophagus. Slide the probe up the dorsal side of the esophagus and pull the gill rakers free from the lower jaw. You should now have the complete alimentary tract separated from the fish.
6. Remove the gut to a clean depression in the slide filled with 3–4 drops of water or (in the case of only one depression) move the gut to a clean area on the slide and remove any tissue fragments from the area.
7. Increase magnification to approximately 25×. Use the tip of one probe to tear off the distal end of the gut where it meets the esophagus. *Gently* tease the gut contents out with the tip of the probe and move them a short distance away, keeping them in order if it is important to you (e.g., what was ingested last). Work from the anterior to the posterior end of the midgut, tearing open the gut wall as needed. Copepods (nauplii and copepodites) will be clear and intact, with the carapace margins visible. Move them gently to avoid fragmentation. Remove pieces of the gut wall from the area as you work. You may also wish to examine the hindgut contents in a similar manner.

Examining Larval Fish Prey

Materials
SPECIMENS
Depression slide containing material dissected from larval fish guts.

CHEMICALS
Rose bengal in 70% ethanol.
Alcian blue.
Glycerin.

EQUIPMENT
Compound microscope.
Hog's eyelash probes.
Small pipette.
Flat microscope slide.

Procedures

1. Place depression slide with dissected prey on microscope stage. Increase magnification for identification and enumeration of prey. Ingested copepods may be clumped together and should be teased apart before counting. Use a hog's eyelash probe to manipulate prey items as they tend to adhere to the metal probe tips. (Note: A human eyelash probe works better.)
2. Prey items may be stained and mounted for viewing at higher magnification using a compound microscope. Stain the prey by adding a few drops of Rose Bengal stain in 70% ethanol or Alcian Blue (200 mg l-1) directly to the prey items in the culture slide. Pipette the stain after several hours and add a few drops of distilled water. Place a drop of 100% glycerin (or a 50:50 mix of water and glycerin) on a regular microscope slide. Transfer the prey items from the culture slide to the glycerin with a pipette, using a minimal amount of water. Arrange the prey items in the glycerin, then place a glass coverslip over the specimens before viewing. Compare organisms from gut to descriptions and illustrations in the literature.

LABORATORY EXERCISE ELEVEN

Histological Condition Index
Midgut Cell Height

Purpose To prepare sections and perform histological examination on larval fish to determine midgut cell height.

Note This exercise requires access to a histological laboratory. Several histological condition indices have been developed (see references). A method to measure midgut cell heights is described here. This method is one of the easiest and most robust. These measurements can be used in an experimental design to evaluate condition of field-caught larvae or larvae from laboratory feeding experiments.

References

O'Connell, C. P. 1976. Histological criteria for diagnosing the starving condition in early post yolk sac larvae of the northern anchovy, *Engraulis mordax* Girard. J. Exp. Mar. Biol. Ecol. 25: 285–312.

Theilacker, G. H. and Y. Watanabe. 1989. Midgut cell height defines nutritional status of laboratory raised larval northern anchovy, *Engraulis mordax*. Fish. Bull. 87: 457–469.

Theilacker, G. H. and S. M. Porter. 1995. Condition of larval walleye pollock, *Theragra chalcogramma*, in the western Gulf of Alaska assessed with histological and shrinkage indices. Fish. Bull. 93: 33–44.

Materials

SPECIMENS
Reared or field-collected larvae of selected species. Field larvae must be collected and preserved quickly (< 15 min) to minimize autolysis.

CHEMICALS
Bouin's solution.
70% ethyl alcohol.
Ethyl n-butyl alcohol dehydrating series.
Histological embedding paraffin (e.g., Pasaplast-plus).
Histological stains: Harris' hematoxylin and eosin-phloxine B.

EQUIPMENT
Microsope slides.
Microtome.
Dissecting and compound microscope with ocular micrometers.

Procedures
1. Immediately after collecting, place larva on microscope slide, measure standard length with ocular micrometer in dissecting microscope, and place in Bouin's solution.
2. After 24 hr transfer larva to 70% ethyl alcohol, dehydrate in ethyl n-butyl series.
3. Embed larva in paraffin.
4. Section larva at 5 μm in sagittal plane. Mount sections in synthetic resin on microscope slide and stain with Harris' hematoxylin and eosin.
5. Examine sections with compound microscope: Select midsagittal section with midgut area clearly visible, measure thickness (using ocular micrometer) of several midgut cells in which the nucleus, brush border, and cell base are clearly defined.

LABORATORY EXERCISE TWELVE

Clearing and Staining Larval Fish

Purpose To prepare larval fish for microscopic examination of developmental osteology.

Note In scheduling this exercise consider that it may take several weeks to complete.

References

Dunn, J. R. 1984. Developmental osteology, pp. 48–50. In: H. G. Moser et al. (eds.), Ontogeny and systematics of fishes. Am. Soc. Ichthyol. Herp., Spec. Publ. No. 1.

Potthoff, T. 1984. Clearing and staining techniques, pp. 35–37. In: H. G. Moser et al. (eds.), Ontogeny and systematics of fishes. Am. Soc. Ichthyol. Herp., Spec. Publ. No. 1.

Taylor, W. R. and G. C. Van Dyke. 1985. Revised procedures for staining and clearing small fishes and other vertebrates for bone and cartilage study. Cybium. 9: 107–119.

Materials

SPECIMENS

Larval fish preserved in marble-chip buffered 10–15% formalin. Alcohol-preserved specimens may disintegrate.

CHEMICALS

Quantities approximate for 1–5 10-mm specimens:
95% ethanol (50 ml).
Distilled water (200 ml).
Absolute ethanol (200 ml).
Acetic acid, 99% glacial (100 ml).
Alcian blue powder (50 mg).
Hydogen peroxide (15 ml of 3% solution).
Potassium peroxide (300 ml of 1% solution).
Sodium borate (70-ml saturated solution).
Trypsin powder (40 mg).
Potassium hydroxide (200 ml of 1% solution).
Alizarin red stain (40 mg).
Glycerin (500 ml).
Thymol (5 mg).

292   Laboratory Exercises

EQUIPMENT

100-ml graduated cylinder.
Balance (1–100 mg).
8–10, 100- or 200-ml glass jars.
5–6 clear glass or plastic watch glasses of suitable size to hold larvae.
Dissecting microscope (about 7–70×) with direct light, microscope tools.

Procedures (from Potthoff [1984], Table 5)

1. Prepare stock solutions listed in table (Table Ex 12.1); store in labeled jars.
2. Place individual specimens in watch dishes and sequentially cover with solutions listed in table for time periods indicated according to the length of the specimen.
3. Record lengths of specimens and amount of time each was in each solution (Table Ex. 12.1).

TABLE EX. 12.1. Solutions and Time Table for Clearing and Staining Larval Fish (from Potthoff 1984)

Procedure	*Solutions*	*Approx. Time*
Fixing	10–15% Formalin	2 Days
Dehydrating	50% Ethanol	1–2 Days
Dehydrating	95% Ethanol	1–2 Days
Staining cartilage	70-ml Ethanol, 30-ml Acetic Acid, 20 mg Alcian Blue	1 Day
Neutralizing	Sodium Borate	1/2 Day
Digesting	35-ml Sodium Borate, 65-ml distilled H_2O, Trypsin	
Bleaching	(Heavily pigmented larvae only) H_2O_2	20 Min[a]
Staining bone	1% KOH Alizarin red-s	1 Day
Destaining	35-ml Sodium Borate, 65-ml distilled H_2O, Trypsin	2 Days
Clearing	30% Glycerin in 1% KOH	1 Week[b]
Clearing	60% Glycerin in 1% KOH	1 Week[b]
Preserving	100% Glycerin with Thymol	1 Week[b]

[a]Up to 40 min for specimens over 20 mm, examine specimens frequently during treatment.
[b]Allow 2 weeks for specimens 20–100 mm.

GLOSSARY

Acellular	Containing no cells; not made of cells.
Actinopterygians	The subclass of vertebrates comprising the bony, ray-finned fishes.
Allometric growth	Growth of one part of the animal which is not in proportion to the overall growth of the animal.
Allometry	The study of the change in proportion of various parts of an organism as a consequence of growth.
Allozyme	A form of enzyme that differs in amino acid sequence, as shown by electrophoretic mobility or some other property, from other forms of the same enzyme.
Anadromous	Migrating from the sea to fresh water to spawn. Pertaining to animals that live their lives in the sea and migrate to fresh water to spawn.
Apomorphic	In cladistics, a character state that occurs only in later descendants is called an apomorphy (meaning "separate form," also called the "derived" state) for that group.
Aquaculture	The science, art, and business of cultivating marine or freshwater food fish or shellfish under controlled conditions.
Asynchronous	Not consecutive, successive, or sequential.
Atresia	The degeneration and resorption of ovarian follicles before maturity has been reached.
Autocorrelated	Two or more time series that vary in the same way over time.
Biogenetic law	The theory that the stages in an organism's embryonic development and differentiation correspond to the stages of evolutionary development characteristic of the species.

Bisexuality	Having both male and female reproductive organs; hermaphroditic.
Blastocoel	The fluid-filled, central cavity of a blastula. Also called segmentation cavity.
Blastodermal cap	The cell mass of an embryo with meroblastic cleavage between the cleavage stage and the morula stage.
Blastodisc	The embryo-forming portion of an egg with discoidal cleavage usually appearing as a small disc on the upper surface of the yolk mass.
Blastomeres	Cells that result from cleavage of a fertilized ovum during early embryonic development.
Blastopore	The opening into the archenteron formed by the invagination of the blastula to form a gastrula.
Breed	To produce offspring by hatching (or gestation).
Brood	To protect, guard, and groom eggs until they hatch.
Catadromous	Living in fresh water but migrating to marine waters to reproduce.
Chondrichthyans	Cartilaginous fishes with well-developed jaws and including the sharks, skates, rays, and chimeras.
Chorion	The outer membrane enclosing the embryo in fishes, reptiles, birds, and mammals.
Cladistics	A system of classification based on the phylogenetic relationships and evolutionary history of groups of organisms.
Cladogram	A tree diagram used to illustrate phylogenetic relationships.
Codend	The rear end (often detachable) of a trawl or other funnel-shaped fishing net where the catch accumulates.
Copepodite	A series of developmental stages in copepods between the nauplius stage and the adult stage.
Cryoprotectant	A substance used to protect cells or tissues from damage during freezing.
Cutaneous	Of, relating to, or affecting the skin.
Cytometry	The study of cells.
Demersal	Dwelling on or near the bottom of the water column.
Depensation	The effect where a decrease in spawning stock leads to reduced survival or production of eggs.
Diel	"Daily" in the sense of a 24-hour period rather than the time between sunrise and sunset.
Dip net	A net or wire mesh bag attached to a handle, used especially to scoop fish from water.
Direct development	A life-history pattern without distinct intermediate stages between hatching or birth and attainment of the adult form.

Glossary

Disconcordant	Time series that vary in opposite ways to each other.
Echosign	The acoustic record produced by echosounders. The return produced by echosounders indicating that organisms are present.
Egg	Animal reproductive body consisting of an ovum or embryo together with nutritive and protective envelopes.
El Niño	An invasion of warm water into the surface of the east Pacific Ocean every 4 to 7 years that causes dramatic changes in local and regional climate and oceanography.
Electrophoresis	A method of separating substances, especially proteins, and analyzing molecular structure based on the rate of movement of each component in a colloidal suspension while under the influence of an electric field.
Embryonic shield	A platelike mass of cells in the blastocyst from which an embryo develops. Also called embryonic disk.
Endonuclease	An enzyme that hydrolyzes nucleic acids.
Energetics	The study of the flow and transformation of energy.
Epiboly	The growth of a rapidly dividing group of cells around a more slowly dividing group of cells, as in the formation of a gastrula.
Epipelagic	Of or relating to the part of the water column into which enough sunlight enters for photosynthesis to take place.
Eutrophication	Water so rich in mineral and organic nutrients that plant life, especially algae, proliferates to the extent that dissolved oxygen content is reduced and other organisms die or are displaced.
Fecundity	The quality or power of producing abundantly; fruitfulness or fertility. Relating to the numbers of eggs produced by fishes.
Fertilization	The fusion of eggs and sperm (creating diploids from haploids).
Finfold	The continuous medial membrane surrounding the trunk of larval fishes before individual fins differentiate.
Flexion stage	The stage during larval fish development when the posterior portion of the notochord bends dorsally.
Flowmeter	A device mounted in the mouths of plankton nets to monitor the flow of water into the net.
Foregut	The anterior part of the embryonic alimentary canal of a vertebrate from which the esophagus, stomach, liver, pancreas, and duodenum develop.
Gastrulation	The process in which a gastrula develops from a blastula by the inward migration of cells.
Germ ring stage	The stage in fish egg development when the blastomeres grow around the yolk.

Gestation	The time young develop within the female. In fishes this applies only to live-bearing species.
Gill rakers	Bony, finger-like projections of the gill arch on the opposite side from the gill filaments which function in retaining food organisms.
Gonad	An organ (testes or ovary) in which gametes (sex cells) are produced.
Gonadotropin	A hormone that stimulates the gonads and controls reproductive activity.
Gonochorism	A sexually reproducing species in which individuals are distinctly male or female. The sex of an individual does not change throughout its lifetime.
Gynogenesis	Female parthenogenesis in which the embryo contains only maternal chromosomes.
Hatch	To emerge from or break out of an egg.
Hermaphroditism	The presence of both male and female reproductive organs in a single animal.
Hindgut	The caudal part of the alimentary canal in vertebrate embryos.
Histology	The study of tissues sectioned as thin slices, using a microscope.
Holoblastic cleavage	A pattern of embryonic cell division in which there is complete cleavage (separation) of cells.
Homocercal	Possessing a symmetrical tail that extends beyond the end of the vertebral column.
Homologous	Similar in structure and evolutionary origin, although not necessarily in function.
Hybridogenesis	A reproductive pattern where fusion occurs but only the haploid female genome is transmitted to the developing ovum.
Hydroacoustics	The study and application of sound in water. Can be used to assess the distribution and abundance of biological particles in the water, including fishes.
Hypochordal	Ventral to the spinal cord.
Ichthyofauna	The fishes occurring in a particular geographic region.
Ichthyoplankton	Eggs and larvae of fish that reside in the plankton.
Illicium	Modified, elongated first ray of the dorsal fin in lophiiform fishes. It is separated and moved anterior from the rest of the fin to function as a "fishing pole" to attract prey.
Incubation time	The time from egg fertilization to hatching.
Indirect development	A life-history pattern with distinct intermediate stages (e.g., larvae) between hatching or birth and attainment of the adult form.

Infauna	Aquatic animals that live in the substrate, especially in a soft sea bottom.
Ingroup	In a cladistic analysis, the set of taxa which are hypothesized to be more closely related to each other than they are to the outgroup.
Interannual	Time variation among years.
Interdecadal	Time variation among decades.
Intertidal	The region between the high tide mark and the low tide mark.
Intraspecific	Arising or occurring within a species; involving the members of one species.
Intromittent organ	The male organ of copulation (penis).
Invaginate	To infold or become infolded so as to form a hollow space within a previously solid structure, as in the formation of a gastrula from a blastula.
Iteroparous	Reproducing more than once in a lifetime.
Juvenile	A fish, usually looking like a miniature adult, between the larval and mature adult stages. A fish before first sexual maturity.
Lactoflavin	A B vitamin.
Larva	The newly hatched, earliest stage of animals that undergo metamorphosis, differing markedly in form and appearance from the adult.
Locomotor	Of or relating to locomotion.
Lognormal	A logarithmic function with a normal distribution.
Macroalgae	Multicellular algae (green, blue-green, and red algae) having filamentous, sheet, or matlike morphology.
Mating	The act of pairing a male and female for fertilizing eggs, copulatory organ present.
Meiotic division	Nuclear division in which there are two successive nuclear divisions without chromosome replication between them.
Melanistic pigment	Black pigment. Pigment that is produced by melanophores.
Melanophore	A pigment cell that contains melanin (black pigment).
Meristic characters	Features of an animal that can be counted; or those that are divided into parts or segments.
Meroblastic cleavage	Incomplete cleavage, characteristic of zygotes with large accumulations of yolk.
Mesocosm	An experimental enclosure for the study of larval fish energetics, predation, etc. that holds several hundred m^3 of water.
Mesopelagic	Of, relating to, or living at ocean depths between about 180 and 900 m.

Mesoscale	The scale of meteorological and oceanographic phenomena that range in size from several kilometers to around 100 km.
Metameric muscles	Muscles that are associated with body segments.
Micronekton	Marine nektonic organisms (living in the open water) about 10–100 mm in length. Micronekton is larger than plankton but smaller than most adult fishes.
Micropyle	A pore in the chorion through which a spermatozoon can enter.
Microvilli	Thin, finger-like protrusions from the surface of a cell, often used to increase absorptive capacity or to trap food particles.
Microzooplankton	Zooplankton smaller than 0.1 mm.
Midgut	The middle part of the alimentary canal in larvae.
Midwater trawls	A tapered net (trawl) designed to be towed in midwater.
Mitosis	The process in cell division by which the nucleus divides and normally results in two new nuclei, each of which contains a complete copy of the parental chromosomes.
Monophyletic	A group of organisms that includes the most recent common ancestor of all of its members and all of the descendants of that most recent common ancestor.
Morula stage	Early cleavage stage of embryo (blastula) resembling a mulberry.
Myomeres	A muscular segment; one of the zones into which the muscles of the trunk are divided (also myotome).
Myosepta	The septa between adjacent myomeres.
Nauplius	A crustacean larva having three pairs of locomotive organs (corresponding to the antennules, antennae, and mandibles), a median eye, and little or no segmentation of the body.
Nearshore	The zone from the splash zone to the start of the offshore zone, typically at water depths of about 20 m.
Neural keel	The first visible indication of the embryo forming in a fish egg consisting of a thickened line on top of the yolk perpendicular to the edge of the germ ring. As the embryo begins to take shape, the neural tube and notochord form along the neural keel.
Neuromast	A group of innervated sensory cells occurring along the lateral line of fishes.
Neuston	Organisms that inhabit the surface layer of the water column.
Ontogeny	The developmental history of an organism.
Oocytes	Female gametocytes that develop into ova (eggs) after two meiotic divisions.

Oogenesis	The formation, development, and maturation of an ovum.
Oogonia	Cells in females that produce primary oocytes by mitosis.
Operculum	The bony structure that serves as a cover for the gill slits in fishes.
Osmoregulation	Adjustment of osmotic concentration of body fluids to changing environmental conditions, for example, when salmon migrate from salt water to fresh water.
Otoliths	Ear bones. Calcium concretions in vertebrates' ears.
Overfishing	A level of fishing that removes so many fish that continued fishing at that level will not let the stock replenish itself.
Overharvest	See *overfishing*. Unsustainable harvest beyond carrying capacity.
Overwintering	To spend winter in a particular place.
Oviparous	Producing eggs that hatch outside the body.
Ovoplasm	The cytoplasm of an unfertilized ovum (egg).
Ovotestes	A hermaphroditic reproductive organ that produces both sperm and eggs.
Ovoviviparous	Producing eggs that hatch within the female's body without obtaining nourishment from it.
Periblast	Protoplasmic matter that surrounds the entoblast, or cell nucleus, and undergoes segmentation.
Perivitelline	Situated around the vitellus, or between the vitellus and zona pellucida of an ovum. The space between the yolk and the chorion.
Photoperiod	The duration of an organism's daily exposure to light, considered especially with regard to the effect of the exposure on growth and development.
Photophore	An organ that emits light from specialized structures or derives light from symbiotic luminescent bacteria; found especially in marine fishes.
Phyletic	Of or relating to the evolutionary descent and development of a species or group of organisms; phylogenetic.
Phylogeny	The evolutionary relationships among organisms; the patterns of lineage branching produced by the evolutionary history of the organisms being considered.
Physoclistous swimbladder	A swimbladder that is closed off from the mouth.
Piscivorous	Habitually feeding on fish; fish-eating.
Planktivore	Habitually feeding on plankton.
Planktonic	Drifting organisms that inhabit the water column.

Pleisiomorphic	A character state that is present in both the outgroups and in the ancestors.
Polarity	In cladistics, characters are either primitive or derived; that is, their polarity is either primitive or derived.
Polymodal	The presence of more than one mode.
Polyspermy	The entry of several sperm into one ovum during fertilization.
Postflexion stage	The stage in larval fish development after notochord flexion is complete.
Preanal	Situated anterior to the anus.
Preanal finfold	The ventral larval finfold in fishes anterior to the anus.
Predation	Capturing and eating prey (other animals) as a means of maintaining life.
Predorsal (supraneural) bones	Medial bones anterior to the dorsal fin in some fishes, some of which may be homologous to dorsal fin pterigyophores.
Preflexion stage	The stage in larval fish development before notochord flexion occurs.
Prejuvenile stage	An ecologically and morphologically distinct stage between the larval stage and the juvenile stage in development of some fishes. Prejuveniles of many fishes are neustonic.
Protandric hermaphrodite	Having male sexual organs while young, and female organs later in life.
Proteolytic enzyme	Enzymes that catalyze the splitting of proteins into smaller peptide fractions and amino acids.
Protogyny	The development of the female organs before the appearance of corresponding male organs.
Resorption	Absorption or, less commonly, adsorption of material by an organism from which the material was previously released.
Semelparity	A life-history pattern in which the organism reproduces only once in its lifetime. In such fishes, they often die shortly after reproduction.
Sequential hermaphroditism	Individuals functioning as males and females at different times during their lives.
Sexual dichromism	Color differences between the sexes.
Sexual dimorphism	Physical differences between males and females that arise as a consequence of sexual maturation, including the secondary sex characteristics.
Shoreside	Located on a shore or along a shore.
Spawner	A female fish at the time of releasing eggs.
Spawning	The depositing and fertilizing of eggs (or roe) by fish. Release of unfertilized eggs into the environment or release

	of larvae into the environment (mating and spawning need not occur simultaneously).
Speciose	An abundance of species.
Squamation	The condition of being scaly.
Subdermal space	The space between the skin and the body in marine fish larvae. The subdermal space is filled with a gelatinous liquid of low specific gravity and aids in maintaining buoyancy.
Sublethal	Less than lethal (deadly).
Swimbladder	An internal organ that contributes to the ability of a fish to control its buoyancy. Also called air bladder or gas bladder.
Synapomorphy	In cladistics, a derived character.
Synchronous (or simultaneous) hermaphroditism	Having functional ovaries and testes in an individual at the same time.
Systematics	The systematic classification of organisms and the evolutionary relationships among them.
Tagesgrade ("Degree-Days")	A measure of the incubation rate of fish eggs. The temperature of incubation multiplied by the days from fertilization to hatching.
Tailbud	The embryonic tail of fishes that becomes raised and free from the yolk.
Taxon (pl., taxa)	A group of genetically similar organisms that are classified together.
Teleconnections	In atmospheric science refer to climate anomalies being related to each other at large distances (typically thousands of kilometers).
Thermocline	A fairly thin zone in a water body with an abrupt temperature change that separates an upper warmer zone from a lower colder zone.
Trophic	Of or involving the feeding habits or food relationships of organisms.
Unisexuality	Relating to only one sex or having only one type of sexual organ; not hermaphroditic.
Vitelline membrane	The membrane that develops around an oocyte. The zona pellucida.
Vitellogenesis	The formation of the yolk of an egg.
Viviparity	Giving birth to living offspring that hatch and develop within the mother. The eggs are fertilized internally and the offspring gain nourishment from the mother during development.

Xanthophores	Pigment cells that contain a yellow pigment.
Yolk sac	A membranous sac enclosing the yolk and attached to the ventral side of an embryo (or yolk sac larva), providing early nourishment in the form of yolk.
Yolk-sac larva	A fish larva that has a yolk sac.
Zona pellucida	The thick, solid, transparent outer membrane of a developed ovum (egg).
Zona radiata	The outer transparent layer, or envelope, of the ovum (egg).
Zonation	The distribution of organisms in biogeographic zones.

LITERATURE CITED

Able, K. W. and M. P. Fahay. 1998. The first year in the life of estuarine fishes in the Middle Atlantic Bight. Rutgers University Press, New Brunswick, NJ.

Able, K.W., D. F. Markle, and M. P. Fahay. 1984. Cyclopteridae: development, pp. 428–437. In: H.G. Moser et al. (eds.), Ontogeny and systematics of fishes. Am. Soc. Ichthyol. Herp., Spec. Publ. No. 1.

Aboussouan, A. 1989. L'identification des larvae de poissons de la mer Mediterranee. Cybium 13: 259–262.

Agardy, M. T. 1997. Marine protected areas and ocean conservation. Academic Press, San Diego, CA.

Agardy, T. 1999. Creating havens for marine life. Issues Sci. Tech. Online 16: 37–44.

Ahlstrom, E. H. 1969. Remarkable movements of oil globules in eggs of bathylagid smelts during embryonic development. J. Mar. Biol. Assoc. India 11: 206–217.

Ahlstrom, E. H. 1972. Kinds and abundance of fish larvae in the eastern tropical Pacific on the second multivessel EASTROPAC survey, and observations on the annual cycle of larval abundance. Fish. Bull. 70: 1153–1242.

Ahlstrom, E. H. 1974. The diverse patterns of metamorphosis in gonostomatid fishes—an aid to classification, pp. 659–674. In: J.H.S. Blaxter (ed.), The early life history of fish. Springer-Verlag, Heidelberg.

Ahlstrom, E. H. and O. P. Ball. 1954. Description of eggs and larvae of jack mackerel (*Trachurus symmetricus*) and distribution and abundance of larvae in 1950 and 1951. Fish. Bull. U.S. 56: 209–245.

Ahlstrom, E. H. and R. C. Counts. 1955. Eggs and larvae of the Pacific hake *Merluccius productus*. Fish. Bull. U.S. 56: 295–329.

Ahlstrom, E. H. and H. G. Moser. 1976. Eggs and larvae of fishes and their role in systematic investigations and in fisheries. Rev. Trav. Inst. Pêches Marit. 40: 379–398.

Ahlstrom, E. H. and H. G. Moser. 1980. Characters useful in identification of pelagic marine fish eggs. Calif. Coop. Oceanic Fish. Invest. Rep. 21: 121–131.

Ahlstrom, E. H., H. G. Moser, and D. M. Cohen. 1984a. Argentinoidei: development and relationships, pp.155–168. In: H. G. Moser et al. (eds.), Ontogeny and systematics of fishes. Am. Soc. Ichthyol. Herp., Spec. Publ. No. 1.

Ahlstrom, E. H., W. J. Richards, and S. H. Weitzman. 1984b. Families Gonostomatidae, and associated stomiiform groups: development and relationships, pp. 184–198. In: H. G. Moser et al. (eds.), Ontogeny and systematics of fishes. Am. Soc. Ichthyol. Herp., Spec. Publ. No. 1.

Ahlstrom, E. H. and E. G. Stevens. 1976. Report of neuston (surface) collections made on an extended CALCOFI cruise during May 1972. Calif. Coop. Oceanic Fish. Invest. Rep. 18: 167–180.

Alderdice, D. F. and C. R. Forrester. 1968. Some effects of salinity and temperature on early development and survival of the English sole (*Parophrys vetulus*). J. Fish. Res. Bd. Can. 25: 495–521.

Alderdice, D. F. and C. R. Forrester. 1971. Effects of salinity and temperature on embryonic development of the petrale sole (*Eopsetta jordani*). J. Fish. Res. Bd. Can. 28: 727–744.

Alderdice, D. F. and C. R. Forrester. 1971. Effects of salinity, temperature, and dissolved oxygen on early development of the Pacific cod (*Gadus macrocephalus*). J. Fish. Res. Bd. Can. 28: 883–902.

Allen, D. M. and D. L. Barker. 1990. Interannual variations in larval fish recruitment to estuarine epibenthic habitats. Mar. Ecol. Prog. Ser. 63: 113–125.

Allen, L. G. 1979. Larval development of *Gobiesox rhessodon* (Gobiesocidae) with notes on the larva of *Rimicola muscarum*. Fish. Bull. 77: 300–304.

Allison, G. W., J. Lubchenco, and M. H. Carr. 1998. Marine reserves are necessary but not sufficient for marine conservation. Ecol. Appl. Suppl. 8: S79–S92.

Almatar, S. M. and E. D. Houde. 1986. Distribution and abundance of sardine *Sardinella fimbriata* (Val.) eggs in Kuwait waters of the Arabian Gulf. Fish. Res. 4: 331–342.

Anderson, J. T. 1988. A review of size dependent survival during pre-recruit stages of fishes in relation to recruitment. J. Northwest Atl. Fish. Sci. 8: 55–66.

Anderson, W. W., J. W. Gehringer, and E. Cohen. 1956. Physical oceanographic, biological, and chemical data. South Atlantic coast of the United States. M/V Theodore N. Gill Cruise 1. U.S. Fish. and Wildl. Serv. Spec. Sci. Rep—Fish. 178.

Angell, C. L., B. S. Miller, and S. R. Wellings. Epizootiology of tumors in a population of juvenile English sole (*Parophrys vetulus*) from Puget Sound, Washington. J. Fish. Res. Bd. Can. 32: 1723–1732.

Apstein, C. 1909. In: Hempel, 1979. Early life history of marine fish: The egg stage.

Arthur, D. K. 1978. Food and feeding of larvae of three fishes occurring in the California Current, *Sardinops sagax, Engraulis mordax*, and *Trachurus symmetricus*. Fish. Bull. 74: 517–530.

Auer, N. A. (ed.). 1982. Identification of larval fishes of the Great Lakes Basin with emphasis on the Lake Michigan drainage. Spec. Publ. 82-3. Great Lakes Fish. Comm., Ann Arbor, MI.

Auster, P. J. and N. L. Shackell. 2000. Marine protected areas for the temperate and boreal Northwest Atlantic: the potential for sustainable fisheries and conservation of biodiversity. Northwest. Natural. 4: 419–434.

Babcock, R. C., S. Kelly, N. T. Shears, J. W. Walker, and T. J. Willis. 1999. Changes in community structure in temperate marine reserves. Mar. Ecol. Prog. Ser. 189: 125–134.

Bagenal, T. B. 1973. Fish fecundity and its relationship with stock and recruitment. Rapp. P.-v. Réun. Cons. Int. Explor. Mer. 164: 186–198.

Bagenal, T. B. 1974. A buoyant net designed to catch freshwater fish larvae quantitatively. Freshwater Biol. 4: 107–109.

Bailey, K. M. 1981. Larval transport and recruitment of Pacific hake *Merluccius productus*. Mar. Ecol. Prog. Ser. 6: 1–9.

Bailey, K. M. 1984. Comparison of laboratory rates of predation on five species of marine larvae by three planktonic invertebrates: effects of larval size and vulnerability. Mar. Biol. 79: 303–309.

Bailey, K. M., R. D. Brodeur, and A. B. Hollowed. 1996. Cohort survival patterns of walleye pollock, *Theragra chalcogramma*, in Shelikof Strait, Alaska: a critical factor analysis. Fish. Oceanogr. 5 (Suppl. 1): 179–188.

Bailey, K. M. and E. D. Houde. 1989. Predation on eggs and larvae of marine fishes and the recruitment problem. Adv. Mar. Biol. 25: 1–83.

Bailey, K. M. and S. A. Macklin. 1994. Analysis of patterns in larval walleye pollock *Theragra chalcogramma* survival and wind mixing events in Shelikof Strait, Gulf of Alaska. Mar. Ecol. Prog. Ser. 113: 1–12.

Bailey, K. M., H. Nakata, and H. Van der Veer. 2005. The planktonic stages of flatfish: Physical and biological interactions in transport processes, pp. 94–119. In: R. Gibson (ed.), Flatfishes: Biology and exploitation. Blackwell Publishing. Oxford, UK.

Bailey, K. M. and C. L. Stehr. 1988. The effects of feeding periodicity and ration on the rate of increment formation in otoliths of larval walleye pollock *Theragra chalcogramma* (Pallas). J. Exp. Mar. Biol. Ecol. 122: 147–161.

Baldwin, C. C. 1990. Morphology of the larvae of American Anthiinae (Teleostei: Serranidae), with comments on relationships within the subfamily. Copeia 1990: 913–955.

Baldwin, C. C. and G. D. Johnson. 1991. A larva of the poorly known serranid fish *Jeboehlkia gladifer* (Teleostei: Serranidae: Epinephelinae). Fish. Bull. 89: 535–537.

Baldwin, C. C. and G. D. Johnson. 1993. Phylogeny of the Epinephelinae (Teleostei: Serranidae). Bull. Mar. Sci. 52: 240–283.

Baldwin, C. C., G. D. Johnson, and P. L. Colin. 1991. Larvae of *Diploprion bifasciatum*, *Belonoperca chabanaudi*, and *Grammistes sexlineatus* (Serranidae: Epinephelinae) with comparisons of known larvae of other epinephelines. Bull. Mar. Sci. 48: 67–93.

Barlow, G. W. 1981. Patterns of parental investment, dispersal and size among coral reef fishes. Environ. Biol. Fish. 6: 65–85.

Battaglene, S. C. 1996. Aquaculture: introduction. Mar. Freshwater Res. 47: 209–210.

Batty, R. S. 1989. Escape responses of herring larvae to visual stimuli. J. Mar. Biol. Assoc. U. K. 69: 647–654.

Baumgartner, T. R., A. Soutar, and V. Ferreira-Bartrina. 1992. Reconstruction of the history of Pacific sardine and northern anchovy populations over the past two millennia from sediments of the Santa Barbara Basin. CalCOFI Rep. 33: 24–40.

Baxter, J. L. 1982. The role of the marine research committee and CalCOFI. CalCOFI Rep. 23: 35–38.

Beamish, R. J., D. J. Noakes, G. A. McFarlane, L. Klyashtorin, V. V. Ivanov, and V. Kurashov. 1999. The regime concept and natural trends in the production of Pacific salmon. Can. J. Fish. Aquat. Sci. 56: 516–526.

Begle, D. P. 1989. Phylogenetic analysis of the cottid genus *Artedius* (Teleostei: Scorpaeniformes). Copeia 1989: 642–652.

Berrien, P. L. 1978. Eggs and larvae of *Scomber scombrus* and *Scomber japonicus* in continental shelf waters between Massachusetts and Florida. Fish. Bull. U.S. 76: 95–115.

Berrien, P. L. and J. D. Sibunka. 1999. Distribution patterns of fish eggs in the U.S. Northeast continental shelf ecosystem, 1977–1987. NOAA Tech. Rep. NMFS. 145.

Beverton, R. J. H. and S. J. Holt. 1957. On the dynamics of exploited fish populations. Fish. Invest., London. Ser. II, 19: 1–533.

Blaxter, J. H. S. 1969. Development: Eggs and larvae, pp. 177–252. In: W. S. Hoar and D. G. Randall (eds.), Fish physiology, Vol. 3. Academic Press, New York.

Blaxter, J. H. S. 1986. Development of sense organs and behaviour of teleost larvae with special reference to feeding and predator avoidance. Trans. Am. Fish. Soc. 115: 98–114.

Blaxter, J. H. S. 1988. Pattern and variety in development, pp. 1–58. In: W. S. Hoar (ed.), Fish physiology, Vol. 11a. Academic Press, San Diego, CA.

Blaxter, J. H. S. 1991. Sensory systems and behaviour of larval fish. In: John Mauchline and Takahisa Nemoto (eds.), Marine biology: its accomplishment and future prospect. Copublished by the University of Tokyo Press, Japan and Elsevier Science Publishers, Amsterdam.

Blaxter, J. H. S. (ed.). 1974. The early life history of fish. Springer-Verlag, Berlin.

Blaxter, J. H. S. and L. A. Fuiman. 1990. The role of the sensory systems of herring larvae in evading predatory fishes. J. Mar. Bio. Assoc. U.K. 70: 413–427.

Blaxter, J. H. S., J. C. Gamble, and H. v. Westernhagen (Eds.). 1989. The Early Life History of Fish. Rapp. P.-v. Réun. Cons. Int. Explor. Mer., 191.

Blom, G., J. T. Nordeide, T. Svåsand, and A. Borge. 1994. Application of two fluorescent chemicals, alizarin complexone and alizarin Red S, to tag otoliths of Atlantic cod (*Gadus morhua* L.). Aquacult. Fish. Manag. 25 (Suppl. 1): 229–243.

Blood, D. M., A. C. Matarese, and M. M. Yoklavich. 1994. Embryonic development of walleye pollock, *Theragra chalcogramma*, from Shelikof Strait, Gulf of Alaska. Fish. Bull. 92: 207–222.

Boehlert, G. W. and B. C. Mundy. 1988. Roles of behavioral and physical factors in larval and juvenile fish recruitment to estuarine nursery areas. Am. Fish. Soc. Sym. 3: 51–67.

Boehlert, G. W. and B. C. Mundy. 1994. Vertical and onshore-offshore distributional patterns of tuna larvae in relation to physical habitat features. Mar. Ecol. Prog. Ser. 107: 1–13.

Boehlert, G. and M. Yoklavich. 1984. Reproduction, embryonic energetics, and the maternal-fetal relationship in the viviparous genus *Sebastes* (Pisces, Scorpaenidae). Biol. Bull. 167: 354–370.

Bortone, S. A. 1977. Osteological notes on the genus *Centropristis* (Pisces: Serranidae). Northeast Gulf Sci. 1: 23–33.

Botsford, L. W., A. Hastings, and S. D. Gaines. 2001. Dependence of sustainability on the configuration of marine reserves and larval dispersal distance. Ecol. Lett. 4: 144–150.

Brander, K. and A. B. Thompson. 1989. Diel differences in avoidance of three vertical profile sampling gears for herring larvae. J. Plankton. Res. 11: 775–784.

Breder, C. and D. Rosen. 1966. Modes of reproduction in fishes. T.F.H. Publications, Neptune City.

Breitburg, D. L., K. A. Rose, and J. H. Cowan. 1999. Linking water quality to larval survival: Predation mortality of fish larvae in an oxygen-stratified water column. Mar. Ecol. Prog. Ser. 178: 39–54.

Bridger, J. P. 1957. On efficiency tests made with a modified Gulf III high-speed tow net. J. Cons. Perm. Int. Explor. Mer. 23: 357–365.

Brodeur, R. D., K. M. Bailey, and S. Kim. 1991. Cannibalism on eggs by walleye pollock (*Theragra chalcogramma*) in Shelikof Strait, Gulf of Alaska. Mar. Ecol. Prog. Ser. 71: 207–218.

Brodeur, R. D., C. E. Mills, J. E. Overland, G. E. Walters, and J. D Schumacher. 1999. Evidence for a substantial increase in gelatinous zooplankton in the Bering Sea, with possible links to climate change. Fish. Oceanogr. 8: 296–306.

Brodeur, R. D. and D. M. Ware. 1992. Interannual and interdecadal changes in zooplankton biomass in the subarctic Pacific Ocean. Fish. Oceanogr. 1: 32–38.

Brodeur, R. D. and M. T. Wilson. 1996. A review of the distribution, ecology and population dynamics of age-0 walleye pollock in the Gulf of Alaska. Fish. Oceanogr. 5 (Suppl.1) 148–166.

Brogan, M. W. 1994. Two methods of sampling fish larvae over reefs: a comparison from the Gulf of California. Mar. Biol. 118: 33–44.

Brothers, E. B., C. P. Mathews, and R. Lasker. 1976. Daily growth increments in otoliths from larval and adult fishes. Fish. Bull. 74: 1–8.

Browman, H. I. and W. J. O'Brien. 1992. The ontogeny of search behavior in the white crappie, *Pomoxis annularis*. Environ. Biol. Fish. 34: 181–195.

Browman, H. I. and A. B. Skiftesvik. 2003 (eds.). The Big Fish Bang. Proceedings of the 26th Annual Larval Fish Conference. Institute of Marine Research, Bergen, Norway.

Brown, D. L. and L. Cheng. 1981. New net for sampling the ocean surface. Mar. Ecol. Prog. Ser. 5: 225–227.

Brownell, C. L. 1985. Laboratory analysis of cannibalism by larvae of the cape anchovy *Engraulis capensis*. Trans. Am. Fish. Soc. 114: 512–518.

Buchanan-Wollaston, H.J. 1926. Plaice-egg production in 1920–21, treated as a statistical problem, with comparison between data from 1911, 1914, and 1921. Fish. Invest. Ser. II, 9(2).

Buckley, L.J. 1980. Changes in ribonucleic acid, deoxyribonucleic acid, and protein content during ontogenesis in winter flounder (*Pseudopleuronectes americanus*) and effect of starvation. Fish. Bull. 77: 703–708.

Buckley, L.J. 1984. RNA-DNA ratio: an index of larval fish growth in the sea. Mar. Biol. 80: 291–298.

Buckley, L.J., A.S. Smigielski, T.A. Halavik, and G.C. Laurence. 1990. Effects of water temperature on size and biochemical composition of winter flounder *Pseudopleuronectes americanus* at hatching and feeding initiation. Fish. Bull. 88: 419–428.

Busby, M.S. 1998. Guide to the identification of larval and early juvenile poachers (Scorpaeniformes: Agonidae) from the northeastern Pacific Ocean and Bering Sea. NOAA Tech. Rep. NMFS 137.

Butler, J.L., H.G. Moser, G.S. Hageman, and L.L. Nordgren. 1982. Developmental stages of three California sea basses (*Paralabrax,* Pisces, Serranidae). Calif. Coop. Oceanic Fish. Invest. Rep. 23: 252–268.

Cada, G. F. and J. M. Loar. 1982. Relative effectiveness of two ichthyoplankton sampling techniques. Can. J. Fish. Aquat. Sci. 39: 811–814.

Cambray, J. A. 2000. 'Threatened Fishes of The World' Series, an update. Environ. Biol. Fish. 59: 353–357.

Campana, S. E. 1996. Year-class strength and growth rate in young Atlantic cod *Gadus morhua*. Mar. Ecol. Prog. Ser. 135: 21–26.

Campana, S. E. and E. Moksness. 1991. Accuracy and precision of age and hatch date estimates from otolith microstructure examination. ICES J. Mar. Sci. 48: 303–316.

Campana, S. E. and J. D. Neilson. 1985. Microstructure of fish otoliths. Can. J. Fish. Aquat. Sci. 42: 1014–1032.

Canino, M. F. 1994. Effects of temperature and food on growth and RNA/DNA ratios of walleye pollock, *Theragra chalcogramma* (Pallas), eggs and larvae. J. Exp. Mar. Biol. Ecol. 175: 1–16.

Canino, M.F. and K.M. Bailey. 1995. Gut evacuation of walleye pollock larvae in response to feeding condition. J. Fish. Biol. 46: 389–403.

Canino, M.F., K.M. Bailey, and L.S. Incze. 1991. Temporal and geographic differences in feeding and nutritional condition of walleye pollock larvae *Theragra chalcogramma* in Shelikof Strait, Gulf of Alaska. Mar. Ecol. Prog. Ser. 79: 27–35.

Caputi, N., W. J. Fletcher, A. Pearce, and C. F. Chubb. 1996. Effect of the Leeuwin Current on the recruitment of fish and invertebrates along the Western Australian coast. Mar. Freshwater Res. 47: 147–155.

Carls, M. G. 1987. Effects of dietary and water-borne oil exposure on larval Pacific herring (*Clupea harengus* Pallasi). Mar. Environ. Res. 22: 253–270.

Carr, M. H. and D. C. Reed. 1993. Conceptual issues relevant to marine harvest refugia: examples from temperate reef fishes. Can. J. Fish. Aquat. Sci. 50: 2019–2028.

Chambers, R. C. and E. A. Trippel. 1997. Early life history and recruitment in fish populations. Chapman and Hall. London.

Chase, J. 1955. Winds and temperatures in relation to the brood-strength of Georges Bank haddock. Cons. Perm. Int. Explor. Mer. J. Cons. 21: 17–24.

Checkley, D. M. 1982. Selective feeding by Atlantic herring (*Clupea harengus*) larvae on zooplankton in natural assemblages. Mar. Ecol. Prog. Ser. 9: 245–253.

Checkley, D. M., Jr., P. B. Ortner, L. R. Settle, and S. R. Cummings. 1997. A continuous underway fish egg sampler. Fish. Oceanogr. 6: 58–73.

Checkley, D. M., Jr., S. Raman, G. L. Maillet, and K. M. Mason. 1988. Winter storm effects on the spawning and larval drift of a pelagic fish. Nature 335: 346–348.

Chen, D., S. E. Zebiak, A. J. Busalacchi, and M. A. Cane. 1995. An improved procedure for El Niño forecasting: implications for predictability. Science 269: 1699–1702.

Choate, J. H., P. J. Doherty, B. A. Kerrigan, and J. M. Leis. 1993. A comparison of towed nets, purse seine, and light-aggregation devices for sampling larvae and pelagic juveniles of coral reef fishes. Fish. Bull. 91: 195–209.

Clark, F. N. 1934. Maturity of the California sardine (*Sardina caerulea*) determined by ova diameter measurements. Calif. Fish Game, Fish. Bull. 42.

Clark, J., W. G. Smith, A. W. Kendall, Jr., and M. P. Fahay. 1969. Studies of estuarine dependence of Atlantic coastal fishes. Data report I: northern section, Cape Cod to Cape Lookout. *R. V. Dolphin* cruises 1965–66: zooplankton volumes, midwater trawl collections, temperatures and salinities. Bur. Sport Fish. Wildl. Tech. Pap. 28, 132 p.

Clark, J., W. G. Smith, A. W. Kendall, Jr., and M. P. Fahay. 1970. Studies of estuarine dependence of Atlantic coastal fishes. Data report II: southern section, New River Inlet, N. C. to Palm Beach, Fla. *R. V. Dolphin* cruises 1967–68: zooplankton volumes, surface-meter net collections, temperatures and salinities. Bur. Sport Fish. Wildl. Tech. Pap. 59.

Clemmesen, C. 1996. Importance and limits of RNA/DNA ratios as a measure of nutritional condition in fish larvae. In: Y. Watanabe, Y. Yamashita, and Y. Oozeki (eds.), Survival strategies in early life stages of marine resources. Proceeding of an international workshop, Yokohama, Japan, 11–14 October 1994. A. A. Balkema, Rotterdam, pp. 67–82.

Clemmesen, C. and T. Doan. 1996. Does otolith structure reflect the nutritional condition of a fish larva? Comparison of otolith structure and biochemical index (RNA/DNA ratio) determined on cod larvae. Mar. Ecol. Prog. Ser. 138: 33–39.

Cohen, D. M. 1984. Ontogeny, systematics, and phylogeny, pp. 7–11. In: H. G. Moser et al. (eds.), Ontogeny and systematics of fishes. Am. Soc. Ichthyol. Herp., Spec. Publ. No. 1.

Collette, B. B., T. Potthoff, W. J. Richards, S. Ueyanagi, J. L. Russo, and Y. Nishikawa. 1984. Scombroidei: development and relationships, pp. 591–619. In: H. G. Moser et al. (eds.), Ontogeny and systematics of fishes. Am. Soc. Ichthyol. Herp., Spec. Publ. No. 1.

Colton, J. B., Jr. 1964. History of oceanography in the offshore waters of the Gulf of Maine. U.S. Fish and Wildl. Serv. Spec. Sci. Rep.—Fish. 496.

Colton, J. B., Jr. and R. F. Temple. 1961. The enigma of Georges Bank spawning. Limnol. Oceanogr. 6: 280–291.

Colton, J. B., Jr., J. R. Green, R. R. Byron, and J. L. Frisella. 1980. Bongo net retention rates as effected by towing speed and mesh size. Can. J. Fish. Aquat. Sci. 37: 606–623.

Conover, D. O., J. Travis, and F. C. Coleman. 2000. Essential fish habitat and marine reserves: an introduction to the second Mote symposium in fisheries ecology. Bull. Mar. Sci. 66: 527–534.

Conway, D. V. P., I. R. B. McFadzen, and P. R. G. Tranter. 1994. Digestion of copepod eggs by larval turbot *Scophthalmus maximus* and egg viability following gut passage. Mar. Ecol. Prog. Ser. 106: 303–309.

Coombs, S. H., C. A. Fosh, and M. A. Keen. 1985. The buoyancy and vertical distribution of eggs of sprat (*Sprattus sprattus*) and pilchard (*Sardina pilchardus*). J. Mar. Biol. Assoc. U.K. 65: 461–474.

Coombs, S. H., J. H. Nicholls, and C. A. Fosh. 1990. Plaice eggs (*Pleuronectes platessa* L.) in the southern North Sea: abundance, spawning area, vertical distribution, and buoyancy. ICES J. Mar. Sci. 47: 133–139.

Cote, I. M., I. Mosqueira, and J. D. Reynolds. 2001. Effects of marine reserve characteristics on the protection of fish populations: a meta analysis. J. Fish. Biol. 59: 178–189.

Cowan, J. H., Jr. and E. D. Houde. 1992. Size-dependent predation on marine fish larvae by ctenophores, scyphomedusae, and planktivorous fish. Fish. Oceanogr. 1: 113–126.

Cowan, J. H., Jr. and R. F. Shaw. 2002. Recruitment, pp. 88–111. In: L. Fuiman and R. Werner, Fishery science: the unique contributions of early life history stages. Blackwell Publishing. Oxford, U.K.

Craik, J. C. A. and S. M. Harvey. 1987. The causes of buoyancy in eggs of marine teleosts. J. Mar. Biol. Assoc. U.K. 67: 169–182.

Crowder, L. B., S. J. Lyman, W. F. Figueira, and J. Priddy. 2000. Source sink population dynamics and the problem of siting marine reserves. Bull. Mar. Sci. 66: 799–820.

Crowder, L. B. and F. E. Werner (eds.). 1999. Fisheries oceanography of the estuarine-dependent fishes of the South Atlantic Bight. Fish. Oceanogr. 8 (Suppl. 2): 1–252.

Cunha, M. E., I. Figueiredo, A. Farinha, and M. Santos. 1992. Estimation of sardine spawning biomass off Portugal by the daily egg production method. Bol. Inst. Esp. Oceanogr. 8: 139–153.

Cury, P. and C. Roy. 1989. Optimal environmental window and pelagic fish recruitment in upwelling areas. Can. J. Fish. Aquat. Sci. 46: 670–680.

Cushing, D. H. 1971. The dependence of recruitment on parent stock in different groups of fishes. J. Cons. Int. Explor. Mer. 33: 340–362.

Cushing, D. H. 1975. Marine ecology and fisheries. Cambridge University Press.

Cushing, D. H. 1980. The decline of the herring stocks and the gadoid outburst. J. Cons. Int. Explor. Mer. 39: 70–81.

Cushing, D. H. 1983. Are fish larvae too dilute to affect the density of their food organisms? J. Plankton Res. 5: 847–854.

Cushing, D. H. 1990. Plankton production and year-class strength in fish populations: an update of the match/mismatch hypothesis. Adv. Mar. Biol. 26: 249–293.

Cushing, D. H. 1995a. Population production and regulation in the sea: a fisheries perspective. Cambridge University Press, Cambridge, UK.

Cushing, D. H. 1995b. The long term relationship between zooplankton and fish. ICES J. Mar. Sci. 52: 611–626.

Cyr, H., J. A. Downing, S. Lalonde, S. B. Baines, and L. M. Pace. 1992. Sampling larval fish populations: choice of sample number and size. Trans. Am. Fish. Soc. 121: 356–368.

D'Vincent, S., H. G. Moser, and E. H. Ahlstrom. 1980. Description of the larvae and early juveniles of the Pacific butterfish, *Peprilus simillimus* (Family Stromateidae). Calif. Coop. Oceanic Fish. Invest. Rep. 21: 172–179.

Daan, N., A. D. Rijnsdorp, and G. R. van Overbeeke. 1985. Predation by North Sea herring *Clupea harengus* on eggs of plaice *Pleuronectes platessa* and cod *Gadus morhua*. Trans. Am. Fish. Soc. 114: 499–506.

Dahlgren, C. P., J. A. Sobel, and D. E. Harper. 2001. Assessment of the reef fish community, habitat, and potential for larval dispersal from the proposed Tortugas South Ecological Reserve. Proc. Gulf Carib. Fish. Inst. 52: 700–712.

Davies, I. E. and E. G. Barham. 1969. The Tucker opening-closing micronekton net and its performance in a study of the deep scattering layer. Mar. Biol. 2: 127–131.

Davis, G. E. 1989. Designated harvest refugia: the next stage of marine fishery management in California. Calif. Coop. Oceanic Fish. Invest. Rep. 30: 53–58.

de Lafontaine, Y. and W. C. Leggett. 1987. Evaluation of in situ enclosures for larval fish studies. Can. J. Fish. Aquat. Sci. 44: 54–65.

DeMartini, E. and P. Sikkel. 2006. Behavioral ecology: reproduction, pp. 483–523. In: L. Allen, D. Pondella, and M. Horn. The ecology of marine fishes: California and adjacent waters. University of California Press, Los Angeles.

Dempsey, C. H. 1988. Ichthyoplankton entrainment. J. Fish. Biol. 33(A): 93–102.

Dennis, G. G., D. Goulet, and J. R. Rooker. 1991. Ichthyoplankton assemblages sampled by night-lighting in nearshore habitats of southwestern Puerto Rico. NOAA-NMFS Tech. Rep. 95: 89–97.

Dight, I. J. 1995. Understanding larval dispersal and habitat connectivity in tropical marine systems. A tool for management. IUCN. 41–46.

Doherty, P. J. 1987. Light-traps: selective but useful devices for quantifying the distributions and abundances of larval fishes. Bull. Mar. Sci. 41: 423–431.

Dorn, M., S. Barbeaux, M. Guttormsen, B. Megrey, A. Hollowed, M. Wilkins, and K. Spalinger. 2003. Assessment of walleye pollock in the Gulf of Alaska, p. 33–148. In: Stock assessment and fishery evaluation report for the groundfish fisheries in the Gulf of Alaska, North Pacific Fisheries Management Council, 605 W 4th Avenue, Suite 306, Anchorage, AK 99501.

Dovel, W. L. 1964. An approach to sampling estuarine macroplankton. Chesapeake Sci. 5: 77–90.

Dower, J. F., T. J. Miller, and W. C. Leggett. 1997. The role of microscale turbulence in the feeding ecology of larval fish. Adv. Mar. Biol. 31: 169–201.

Doyle, M. J., K. L. Mier, M. S. Busby, and R. D. Brodeur. 2002. Regional variation in springtime ichthyoplankton assemblages in the northeast Pacific Ocean. Prog. Oceanogr. 53: 247–281.

Dugan, J. E. and G. E. Davis. 1993a. Introduction to the international symposium on marine harvest refugia. Can. J. Fish. Aquat. Sci. 50: 1991–1992.

Dugan, J. E. and G. E. Davis. 1993b. Applications of marine refugia to coastal fisheries management. Can. J. Fish. Aquat. Sci. 50: 2029–2042.

Dunn, J., R. B. Mitchell, G. G. Urquhart, and B. J. Ritchie. 1993. LOCHNESS a new multi-net midwater sampler. J. du Cons. 50: 203–212.

Dunn, J. R. 1984. Developmental osteology, pp. 48–50. In: H. G. Moser et al. (eds.), Ontogeny and systematics of fishes. Am. Soc. Ichthyol. Herp., Spec. Publ. No. 1.

Dunn, J. R. and B. M. Vinter. 1984. Development of larvae of saffron cod, *Eleginus gracilus*, with comments on the identification of gadid larvae in Pacific and Arctic waters contiguous to Canada and Alaska. Can. J. Fish. Aquat. Sci. 41: 304–318.

Eckmann, R., U. Gaedke, and H. J. Wetzlar. 1988. Effects of climatic and density-dependent factors on year-class strength of *Coregonus lavaretus* in Lake Constance. Can. J. Fish. Aquat. Sci. 45: 1088–1093.

Economou, A. N. 1991. Food and feeding ecology of five gadoid larvae in the northern North Sea. ICES J. Mar. Sci. 47: 339–351.

Ehrenbaum, E. 1905–1909. Eier und Larven von Fischen. Nord. Plankton 1 and 2.

Ellertsen, B., P. Fossum, P. Solemdal, and S. Sundby. 1989. The relationship between temperature and survival of eggs and first-feeding larvae of northeast Arctic cod (*Gadus morhua* L.) Rapp. P.-v. Reun. Cons. Int. Explor. Mer. 191: 209–219.

Ellertsen, B., P. Fossum, P. Solemdal, S. Sundby, and S. Tilseth. 1984. A case study on the distribution of cod larvae and availability of prey organism in relation to physical processes in Lofoten, pp. 453–477. In: E. Dahl, D. S. Danielssen, E. Moksness, and P. Solemdal (eds.), The propagation of cod *Gadus morhua*. Institute of Marine Research, Arendal, Norway.

Fahay, M. P. 1983. Guide to the early stages of marine fishes occurring in the western North Atlantic Ocean, Cape Hatteras to the southern Scotian shelf. J. Northwest Atl. Fish. Sci. 4.

Fahay, M. P. 2007a. Early stages of fishes in the western North Atlantic Ocean (Davis Strait, southern Greenland and Flemish Cap to Cape Hatteras). Volume One: Acipenseriformes through Syngnathiformes. Monograph No. 1, North Atlantic Fisheries Organization. Dartmouth.

Fahay, M. P. 2007b. Early stages of fishes in the western North Atlantic Ocean (Davis Strait, southern Greenland and Flemish Cap to Cape Hatteras). Volume Two: Scorpaeniformes through Tetraodontiformes. Monograph No. 1, North Atlantic Fisheries Organization. Dartmouth.

Fahay, M. P. and D. F. Markle. 1984. Gadiformes: development and relationships, pp. 265–282. In: H. G. Moser et al. (eds.), Ontogeny and systematics of fishes. Am. Soc. Ichthyol. Herp., Spec. Publ. No. 1.

Fay, L. A., M. L. Gessner, and P. C. Stromberg. 1985. Programs and abstracts of the 28th Conf. on Great Lakes Research, University of Wisconsin, 38.

Fitch, J. E. 1982. Revision of the eastern North Pacific anthiine basses (Pisces: Serranidae). Contrib. Sci. (Los Angeles) 339.

Fogarty, M. J. (ed.) 2000. Recruitment dynamics of exploited marine populations: physical-biological interactions. Part 1. ICES J. Mar. Sci. 57: 189–464.

Fogarty, M. J. (ed.) 2001. Recruitment dynamics of exploited marine populations: physical-biological interactions. Part 2. ICES J. Mar. Sci. 58: 935–1114.

Fogarty, M. J., M. P. Sissenwine, and E. B. Cohen. 1991. Recruitment variability and the dynamics of exploited marine populations. Trends Ecol. Evol. 6: 241–246.

Fontaine, C. T. and D. B. Revera. 1980. The mass culture of the rotifer, *Brachionus plicatilis*, for use as foodstuff in aquaculture. Proc. World Maric. Soc. 11: 211–218.

Fortier, L. and J. A. Gagne. 1990. Larval herring (*Clupea harengus*) dispersion, growth and survival in the St. Lawrence estuary: match/mismatch or membership/vagrancy? Can. J. Fish. Aquat. Sci. 47: 1898–1912.

Fortier, L. and W. C. Leggett. 1984. Small scale covariability on the abundance of fish larvae and their prey. Can. J. Fish. Aquat. Sci. 41: 502–512.

Fortier, L. and A. Villeneuve. 1996. Cannibalism and predation on fish larvae by larvae of Atlantic mackerel, *Scomber scombrus*: trophodynamics and potential impact on recruitment. Fish. Bull. 94: 268–281.

Foucher, R. and R. Beamish. 1977. Production of nonviable oocytes by Pacific hake, *Merluccius productus*. Can. J. Fish. Aquat. Sci. 37: 41–48.

Francis, M. P. 1993. Does water temperature determine year class strength in New Zealand snapper (*Pagurus auratus*, Sparidae)? Fish. Oceanogr. 2: 65–72.

Francis, R. C., S. R. Hare, A. B. Hollowed, and W. S Wooster. 1998. Effects of interdecadal climate variability on the oceanic ecosystems of the NE Pacific. Fish. Oceanogr. 7: 1–21.

Frank, K. T. and W. C. Leggett. 1982. Selective exploitation of capelin (*Mallotus villosus*) eggs by winter flounder (*Pseudopleuronectes americanus*): Capelin egg mortality rates, and contribution of egg energy to the annual growth of flounder. Can. J. Fish. Aquat. Sci. 41: 1294–1302.

Franzin, W. G. and S. M. Harbicht. 1992. Test of drift samplers for estimating abundance of recently hatched walleye larvae in small rivers. N. Am. J. Fish. Manag. 12: 396–405.

Freeberg, M. H., W. W. Taylor, and R. W. Brown. 1990. Effect of egg and larval survival on year-class strength of lake whitefish in Grand Traverse Bay, Lake Michigan. Trans. Am. Fish. Soc. 119: 92–100.

Fritzsche, R. A. 1978. Development of fishes of the mid-Atlantic bight, an atlas of egg, larval, and juvenile stages. Volume V. Chaetodontidae through Ophidiidae. U.S. Fish. Wildl. Serv., Biol. Prog. FWS/OBS-78/12.

Fuiman, L. A. (ed.). 1993. Water quality and the early life stages of fishes. Am. Fish. Soc. Sym. 14.

Fuiman, L. A. and J. C. Gamble. 1988. Influence of experimental manipulations on predation of herring larvae by juvenile herring in large enclosures. Rapp. Proc. Verb. 191: 359–365.

Fuiman, L. A. and R. G. Werner. (eds.) 2002. Fishery Science: The unique contributions of early life stages. Blackwell science. Oxford, UK.

Fukuhara, O. 1990. Effects of temperature on yolk utilization, initial growth, and behavior of unfed marine fish-larvae. Mar. Biol. 106: 169–174.

Furuyu, K. 1995. Current condition and subject of marine fish culture in Japan, pp. 219–230. In: K. L. Main and C. Rosenfeld (eds.), Culture of high-value marine fishes in Asia and the United States. Oceanic Inst., Honolulu.

Gallagher, R. P. and J. V. Conner. 1983. Comparison of two ichthyoplankton sampling gears with notes on microdistribution of fish larvae in a large river. Trans. Am. Fish. Soc. 112: 280–286.

Gehringer, J. W. 1962. The Gulf III and other high-speed plankton samplers. Rapp. P.-v. Réun. Cons. Int. Explor. Mer. 153: 19–22.

Goksoyr, A., T. S. Solberg, and B. Serigstad. 1991. Immunochemical detection of cytochrome P450IA1 induction in cod (*Gadus morhua*) larvae and juveniles exposed to a water soluble fraction of North Sea crude oil. Mar. Pollut. Bull. 22: 122–127.

Gong, K., L. Yumei, and L. Hou. 1987. Effects of six heavy metals on hatching eggs and survival of marine fish. Oceanol. Limnol. 18: 138–144.

Goode, G. B. 1883. The first decade of the United States Fish Commission: Its plan of work and accomplished results, scientific and economical. Rep. U.S. Comm. Fish and Fish. for 1880: 53–62. (reprinted in Mar. Fish. Rev. 50: 130–134, 1988).

Gordon, D. J., D. F. Markle, and J. E. Olney. 1984. Ophidiiforms: development and relationships, pp. 308–319. In: H. G. Moser et al. (eds.), Ontogeny and systematics of fishes. Am. Soc. Ichthyol. Herp., Spec. Publ. No. 1.

Gosline, W. A. 1966. The limits of the fish family Serranidae, with notes on other lower percoids. Proc. Calif. Acad. Sci. 33: 91–111.

Gould, S. J. 1977. Ontogeny and phylogeny. Belknap Press, Cambridge, MA.

Govoni, J. J. (ed.) 2004. Development of form and function in fishes, and the question of larval adaptation. Am. Fish. Soc. Symp. 40.

Govoni, J. J., G. W. Boehlert, and Y. Watanabe. 1986. The physiology of digestion in fish larvae. Environ. Biol. Fish. 16: 59–77.

Govoni, J. J. and C. B. Grimes. 1992. The surface accumulation of larval fishes by hydrodynamic convergence within the Mississippi River Plume front. Cont. Shelf Res. 12: 1265–1276.

Graham, J. J., S. R. Chenoweth, and C. W. Davis. 1972. Abundance, distribution movements and lengths of larval herring along the western coast of the Gulf of Maine. Fish. Bull. 70: 307–321.

Granmo, A., R. Ekelund, J. A. Sneli, M. Berggren, and J. Svavarsson. 2002. Effects of antifouling paint components (TBTO, copper and triazine) on the early development of embryos in cod (*Gadus morhua* L.). Mar. Poll. Bull. 44: 1142–1148.

Grant, A. (ed.) 1996. Papers from the International Larval Fish Conference, Sydney, 1995, held under the auspices of the Early Life History Section of the American Fisheries Society and the Australian Society for Fish Biology. Mar. Freshw. Res. 47: 97–482.

Grave, H. 1981. Food and feeding of mackerel larvae and early juveniles in the North Sea. Rapp. P.-v. Réun. Cons. Int. Explor. Mer. 178: 454–459.

Graves, J. E., M. J. Curtis, P. A. Oeth, and R. S. Waples. 1989. Biochemical genetics of southern California basses of the genus *Paralabrax*: specific identification of fresh and ethanol-preserved individual eggs and early larvae. Fish. Bull. 88: 59–66.

Gregory, R. S. and P. M. Powles. 1988. Relative selectivities of Miller high-speed samplers and light traps for collecting ichthyoplankton. Can. J. Fish. Aquat. Sci. 45: 993–998.

Gross, M. R. and R. Shine. 1981. Parental care and mode of fertilization in ectothermic vertebrates. Evolution 35: 775–793.

Guenette, S., T. Lauk, and C. Clark. 1998. Marine reserves from Beverton and Holt to the present. Rev. Fish. Biol. Fish. 8: 251–272.

Gulec, I. and D. A. Holdway. 2000. Toxicity of crude oil and dispersed crude oil to ghost shrimp, *Palaemon serenus*, and larvae of Australian bass, *Macquaria novemaculeata*. Environ. Toxicol. 15: 91–98.

Gunderson, D. R. 1993. Surveys of fisheries resources. John Wiley & Sons, New York.

Gunderson, D. R., D. A. Armstrong, Y-B. Shi, and R. A. McConnaughey. 1990. Patterns of estuarine use by juvenile English sole (*Parophrys vetulus*) and Dungeness crab (*Cancer magister*). Estuaries. 13: 59–71.

Gunderson, D. R. and R. D. Vetter. 2006. Temperate rocky reef fishes, pp. 69–117. In: J. P. Kritzer and P. F. Sale, Marine Metapopulations. Elsevier, San Francisco.

Hales, L. S. and K. W. Able. 2001. Winter mortality of young-of-the-year of four coastal fishes in New Jersey (USA). Mar. Biol. 139: 45–54.

Halpern, B. S. and R. R. Warner. 2002. Marine reserves have rapid and lasting effects. Ecol. Lett. 5: 361–366.

Hardy, J. D., Jr. 1978a. Development of fishes of mid-Atlantic bight, an atlas of egg, larval, and juvenile stages. Vol. II. Anguillidae through Syngnathidae. U.S. Fish Wildl. Serv. Biol. Serv. Program FWS/OBS-78/12.

Hardy, J. D., Jr. 1978b. Development of fishes of the mid-Atlantic bight, an atlas of egg, larval, and juvenile stages. Volume III. Aphredoderidae through Rachycentridae. U.S. Fish. Wildl. Serv., Biol. Prog. FWS/OBS-78/12.

Hare, J. A. and R. K. Cowen. 1996. Transport mechanisms of larval and pelagic juvenile bluefish (*Pomatomus saltatrix*) from South Atlantic Bight spawning grounds to Middle Atlantic Bight nursery habitats. Limnol. Oceanogr. 41: 1264–1280.

Hare, J. A. and R. K. Cowen. 1997. Size, growth, development and survival of the planktonic larvae of *Pomatomus saltatrix* (Pisces: Pomatomidae). Ecology 78: 2415–2431.

Hare, J. A., M. P. Fahay, and R. K. Cowan. 2001. Springtime ichthyoplankton of the slope region of the northeastern United States of America: larval assemblages, relation to hydrography and implications for larval transport. Fish. Oceanogr. 10: 164–192.

Harris, R. P., L. Fortier, and R. P. Young. 1986. A large-volume pump system for studies of the vertical distribution of fish larvae under open sea conditions. J. Mar. Biol. Assoc. U.K. 66: 846–854.

Harris, S. A., D. P. Cyrus, and L. E. Beckley. 2001. Horizontal trends in larval diversity and abundance along an ocean-estuarine gradient on the northern KwaZulu-Natal Coast, South Africa. Est., Coast. Shelf Sci. 53: 221–235.

Haryu, T. and T. Nishiyama. 1981. Larval form of zaprorid fish, *Zaprora silenus* from the Bering Sea and the Northern Pacific. Jpn. J. Ichthyol. 28: 313–318.

Haury, L. R., J. A. McGowan, and P. H. Weibe. 1978. Patterns and process in the time space scales of plankton distribution, pp. 277–328. In: J. H. Steele (ed.), Spatial pattern in plankton communities. Plenum Press, New York.

Haury, L. R., P. H. Weibe, and S. H. Boyd. 1976. Longhurst-Hardy plankton recorders: their design and use to minimize bias. Deep-Sea Res. 23: 1217–1229.

Hauser, J. W. and M. P. Sissenwine. 1991. The uncertainty in estimates of the production of larval fish derived from samples of larval abundance. ICES J. Mar. Sci. 48: 23–32.

Hay, D. E. 1981. Effects of capture and fixation on gut contents and body size of Pacific herring larvae. Rapp. P.-v. Réun. Cons. Int. Explor. Mer. 178: 395–400.

Heath, M. W. 1992. Field investigations of the early life stages of marine fish. Adv. Mar. Biol. 28: 1–174.

Hempel, G. 1979. Early life history of marine fish: the egg stage. Washington Sea Grant Publication, Seattle.

Hermes, R., N. N. Navaluna, and A. C. del Norte. 1984. A push-net ichthyoplankton sampler attachment to an outrigger boat. Prog. Fish-Cult. 46: 67–70.

Hewitt, R. P. 1988. Historical review of the oceanographic approach to fishery research. CalCOFI Rep. 29: 27–41.

Hewitt, R. P., G. H. Theilacker, and N. C. Lo. 1985. Causes of mortality in young jack mackerel. Mar. Ecol. Prog. Ser. 26: 1–10.

Hill, P. S. and D. P. Demaster. 1998. Alaska marine mammal stock assessments, 1998. U.S. Dep. Commer., NOAA Tech. Memo. NMFS-AFSC-97.

Hill, R. L. and R. L. Creswell. 1998. Using knowledge of microhabitat selection to maximize recruitment to marine fishery reserves (MFR). Proc. Gulf Carib. Fish. Inst. 50: 417–426.

Hinckley, S. 1987. The reproductive biology of walleye pollock, *Theragra chalcogramma*, in the Bering Sea, with reference to spawning stock structure. Fish. Bull. 85: 481–498.

Hinckley, S., A. J. Hermann, K. L. Meir, and B. A. Megrey. 2001. The importance of spawning location and timing to successful transport to nursery areas: a simulation modeling study of Gulf of Alaska walleye pollock. ICES J. Mar. Sci. 58: 1042–1052.

Hislop, J. R. G. 1996. Changes in North Sea gadoid stocks. ICES J. Mar. Sci. 53: 1146–1156.

Hissmann, K., H. Fricke, and J. Schauer 1998. Population monitoring of the coelacanth (*Litimeria chalumnae*). Conservation Biology 12: 759–765.

Hjort, J. 1914. Fluctuations in the great fisheries of northern Europe viewed in the light of biological research. Rapp. P.-v. Réun. Cons. Int. Explor. Mer. 20: 1–228.

Hjort, J. 1926. Fluctuations in the year classes of important food fishes. J. Cons. Int. Explor. Mer. 1: 5–38.

Hodson, R. G., C. R. Bennett, and R. J. Monroe. 1981. Ichthyoplankton samplers for simultaneous replicate samples at surface and bottom. Estuaries 4: 176–184.

Hollowed, A. B., K. M. Bailey, and W. S. Wooster. 1987. Patterns in recruitment of marine fishes in the Northeast Pacific Ocean. Biol. Oceanogr. 5: 99–131.

Holm, E. R. 1990. Effects of density-dependent mortality on the relationship between recruitment and larval settlement. Mar. Ecol. Prog. Ser. 60: 141–146.

Horn, M. H. and L. G. Allen. 1981. Comparison of the structure and function of estuarine and harbor fish communities in southern California. Estuaries 4: 243.

Houde, E. D. 1978. Critical food concentrations for larvae of three species of subtropical marine fishes. Bull. Mar. Sci. 28: 395–411.

Houde, E. D. 1984a. Bregmacerotidae: development and relationships, pp. 300–307. In: H. G. Moser et al. (eds.), Ontogeny and systematics of fishes. Am. Soc. Ichthyol. Herp., Spec. Publ. No. 1.

Houde, E. D. 1984b. Callionymidae: development and relationships, pp. 637–639. In: H. G. Moser et al. (eds.), Ontogeny and systematics of fishes. Am. Soc. Ichthyol. Herp., Spec. Publ. No. 1.

Houde, E. D. 1987. Fish early life dynamics and recruitment variability. Am. Fish. Soc. Symp. 2: 17–29.

Houde, E. D. 1994. Differences between marine and freshwater fish larvae: implications for recruitment. ICES J. Mar. Sci. 51: 91–97.

Houde, E. D., P. B. Ortner, L. Lubbers, and S. R. Cummings. 1989. Test of a cameranet system to determine abundance and heterogeneity in anchovy egg distributions. Rapp. P.-v. Réun. Cons. Int. Explor. Mer. 191: 112–118.

Houde, E. D. and R. C. Schekter. 1980. Feeding by marine fish larvae: development and functional responses. Environ. Biol. Fish. 5: 315–334.

Hourston, A. and C. Haegele. 1980. Herring on Canada's Pacific Coast. Dept. Fisheries and Oceans, Ottawa.

Howell, B. R. 1994. Fitness of hatchery-reared fish for survival in the sea. Aquacult. Fish. Manag. 25 (Suppl. 1) 3–17.

Howell, B. R., E. Moksness, and T. Svåsand (eds.). 1999. Stock enhancement and sea ranching. Blackwell Science. Oxford, UK.

Hubbs, C. L. 1943. Terminology of early stages of fishes. Copeia. 1943: 260.

Hubbs, C. L. and Y. T. Chu. 1934. Asiatic fishes (*Diploprion* and *Laeops*) having a greatly elongated dorsal ray in very large post-larvae. Occas. Pap. Mus. Zool. Univ. Mich. 299: 1–7.

Hunter, J. R. 1972. Swimming and feeding behavior of larval anchovy, *Engraulis mordax*. Fish. Bull. 70: 821–838.

Hunter, J. R. 1981. Feeding ecology and predation of marine fish larvae. In: R. Lasker (ed.), Marine fish larvae. Washington Sea Grant Program, University of Washington Press.

Hunter, J. R. 1984. Synopsis of culture methods for marine fish larvae, pp. 24–27. In: H. G. Moser et al. (eds.), Ontogeny and systematics of fishes. Am. Soc. Ichthyol. Herp., Spec. Publ. No. 1.

Hunter J. R. and N. C. H. Lo. 1993. Ichthyoplankton methods for estimating fish biomass introduction and terminology. Bull. Mar. Sci. 53: 723–727.

Hunter J. R., N. C. H. Lo, and L. A. Fuiman. (eds.) 1993. Ichthyoplankton methods for estimating fish biomass. Bull. Mar. Sci. 53: 723–935.

Hunter, J. R. and B. J. Macewicz. 1985. Measurement of spawning frequency in multiple spawning fishes, pp. 67–78. In: R. Lasker (ed.), An egg production method for estimating spawning biomass of pelagic fishes: application to the northern anchovy, *Engraulis mordax*. U.S. Dep. Commer., NOAA Tech Rep. NMFS 36.

Hunt von Herbing, I. 2002. Effects of temperature on larval fish swimming performance: the importatnce of physics to physiology. J. Fish Biol. 60: 865–876.

Hurst, T. P. and D. O. Conover. 1998. Winter mortality of young-of-the-year Hudson River striped bass (*Morone saxatilis*): size-dependent patterns and effects on recruitment. Can. J. Fish. Aquat. Sci. 55: 1122–1130.

Iles, T. C. 1994. A review of stock-recruitment relationships with reference to flatfish populations. Neth. J. Sea Res. 32: 399–420.

Isaacs, J. D. and L. W. Kidd. 1953. Isaacs-Kidd mid-water trawl. Scripps Inst. Oceanogr. Equipment Report. 1: 1–18.

Iwata, N. and K. Kikuchi. 1998. Effects of rearing conditions on the ambicoloration of cultured Japanese flounder. Aquaculture, '98 Book of Abstracts. p. 265.

Jahn, A. E. 1987. Precision of estimates of abundance of coastal fish larvae. Am. Fish Soc. Symp. 2: 30–38.

Jameson, G. S. and C. O. Levings. 2001. Marine protected areas in Canada implications for both conservation and fisheries management. Can. J. Fish. Aquat. Sci. 58: 138–156.

Jennings, S. 2000. Patterns and prediction of population recovery in marine reserves. Rev. Fish. Biol. Fish. 10: 209–231.

Jesperson, P. and Å. V. Taning. 1934. Introduction to the reports from the Carlsberg Foundation's oceanographical expedition round the world 1928–1930. Dana Rep. 1. Copenhagen.

Johns, D. M. and W. H. Howell. 1980. Yolk utilization in summer flounder (*Paralichthys dentatus*) embryos and larvae reared at two temperatures. Mar. Ecol. Prog. Ser. 2: 1–8.

Johnson, G. D. 1978. Development of fishes of the mid-Atlantic bight, an atlas of egg, larval, and juvenile stages. Volume IV. Carangidae through Ephippidae. U.S. Fish.Wildl. Serv., Biol. Prog. FWS/OBS-78/12.

Johnson, G. D. 1983. *Niphon spinosus*: a primitive epinepheline serranid, with comments on the monophyly and interrelationships of the Serranidae. Copeia 1983: 777–787.

Johnson, G. D. 1988. *Niphon spinosus:* a primitive epinepheline serranid: corroborative evidence from the larvae. Japanese J. Ichthyol. 35: 7–18.

Johnson, G. D. and P. Keener. 1984. Aid to the identification of American grouper larvae. Bull. Mar. Sci. 34: 106–134.

Johnson, M. W., J. R. Rooker, D. M. Gatlin III, and G. J. Holt. 2002. Effects of variable ration levels on direct and indirect measures of growth in juvenile red drum (*Sciaenops ocellatus*). J. Exp. Mar. Biol. Ecol. 274: 141–157.

Johnson, R. and M. Barnett. 1975. An inverse correlation between meristic characters and food supply in mid-water fishes: evidence and possible explanations. Fish. Bull. 73: 284–299.

Jones, C. M. 1986. Determining age of larval fish with the otolith increment technique. Fish. Bull. 84: 91–103.

Jones, P. J. S. 1994. A review and analysis of the objectives of marine nature reserves. Ocean Coast. Manage. 24: 149–178.

Jones, P. W., F. D. Martin, and J. D. Hardy, Jr. 1978. Development of fishes of the mid-Atlantic bight, an atlas of egg, larval, and juvenile stages. Volume I. Acipenseridae through Ictaluridae. U.S. Fish. Wildl. Serv., Biol. Prog. FES/OBS-78/12.

Jones, S. and M. Kumaran. 1962. Notes on eggs, larvae and juveniles of fishes from Indian waters. XII. *Myripristis murdjan*. XIII. *Holocentrus* sp. Indian J. Fish, Sect. A 9(1): 155–167.

Jørstad, K. E., V. Øiestad, O. I. Paulsen, K. Naas, and O. Skaala. 1987. A genetic marker for artificially reared cod (*Gadus morhua* L.). ICES CM 1987/F:22.

Juanes, F. and D. O. Conover. 1995. Size-structured piscivory: advection and the linkage between predator and prey recruitment in young-of-the-year bluefish. Mar. Ecol. Prog. Ser. 128: 287–304.

Kalmer, E. 1992. Early life history of fish: an energetic approach. Chapman & Hall, London.

Kamba, M. 1977. Feeding habits and vertical distribution of walleye pollock, *Theragra chalcogramma* (Pallas), in early life stage in Uchiura Bay, Hokkaido. Res. Inst. N. Pacific Fish., Hokkaido Univ., Spec. Vol., 175–197.

Kaufman, L., J. Ebersole, J. Beets, and C. C. McIvor. 1992. A key phase in the recruitment dynamics of coral reef fishes: post-settlement transition. Environ. Biol. Fishes. 34: 109–118.

Kawaguchi, K. and H. G. Moser. 1984. Stomiatoidea: development, pp. 169–180. In: H. G. Moser et al. (eds.), Ontogeny and systematics of fishes. Am. Soc. Ichthyol. Herp., Spec. Publ. No. 1.

Keene, M. J. and K. A. Tighe. 1984. Beryciformes: development and relationships, pp. 383–392. In: H. G. Moser et al. (eds.), Ontogeny and systematics of fishes. Am. Soc. Ichthyol. Herp., Spec. Publ. No. 1.

Keener, P., G. D. Johnson, B. W. Stender, E. B. Brothers, and H. R. Beatty. 1988. Ingress of post larval gag, *Mycteroperca microlepis* (Pisces: Serranidae), through a South Carolina barrier island inlet. Bull. Mar. Sci. 42: 376–396.

Kellermann, A. (ed.). 1989. Identification key and catalogue of larval Antarctic fishes. Biomass Scientific Ser. No. 10.

Kelso, W. E. and D. A. Rutherford. 1996. Collection, preservation, and identification of fish eggs and larvae, pp. 255–302. In: B. R. Murphy and D. W. Willis (eds.), Fisheries techniques, 2nd ed. Am. Fish. Soc., Bethesda, MD.

Kendall, A. W., Jr. 1976. Predorsal and associated bones in serranid and grammistid fishes. Bull. Mar. Sci. 26: 585–592.

Kendall, A. W., Jr. 1977. Relationships among American serranid fishes based on the morphology of their larvae. Unpublished PhD dissertation. Univ. Calif. San Diego, San Diego, CA.

Kendall, A. W., Jr. 1979. Morphological comparisons of North American sea bass larvae (Pisces: Serranidae). NOAA Tech. Rep. NMFS Circ. 428.

Kendall, A. W., Jr. 1984. Serranidae: development and relationships, pp. 499–510. In: H. G. Moser et al. (eds.), Ontogeny and systematics of fishes. Am. Soc. Ichthyol. Herp., Spec. Publ. No. 1.

Kendall, A. W., Jr., E. H. Ahlstrom, and H. G. Moser. 1984. Early life history stages of fishes and their characters, pp. 11–22. In: H. G. Moser et al. (eds.), Ontogeny and systematics of fishes. Am. Soc. Ichthyol. Herp., Spec. Publ. No. 1.

Kendall, A. W., Jr. and G. J. Duker. 1998. The development of recruitment fisheries oceanography in the United States. Fish. Oceanogr. 7: 69–88.

Kendall, A. W., Jr. and M. P. Fahay. 1979. Larva of the serranid fish *Gonioplectrus hispanus* with comments on its relationships. Bull. Mar. Sci. 29: 117–121.

Kendall, A. W., Jr., L. S. Incze, P. B. Ortner, S. R. Cummings, and P. K. Brown. 1994. The vertical distribution of eggs and larvae of walleye pollock, *Theragra chalcogramma*, in Shelikof Strait, Gulf of Alaska. Fish. Bull. 92: 540–554.

Kendall, A. W., Jr. and A. J. Mearns. 1996. Egg and larval development in relation to systematics of *Novumbra hubbsi*, the Olympic mudminnow. Copeia 1996: 684–695.

Kendall, A. W., Jr., J. D. Schumacher, and S. Kim. 1996. Walleye pollock recruitment in Shelikof Strait: applied fisheries oceanography. Fish. Ocean. 5 (Suppl. 1): 4–18.

Kendall, A. W., Jr., J. D. Schumacher, and S. Kim. (eds.) 1996. Fisheries oceanography of walleye pollock in Shelikof Strait, Alaska. Fish. Oceanogr. 5 (Suppl. 1): 1–203.

Kendall, A. W., Jr. and B. Vinter. 1984. Development of hexagrammids (Pisces: Scorpaeniformes) in the Northeastern Pacific Ocean. NOAA Tech. Rep. NMFS 2.

Kim, S. and C.I. Zhang. 1994. Fish Ecology: Spawning and early life history of fish. Seoul Press.

Kingsford, M.J., I.M. Suthers, and C.A. Gray. 1997. Exposure to sewage plumes and the incidence of deformities in larval fishes. Mar. Pollut. Bull. 33: 201–212.

Klyashtorin, L. and B. Smirnov. 1995. Climate-dependent salmon and sardine stock fluctuations in the North Pacific. Can. J. Fish. Aquat. Sci., Special Publ. 121: 687–689.

Knutsen, G.M. and S. Tilseth. 1985. Growth, development, and feeding success of Atlantic cod larvae *Gadus morhua* related to egg size. Trans. Am. Fish. Soc. 114: 507–511.

Koening, C.C., F.C. Coleman, C.B. Grimes, G.R. Fitzhugh, K.M. Scanlon, C.T. Gledhill, and M. Grace. 2000. Protection of fish spawning habitat for the conservation of warm temperate reef fish fisheries of shelf edge reefs of Florida. Bull. Mar. Sci. 66: 593–616.

Koslow, J.A. 1984. Recruitment patterns in northwest Atlantic fish stocks. Can. J. Fish. Aquat. Sci. 41: 1722–1729.

Koslow, J.A. 1992. Fecundity and the stock-recruitment relationship. Can. J. Fish. Aquat. Sci. 49: 210–217.

Koslow, J.A., S. Brault, J. Dugas, and F. Page. 1985. Anatomy of an apparent year-class failure: the early life history of the 1983 Browns Bank haddock *Melanogrammus aeglefinus*. Trans. Am. Fish. Soc. 114: 478–489.

Koslow, J.A., K.R. Thompson, and W. Silvert. 1987. Recruitment to northwest Atlantic Cod (*Gadus morhua*) and haddock (*Melanogrammus aeglefinus*) stocks: influence of stock size and climate. Can. J. Fish. Aquat. Sci. 44: 26–39.

Kotthaus, A. 1970. *Flagelloserranus* a new genus of serranid fishes with the description of two new species (Pisces, Percomorphi). Dana-Rep. Carlsberg Found. 78.

Kramer, D. and E.H. Ahlstrom. 1968. Distributional atlas of fish larvae of the California Current region: northern anchovy *Engraulis mordax* (Girard), 1951 through 1965. CalCOFI Atlas 9.

Kramer, D.L. and M.R. Chapman. 1999. Implications of fish home range size and relocation for marine reserve function. Environ. Biol. Fish. 55: 65–79.

Kuntz, Y. 2004. Developmental biology of teleost fishes. Springer, The Netherlands.

La Bolle, L.D., Jr., H.W. Li, and B.C. Mundy. 1985. Comparison of two samplers for quantitatively collecting larval fishes in upper littoral habitats. J. Fish. Biol. 26: 139–146.

Landry, F., T.J. Miller, and W.C. Leggett. 1995. The effects of small-scale turbulence on the ingestion rate of fathead minnow (*Pimephales promelas*). Can. J. Fish. Aquat. Sci. 52: 1714–1719.

Laprise, R. and P. Pepin. 1995. Factors influencing the spatio-temporal occurrence of fish eggs and larvae in a northern, physically dynamic coastal environment. Mar. Ecol. Prog. Ser 122: 1–3.

Laroche, J. and S. Richardson 1983. Reproduction of northern anchovy, *Engraulis mordax*, off Oregon and Washington, USA. Fish. Bull. 78: 603–618.

Laroche, W. A. 1977. Description of larval and early juvenile vermilion snapper, *Rhomboplites aurorubens*. Fish. Bull. 75: 547–554.

Lasker, R. 1965. Symposium on larval fish biology. Preface. CalCOFI Rep. 10: 12.

Lasker, R. 1975. Field criteria for survival of anchovy larvae: the relation between inshore chlorophyll maximum layers and successful first feeding. Fish. Bull. 73: 453–462.

Lasker, R. 1981. Factors contributing to variable recruitment of the northern anchovy (*Engraulis mordax*) in the California Current: contrasting years 1975 through 1978. Rapp. P.-v. Reun. Cons. Int. Explor. Mer. 178: 375–388.

Lasker, R. 1987. Use of fish eggs and larvae in probing some major problems in fisheries and aquaculture. Am. Fish. Soc. Symp. 2: 1–16.

Lasker, R. (ed.). 1985. An egg production method for estimating spawning biomass of pelagic fish: application to the northern anchovy, *Engraulis mordax*. NOAA Tech. Rep. NMFS 36: 1–99.

Lasker, R., H. M. Feder, G. H. Theilacker, and R. C. May. 1970. Feeding, growth, and survival of (*Engraulis mordax*) larvae reared in the laboratory. Mar. Biol. 5: 345–353.

Lasker, R. and K. Sherman (eds.). 1981. The early life history of fish: recent studies. Rapp. P.-v. Réun. Cons. Int. Explor. Mer., 178.

Last, J. M. 1978a. The food of four species of pleuronectiform larvae from the English Channel and southern North Sea. Mar. Biol. 45: 359–368.

Last, J. M. 1978b. The food of three species of gadoid larvae in the English Channel and southern North Sea. Mar. Biol. 48: 377–386.

Last, J. M. 1980. The food of twenty species of fish larvae in the west-central North Sea. Fish. Res. Tech. Pap. MAFF Direct. Fish Res. Lowestoft, 60.

Lauck, T., C. W. Clark, M. Mangel, and G. R. Munro. 1998. Implementing the precautionary principle in fisheries management through marine reserves. Ecol. Appl. Suppl. 8: S72–S78.

Laurence, G. C. 1974. Growth and survival of haddock (*Melanogrammus aeglefinus*) larvae in relation to planktonic prey concentration. J. Fish. Res. Bd. Canada. 31: 1415–1419.

Laurence, G. C. and W. H. Howell. 1981. Embryology and influence of temperature and salinity on early development and survival of yellowtail flounder *Limanda ferruginea*. Mar. Ecol. Prog. Ser. 6: 11–18.

Leggett, W. C. and E. Deblois. 1994. Recruitment in marine fishes: is it regulated by starvation and predation in the egg and larval stages? Neth. J. Sea Res. 32: 119–134.

Lehtinen, K. J., K. Mattsson, J. Tana, C. Engstrom, O. Lerche, and J. Hemming. 1999. Effects of wood-related sterols on the reproduction, egg survival, and offspring of brown trout (*Salmo trutta lacustris* L.). Ecotoxicol. Environ. Safety 42: 40–49.

Leis, J. M. 1977. Development of the eggs and larvae of the slender mola, *Ranzania laevis* (Pisces: Molidae). Bull. Mar. Sci. 27: 448–466.

Leis, J. M. 1986. Larval development in four species of Indo-Pacific coral trout *Plectropomus* (Pisces: Serranidae: Epinephelinae) with an analysis of the relationships of the genus. Bull. Mar. Sci. 38: 525–552.

Leis, J. M. and B. M. Carson-Ewart. (eds.) 2000. The larvae of Indo-Pacific coastal fishes. Brill. Leiden.

Leis, J. M., J. E. Olney, and M. Okiyama. 1997. Introduction to the proceedings of the symposium fish larvae and systematics: ontogeny and relationships. Bull. Mar. Sci. 60: 1–5.

Leis, J. M. and D. S. Rennis. 1983. The larvae of Indo-Pacific Coral reef fishes. New South Wales University Press, Sydney.

Leis, J. M. and W. J. Richards. 1984. Acanthuroidei: development and relationships, pp. 547–550. In: H. G. Moser et al. (eds.), Ontogeny and systematics of fishes. Am. Soc. Ichthyol. Herp., Spec. Publ. No. 1.

Leis, J. M. and T. Trnski. 1989. The larvae of Indo-Pacific shorefishes. University of Hawaii Press, Honolulu.

Leithiser, R. L., K. F. Ehrlich, and A. B. Thum. 1979. Comparison of a high volume pump and conventional plankton nets for collecting fish larvae entrained in power plant cooling systems. Can. J. Fish. Aquat. Sci. 36: 81–84.

Leslie, J. K. and J. R. M. Kelso. 1977. Influence of a pulp and paper mill effluent on aspects of distribution, survival and feeding of Nipigon Bay, Lake Superior, larval fish. Bull. Environ. Contam. Toxicol. 18: 602–610.

Levin, P. 1996. Recruitment in a temperate demersal fish: does larval supply matter? Limnol. Oceanogr. 41: 672–679.

Lewis, R. M., W. F. Hettler, Jr., E. P. H. Wilkins, and G. N. Johnson. 1970. A channel net for catching larval fishes. Chesapeake Sci. 11: 196–197.

Lindeman, K. C., R. Pugliese, G. T. Waugh, and J. S. Ault. 2000. Developmental patterns within a multispecies reef fishery: management applications for essential fish habitats and protected areas. Bull. Mar. Sci. 66: 929–956.

Lindholm, J. B., P. J. Auster, M. Ruth, and L. Kaufman. 2001. Modeling the effects of fishing and implications for the design of marine protected areas: juvenile fish responses to variations in seafloor habitat. Conserv. Biol. 15: 424–437.

Lindquist, A. 1970. The distribution of different types of eggs in the various zones of the Skagerak. Fig. 19, p. 34. In: G. Hempel, 1979, Early life history of marine fish: the egg stage. Washington Sea Grant Publication.

Litvak, M. K. and W. C. Leggett. 1992. Age and size-selective predation on larval fishes: the bigger-is-better hypothesis revisited. Mar. Ecol. Prog. Ser. 81: 13–24.

Liu, H. W., R. R. Stickney, W. W. Dickhoff, and D. A. McCaughran. 1994. Effects of environmental factors on egg development and hatching of Pacific halibut *Hippoglossus stenolepis*. J. World Aquacult. Soc. 25: 317–321.

Lluch-Belda, D., R. A. Schwartzlose, R. Serra, R. Parrish, T. Kawasaki, D. Hedgecock, and R. J. M. Crawford. 1992. Sardine and anchovy regime fluctuations of abundance in four regions of the world oceans: a workshop report. Fish. Oceanogr. 1: 339–347.

Lo, N. C. H. 1985. Egg production of the central stock of northern anchovy, *Engraulis mordax*, 1951-82. Fish. Bull., U.S. 83: 137–150.

Lo, N.C.H. 1986. Modelling life-stage-specific instantaneous mortality rates, and application to northern anchovy, *Engraulis mordax*, eggs and larvae. Fish. Bull., U.S. 84: 395–407.

Lo, N.C.H., J.R. Hunter, H.G. Moser, P.E. Smith, and R.D. Methot. 1992. A daily fecundity reduction method: a new procedure for estimating adult biomass. ICES J. Mar. Sci. 49: 209–215.

Longhurst, A. 2002. Murphy's law revisited: longevity as a factor in recruitment to fish populations. Fish. Res. 56: 125–131.

Longhurst, A.R., A.D. Reith, R.E. Bower, and D.L.R. Seibert. 1966. A new system for the collection of multiple serial plankton samples. Deep-Sea Res. 13: 213–222.

Lough, R.G. and D.G. Mountain. 1996. Effect of small-scale turbulence on feeding rates of larval cod and haddock in stratified water on Georges Bank. Deep-Sea Res. 43: 1745–1772.

Love, M.S., M. Yoklavich, and L. Thorsteinson. 2002. The rockfishes of the northeast Pacific. Univ. California Press, Berkeley.

Mabee, P.M. 1988. Supraneural and predorsal bones in fishes: Development and homologies. Copeia 1988: 827–838.

Mabee, P.M. 1989. Assumptions underlying the use of ontogenetic sequences for determining character state order. Trans. Am. Fish. Soc. 118: 151–158.

MacCall, A.D. 1990. Dynamic geography of marine fish population. Washington Sea Grant, Univ. Washington Press, Seattle.

MacKenzie, B.R. and T. Kiørboe. 1995. Encounter rates and swimming behavior of pause-travel and cruise larval fish predators in calm and turbulent laboratory environments. Limnol. Oceanogr. 40: 1278–1289.

MacKenzie, B.R., W.C. Leggett, and R.H. Peters. 1990. Estimating larval fish ingestion rates: can laboratory derived values be reliably extrapolated to the wild? Mar. Ecol. Prog. Ser. 67: 209–225.

MacKenzie, B.R., T.J. Miller, S. Cyr, and W.C. Leggett. 1994. Evidence for a dome-shaped relationship between turbulence and larval fish ingestion rates. Limnol. Oceanogr. 39: 1790–1799.

Madenjian, C.P. and D.J. Jude. 1985. Comparison of sleds versus plankton nets for sampling fish larvae and eggs. Hydrobiologica. 124: 275–281.

Man, A., R. Law, and N.V.C. Polunin. 1995. Role of marine reserves in recruitment to reef fisheries: a metapopulation model. Biol. Conserv. 71: 197–204.

Mangor-Jensen, A. 1986. Osmoregulation in eggs and larvae of cod (*Gadus morhua* L.): basic studies and effects of oil exposure. Univ. Bergen Inst. of Mar. Res. Rep., H.J. Fyhn ed.: 117–166.

Mantua, N.J., S.R. Hare, Y. Zhang, J.M. Wallace, and R.C. Francis. 1997. A Pacific interdecadal climate oscillation with impacts on salmon production. Bull. Am. Meteorol. Soc. 78: 1069–1079.

Marcy, B.C., Jr. and M.D. Dahlberg. 1980. Sampling problems associated with ichthyoplankton field-monitoring studies with emphasis on entrainment, pp. 233–252.

In: C. H. Hocutt and J. R. Stauffer, Jr. (eds.), Biological monitoring of fish. Lexington Books, Lexington, MA.

Margulies, D. 1989. Effects of food concentrations and temperature on development, growth, and survival of white perch (*Morone americana*) eggs and larvae. Fish. Bull. 87: 63–72.

Marliave, J. 1975. Seasonal shifts in the spawning site of a Northeast Pacific intertidal fish. J. Fish. Res. Bd. Canada 32: 1687–1691.

Marliave, J. B. 1986. Lack of planktonic dispersal of rocky intertidal fish larvae. Trans. Am. Fish. Soc. 115: 149–154.

Marr, J. C. 1960. The causes of major variations in the catch of the Pacific sardine *Sardinops caerulea* (Girard), pp. 667–791. In: H. Rosa, Jr. and G. Murphy (eds.), Proceedings of the world scientific meeting on the biology of sardines and related species. 3. FAO.

Marsden, J. E., C. C. Krueger, and H. M. Hawkins. 1991. An improved trap for passive capture of demersal eggs during spawning: an efficiency comparison with egg nets. N. Am. J. Fish. Manag. 11: 364–368.

Martell, S. J. D., C. J. Walters, and S. S. Wallace. 2000. The use of marine protected areas for conservation of lingcod (*Ophiodon elongatus*). Bull. Mar. Sci. 66: 729–743.

Martin, F. D. and G. E. Drewry. 1978. Development of fishes of the mid-Atlantic bight, an atlas of egg, larval, and juvenile stages. Volume VI. Stromateidae through Ogcocephalidae. U.S. Fish. Wildl. Serv., Biol. Prog. FES/OBS-78/12.

Masuda, R. and K. Tsukamoto. 1998. Stock enhancement in Japan: review and perspective. Bull. Mar. Sci. 62: 337–358.

Matarese, A. C., A. W. Kendall, Jr., D. M. Blood, and B. M. Vinter. 1989. Laboratory guide to early life history stages of Northeast Pacific fishes. NOAA Tech. Rep. NMFS 80.

Matarese, A. C., S. L. Richardson, and J. R. Dunn. 1981. Larval development of Pacific tomcod, *Microgadus proximus*, in the northeast Pacific Ocean with comparative notes on larva of walleye pollock, *Theragra chalcogramma*, and Pacific cod, *Gadus macrocephalus* (Gadidae). Fish. Bull. U.S. 78: 923–940.

Matarese, A. C. and E. M. Sandknop. 1984. Identification of fish eggs, pp. 27–30. In: H. G. Moser et al. (eds.), Ontogeny and systematics of fishes. Am. Soc. Ichthyol. Herp., Spec. Publ. No. 1.

Mathias, J. A., and S. Li. 1982. Feeding habits of walleye larvae and juveniles: comparative laboratory and field studies. Trans. Am. Fish. Soc. 111: 722–735.

May, R. C. 1974. Larval mortality in marine fishes and the critical period concept, pp. 3–19. In: J. H. S. Blaxter (ed.), The early life history of fish. Springer-Verlag, New York.

McDermott, S. F. and S. A. Lowe. 1997. The reproductive cycle and sexual maturity of Atka mackerel, *Pleurogrammus monopterygius*, in Alaskan Waters. Fish. Bull. 95: 321–333.

McEachron, L. W., R. L. Colura, B. W. Bumguardner, and R. Ward. 1998. Survival of stocked red drum in Texas. Bull. Mar. Sci. 62: 359–368.

McEvoy, L. A. and J. McEvoy. 1991. Size fluctuation in the eggs and newly hatched larvae of captive turbot (*Scophthalmus maximus*). J. Mar. Biol. Assoc. U.K. 71: 679–690.

McGowan, J. A. and Brown, D. M. 1966. A new opening-closing paired zooplankton net. SIO Ref. Pap. (66–23).

McGurk, M. S. 1984. Effects of delayed feeding and temperature on the age of irreversible starvation and on the rates of growth and mortality of Pacific herring larvae. Mar. Biol. 84: 13–26.

McClain, C. R., M. F. Fougerolle, M. A. Rex, and J. Welch. 2001. MOCNESS estimates of the size and abundance of a pelagic gonostomatid fish *Cyclothone pallida* off the Bahamas. J. Mar. Biol. Assoc. U.K. 81: 869–871.

Meador, M. W. and J. S. Bulak. 1987. Quantifiable ichthyoplankton sampling in congested shallow-water areas. J. Freshw. Ecol. 4: 65–69.

Megrey, B. A., A. B. Hollowed, S. R. Hare, S. A. Macklin, and P. J. Stabeno. 1996. Contributions of FOCI research to forecasts of year-class strength of walleye pollock in Shelikof Strait, Alaska. Fish. Oceanogr. 5 (Suppl. 1) 189–203.

Merrett, N. R. and S. H. Barnes. 1996. Preliminary survey of egg envelope morphology in the Macrouridae and the possible implications of its ornamentation. J. Fish. Biol. 48: 101.

Mertz, G. and R. A. Myers. 1994a. Match/mismatch predictions of spawning duration versus recruitment variability. Fish. Oceanogr. 3: 236–245.

Mertz, G. and R. A. Myers. 1994b. The ecological impact of the Great Salinity Anomaly in the northern North-West Atlantic. Fish. Oceanogr. 3: 1–14.

Messieh, S. N. and H. Rosenthal. 1989. Mass mortality of herring eggs on spawning beds on and near Fisherman's Bank, Gulf of St. Lawrence Canada. Aquat. Liv. Res. 2: 1–8.

Methot, R. D., Jr. 1983. Seasonal variation in survival of larval northern anchovy, (*Engraulis mordax*), estimated from the age distribution of juveniles. Fish. Bull. 81: 741–750.

Methot, R. D. 1986. Frame trawl for sampling pelagic juvenile fish. CalCOFI Rep. 27: 267–278.

Methot, R. D. and N. C. H. Lo. 1987. Spawning biomass of the northern anchovy in 1987. SWFC Admin. Rep., La Jolla. LJ-87-14.

Meyers, R. A., J. Bridson, and J. Barrowman. 1995. Summary of world-wide spawner and recruitment data. Can. Tech. Rep. Fish. Aquat. Sci. 2024.

Milicich, M. J. and P. J. Doherty. 1994. Larval supply of coral reef fish populations: magnitude and synchrony of replenishment to Lizard Island, Great Barrier Reef. Mar. Ecol. Prog. Ser. 110: 121–134.

Miller, A. J., D. R. Cayan, T. P. Barnett, N. E. Graham, J. M. Oberhuber. 1994. The 1976–77 climate shift in the Pacific Ocean. Oceanography 7: 21–26.

Miller, D. 1961. A modification of the small Hardy plankton sampler for simultaneous high-speed plankton hauls. Bull. Mar. Ecol. 5: 165–172.

Miller, J. M. 1973. A quantitative push-net system for transect studies of larval fish and macrozooplankton. Limnol. Oceanogr. 18: 175–178.

Miller, J. M. 1974. Nearshore distribution of Hawaiian marine fish larvae: effects of water quality, turbidity and currents, pp. 217–231. In: J. H. S. Blaxter (ed.), The early life history of fish. Springer-Verlag, Berlin.

Miller, J. M., W. Watson, and J. M. Leis. 1979. An atlas of common nearshore marine fish larvae of the Hawaiian Islands. Sea Grant Misc. Rep. UNIHI-Seagrant-MR-80-02.

Miller, T. J., L. B. Crowder, J. A. Rice, and E. A. Marschall. 1988. Larval size and recruitment mechanisms in fishes: toward a conceptual framework. Can. J. Fish. Aquat. Sci 45: 1657–1670.

Misitano, D. A., E. Casillas, and C. R. Haley. 1994. Effects of contaminated sediments on viability, length, DNA and protein content of larval surf smelt, *Hypomesus pretiosus*. Mar. Environ. Res. 37: 1–21.

Monteleone, D. M. and L. E. Duguay. 1988. Laboratory studies of predation by the ctenophore *Mnemiopsis leidyi* on the early stages in the life history of the bay anchovy, *Anchoa mitchilli*. J. Plankton Res. 10: 359–372.

Morgan, R. P. 1975. Distinguishing larval white perch and striped bass by electrophoresis. Chesapeake Sci. 16: 68–70.

Morgan, R. P., Jr. and R. D. Prince. 1978. Chlorine effects on larval development of striped bass (*Morone saxatilis*), white perch (*M. americana*) and blueback herring (*Alosa aestivalis*). Trans. Am. Fish. Soc. 107: 636–641.

Mork, J., P. Solemdal, and G. Sundnes. 1983. Identification of marine fish eggs: a biochemical approach. Can. J. Fish. Aquat. Sci. 40: 361–369.

Morse, W. W. 1989. Catchability, growth, and mortality of larval fishes. Fish. Bull. 87: 417–446.

Moser, H. G. 1967. Reproduction and development of *Sebastodes paucispinus* and comparison with other rockfishes off southern California. Copeia 1967: 773–797.

Moser, H. G. (ed.). 1996. The early stages of fishes of the California Current region. CalCOFI Atlas 33.

Moser, H. G. and E. H. Ahlstrom. 1970. Development of lanternfishes (family Myctophidae) in the California Current. Part I. Species with narrow-eyed larvae. Bull. Los Ang. Cty. Mus. Nat. Hist. 7.

Moser, H. G. and E. H. Ahlstrom. 1974. The role of larval stages in systematic investigations of marine teleosts: the Myctophidae, a case study. Fish. Bull. 72: 391–413.

Moser, H. G., E. H. Ahlstrom, D. Kramer, and E. G. Stevens. 1974. Distribution and abundance of fish eggs and larvae in the Gulf of California. CalCOFI Rep. 17: 112–128.

Moser, H. G., E. H. Ahlstrom, and J. R. Paxton. 1984b. Myctophidae: development, pp. 218–239. In: H. G. Moser et al. (eds.), Ontogeny and systematics of fishes. Am. Soc. Ichthyol. Herp., Spec. Publ. No. 1.

Moser, H. G., E. H. Ahlstrom, and E. M. Sandknop. 1977. Guide to the identification of scorpionfish larvae (family Scorpaenidae) in the eastern Pacific with comparative notes on species of *Sebastes* and *Helicolenus* from other oceans. NOAA Tech. Rep., NMFS Cir. 402.

Moser, H. G., R. L. Charter, P. E. Smith, D. A. Ambrose, S. R. Charter, C. A. Myers, E. M. Sandknop, and W. Watson. 1993. Distributional atlas of fish larvae and eggs in the California Current region: taxa with 1000 or more total larvae, 1951 through 1984. CalCOFI Atlas 31.

Moser, H. G., R. L. Charter, P. E. Smith, D. A. Ambrose, S. R. Charter, C. A. Myers, E. M. Sandknop and W. Watson. 1994. Distributional atlas of fish larvae and eggs in the California Current region: Taxa with less than 1000 total larvae, 1951 through 1984. CalCOFI Atlas. 32.

Moser, H. G., W. J. Richards, D. M. Cohen, M. P. Fahay, A.W. Kendall, Jr., and S. L. Richardson (eds.). 1984a. Ontogeny and systematics of fishes. Am. Soc. Ichthyol. Herp., Spec. Publ. No. 1.

Moser, H. G. and P. E. Smith. 1993. Larval fish assemblages and oceanic boundaries. Bull. Mar. Sci. 53: 283–289.

Moser, H. G. and W. Watson. 2006. Ichthyoplankton. In: L. Allen, D. Pondella II, and M. Horn (eds.), The ecology of marine fishes: California and adjacent waters. University of California Press, Berkeley.

Moser et al. (eds.). 1984a. www.biodiversitylibrary.org/bibliography/4334.

Mosquera, I., I. M. Cote, S. Jennings, and J. D. Reynolds. 2000. Conservation benefits of marine reserves for fish populations. Anim. Conserv. 3: 321–332.

Moulton, L. L. 1977. An ecological analysis of fishes inhabiting the rocky nearshore regions of northern Puget Sound, Washington. Ph.D. Dissertation, University of Washington, Seattle, WA.

Müller, U. K. and J. J. Videler. 1996. Inertia as a "safe harbour": do fish larvae increase length growth to escape viscous drag? Rev. Fish. Biol. Fish. 6: 353–360.

Mulligan, T. J., F. D. Martin, R. A. Smucker, and D. A. Wright. 1987. A method of stock identification based on the elemental composition of striped bass *Morone saxatilis* (Walbaum) otoliths. J. Exp. Mar. Biol. Ecol. 114: 241–248.

Mullin, M. M. 1993. Webs and scales: physical and ecological processes in marine fish recruitment. WSG-IS 93-01. Washington Sea Grant, Univ. Washington Press, Seattle.

Munk, P. 1988. Catching large herring larvae: gear applicability and larval distribution. J. Cons. Int. Explor. Mer. 45: 97–114.

Munk, P. and T. Kiørboe. 1985. Feeding behavior and swimming activity of larval herring (*Clupea harengus*) in relation to density of copepod nauplii. Mar. Ecol. Prog. Ser. 24: 15–21.

Munk, P., T. Kiørboe, and V. Christensen. 1989. Vertical migrations of herring, *Clupea harengus*, larvae in relation to light and prey distribution. Environ. Biol. Fish. 26: 87–96.

Munk, P., and J. G. Nielsen. 2005. Eggs and larvae of North Sea fishes. Biofolia Press, Fredericksberg, Denmark.

Murawski, S. A., R. Brown, H. L. Lai, P. J. Rago, and L. Hindrickson. 2000. Large scale closed areas as a fishery management tool in temperate marine systems: the Georges Bank experience. Bull. Mar. Sci. 66: 775–798.

Murphy, G. I. 1961. Oceanography and variations in the Pacific sardine population. CalCOFI Rep. 8: 55–64.

Myers, R. A. 2001. Stock and recruitment: generalizations about maximum reproductive rate, density dependence, and variability using meta-analysis approaches. ICES J. Mar. Sci. 58: 937–951.

Myers, R. A. and N. J. Barrowman. 1996. Is fish recruitment related to spawner abundance? Fish. Bull. 94: 707–724.

Myers, R. A. and N. G. Cadigan. 1993. Density-dependent juvenile mortality in marine demersal fish. Can. J. Fish. Aquat. Sci. 50: 1576–1590.

Myers, R. A., G. Mertz, and N. J. Barrowman. 1995. Spatial scales of variability in cod recruitment in the North Atlantic. Can. J. Fish. Aquat. Sci. 52: 1849–1862.

Myers, R. A. and P. Pepin. 1994. Recruitment variability and oceanographic stability. Fish. Oceanogr. 3: 246–255.

Nagahama, Y., M. Yoshikuni, M. Yamashita, T. Tokumoto, and Y. Katsu. 1995. Regulation of oocyte growth and maturation in fish. Current Topics in Developmental Biology 30: 103–145.

National Research Council (NRC). 2001. Marine protected areas: tools for sustaining ocean ecosystems. Nat. Acad. Press, Washington, D.C.

Nedreaas, K. and G. Naevdal. 1991. Identification of 0- and 1-group redfish (genus *Sebastes*) using electrophoresis. ICES J. Mar. Sci. 48: 91.

Neilson, J. D. and R. I. Perry. 1990. Diel vertical migrations in marine fishes: an obligate or facultative process? Adv. Mar. Biol. 26: 115–168.

Neira, F. J., A. G. Miskiewicz, and T. Trnski. 1998. Larvae of temperate Australian fishes. University of Western Australia Press, Nedlands.

Nelson, J. S. 1994. Fishes of the world, 3rd ed. John Wiley & Sons, New York.

Nichol, D. G. and E. I. Acuna. 2001. Annual and batch fecundities of yellowfin sole, *Limanda aspera*, in the eastern Bering Sea. Fish. Bull. 99: 108–122.

Nissling, A. and L. Vallin. 1996. The ability of Baltic cod eggs to maintain neutral buoyancy and the opportunity for survival in fluctuating conditions in the Baltic Sea. J. Fish. Biol. 48: 217–227.

Noble, R. L. 1970. Evaluation of the Miller high-speed sampler for sampling yellow perch and walleye fry. J. Fish. Res. Bd. Can. 27: 1033–1044.

Norcross, B. L. and M. Frandsen. 1996. Distribution and abundance of larval fishes in Prince William Sound, Alaska, during 1989 after the Exxon Valdez oil spill. Proc. Exxon Valdez Oil Symposium (S. D. Rice, R. B. Spies, D. A. Wolfe, and B. A. Wright, eds.) 18: 463–486.

Nordeide, J. T., J. H. Fossa, A. G. V. Salvanes, and O. M. Smedstad. 1994. Testing if year-class strength of coastal cod, *Gadus morhua* L., can be determined at the juvenile stage. Aquacult. Fish. Manage. 25: 101–116.

Nordeide, J. T., J. C. Holm, H. Otteraa, G. Blom, and A. Borge. 1992. Use of oxytetracycline as a marker for juvenile cod (*Gadus morhua* L.). J. Fish. Biol. 41: 21–30.

O'Connell, C. P. 1976. Histological criteria for diagnosing the starving condition in early post yolk sac larvae of the northern anchovy, *Engraulis mordax* Girard. J. Exp. Mar. Biol. Ecol. 25: 285–312.

O'Connell, C. P. and L. P. Raymond. 1970. The effect of food density on survival and growth of early post yolk-sac larvae of northern anchovy (*Engraulis mordax* Girard) in the laboratory. J. Exp. Mar. Biol. Ecol. 5: 187–197.

Øiestad, V. and E. Moksness. 1981. Study of growth and survival of herring larvae (*Clupea harengus* L.) using plastic bag and concrete basin enclosures. Rapp. Proc. Verb. 178: 144–149.

Okiyama, M. 1984. Myctophiformes: development, pp. 206–217. In: H. G. Moser et al. (eds.), Ontogeny and systematics of fishes. Am. Soc. Ichthyol. Herp., Spec. Publ. No. 1.

Okiyama, M. (ed.). 1988. An atlas of the early stage fishes in Japan. Tokai University Press, Tokyo.

Olivar, M. P. and J. M. Fortuño. 1991. Guide to ichthyoplankton of the southeast Atlantic (Benguela current region). Sci. Mar. 55 (1): 1–383.

Olla, B. L., M. Davies, and C. H. Ryer. 1998. Understanding how the hatchery environment represses or promotes the development of behavioral survival skills. Bull. Mar. Sci. 62: 531–550.

Olney, J. E. and G. W. Boehlert. 1988. Nearshore ichthyoplankton associated with seagrass beds in the lower Chesapeake Bay. Mar. Ecol. Prog. Ser. 45: 33–43.

Oozeki, Y. and K. M. Bailey. 1995. Ontogenetic development of digestive enzyme activities in larval walleye pollock, *Theragra chalcogramma*. Mar. Biol. 122: 177–186.

Orton, G. L. 1953. The systematics of vertebrate larvae. Syst. Zool. 2: 63–75.

Ozawa, T. 1986. Studies on the oceanic ichthyoplankton in the western North Pacific. Kyushu Univ. Press.

Palsson, W. 2001. Marine refuges offer haven for Puget Sound fish. http://www.wa.gov/wdfw/science/current/marine_sanctuary.html.

Palsson, W. A. 1984. Egg mortality upon natural and artificial substrata within Washington state spawning grounds of Pacific herring (*Clupea harengus pallasi*). University of Washington M. S. Thesis.

Palsson, W. A. 1998. Monitoring the response of rockfishes to protected areas, pp. 64–73. In: M. M. Yoklavich (ed.), Marine harvest refugia for west coast rockfish: a workshop. NOAA TM NMFS SWFSC 255.

Paul, A. J. 1983. Light, temperature, nauplii concentrations, and prey capture by first feeding pollock larvae *Theragra chalcogramma*. Mar. Ecol. Prog. Ser. 13: 175–179.

Paxton, J. R., E. H. Ahlstrom, and H. G. Moser. 1984. Myctophidae: relationships, pp. 239–244. In: H. G. Moser et al. (eds.), Ontogeny and systematics of fishes. Am. Soc. Ichthyol. Herp., Spec. Publ. No. 1.

Pearson K. E. and D. R. Gunderson. 2003. Reproductive biology and ecology of shortspine thornyhead rockfish, *Sebastolobus alascanus*, and longspine thornyhead rockfish, *S. altivelis*, from the northeastern Pacific Ocean. Environ. Biol. Fish. 67: 117–136.

Pepin, P. 1990. Biological correlates of recruitment variability in North Sea fish stocks. ICES J. Mar. Sci. 47: 89–98.

Pepin, P. and R. A. Myers. 1991. Significance of egg and larval size to recruitment variability of temperate marine fish. Can. J. Fish. Aquat. Sci. 48: 1820–1828.

Pepin, P., S. Pearre, and J. A. Koslow. 1987. Predation on larval fish by Atlantic mackerel, *Scomber scombrus*, with a comparison of predation by zooplankton. Can. J. Fish. Aquat. Sci. 44: 2012–2018.

Pepin, P. and T. H. Shears. 1995. Influence of body size and alternate prey abundance on the risk of predation to fish larvae. Mar. Ecol. Prog. Ser. 128: 279–285.

Peterson, C. H. 2001. The "Exxon Valdez" oil spill in Alaska: acute, indirect and chronic effects on the ecosystem. Adv. Mar. Biol. 39: 1–103.

Phillips, A. C. and J. C. Mason. 1986. A towed, self-adjusting sled sampler for demersal fish eggs and larvae. Fish. Res. 4: 235–242.

Picquelle, S. J. and B. A. Megrey. 1993. A preliminary spawning biomass estimate of walleye pollock, *Theragra chalcogramma*, in the Shelikof Strait, Alaska, based on the annual egg production method. Bull. Mar. Sci. 53: 728–749.

Pinder, A. C. 2001. Keys to larval and juvenile stages of coarse fishes from fresh waters in the British Isles. Freshwater Biological Association. The Ferry House, Far Sawrey, Ambleside, Cumbria, UK. Scientific Publication No. 60.

Pipe, R. P., S. H. Coombs, and K. F. Clarke. 1981. On the sample validity of the Longhurst-Hardy Plankton Recorder for fish eggs and larvae. J. Plankton Res. 3: 675–684.

Planes, S., R. Galzin, A. Garcia Rubies, R. Goñi, J.-G. Harmelin, L. Le Diréach, P. Lenfant, and A. Quetglas. 2000. Effects of marine protected areas on recruitment processes with special reference to Mediterranean littoral ecosystems. Environ. Conserv. 27: 126–143.

Polacheck, T., D. Mountain, D. McMillan, W. Smith, and P. Berrien. 1992. Recruitment of the 1987 year class of Georges Bank haddock (*Melanogrammus aeglefinus*): the influence of unusual larval transport. Can. J. Fish. Aquat. Sci. 49: 484–496.

Polovina, J. J., G. T. Mitchum, N. E. Graham, M. P. Craig, E. E. Demartini, and E. N. Flint. 1994. Physical and biological consequences of a climate event in the central North Pacific. Fish. Oceanogr. 3: 15–21.

Porter, S. M. and G. H. Theilacker. 1996. Larval walleye pollock, *Theragra chalcogramma*, rearing techniques used at the Alaska Fisheries Science Center, Washington. AFSC Proc. Rep. 96-06.

Posgay, J. A. and R. R. Marak. 1980. The MARMAP bongo zooplankton samplers. J. Northwest Atl. Fish. Sci. 1: 91–99.

Potthoff, T. 1984. Clearing and staining techniques, pp. 35–37. In: H. G. Moser et al. (eds.), Ontogeny and systematics of fishes. Am. Soc. Ichthyol. Herp., Spec. Publ. No. 1.

Potthoff, T., W. J. Richards, and S. Ueyanagi. 1980. Development of *Scombrolabrax heterolepis* (Pisces, Scombrolabracidae) and comments on familial relationships. Bull. Mar. Sci. 30: 329–357.

Powell, P. 1972. Oscar Elton Sette: fishery biologist. Fish. Bull. 70: 525–535.

Powell, P. P. 1982. Personalities in California fishery research. CalCOFI Rep. 23: 43–47.

Powles, H. and D. F. Markle. 1984. Identification of larvae, pp. 31–33. In: H. G. Moser et al. (eds.), Ontogeny and systematics of fishes. Am. Soc. Ichthyol. Herp., Spec. Publ. No. 1.

Pryor-Connaughton, V., C. E. Epifanio, and R. Thomas. 1994. Effects of varying irradiance on feeding in larval weakfish (*Cynoscion regalis*). J. Exp. Mar. Biol. Ecol. 180: 151–163.

Purcell, J. E. 1985. Predation on fish larvae by pelagic cnidarians and ctenophores. Bull. Mar. Sci. 37: 739–755.

Purcell, J. E., T. D. Siferd, and J. B. Marliave. 1987. Vulnerability of larval herring (*Clupea harengus pallasi*) to capture by the jellyfish *Aequorea victoria*. Mar. Biol. 94: 157–162.

Quantz, G. 1985. Use of endogenous energy sources by larval turbot *Scopthalmus maximus*. Trans. Am. Fish. Soc. 114: 558–563.

Radovich, J. 1960. Some causes of fluctuations in catches of the Pacific sardine *Sardinops caerulea* (Girard), pp. 1081–1093. In: H. Rosa, Jr. and G. Murphy (eds.), Proceedings of the world scientific meeting on the biology of sardines and related species. 3. FAO.

Radovich, J. 1982. The collapse of the California sardine fishery: what have we learned? CalCOFI Rep. 23: 56–78.

Randall, J. E. 2007. Reef and shore fishes of the Hawaiian Islands. Hawaii Sea Grant, Honolulu.

Rice, J. A., L. B. Crowder, and F. P. Binkowski. 1987. Evaluating potential sources of mortality for larval bloater (*Coregonus hoyi*): starvation and vulnerability to predation. Can. J. Fish. Aquat. Sci. 44: 467–472.

Rice, J. A., L. B. Crowder, and M. E. Holey. 1987. Exploration of mechanisms regulating larval survival in Lake Michigan bloater: a recruitment analysis based on characteristics of individual larvae. Trans. Am. Fish. Soc. 116: 703–718.

Richards, W. J. 1969. Elopoid leptocephali from Angolan waters. Copeia 1969: 515–518.

Richards, W. J. 1999. Preliminary guide to the identification of the early life history stages of serranid fishes of the western central Atlantic. NOAA Tech. Memo. NMFS-SEFSC-419.

Richards, W. J. (ed.). 2006. Early stages of Atlantic fishes: an identification guide for the Western Central Atlantic. CRC Press. ISBN 0849319161.

Richards, W. J. and J. M. Leis. 1984. Labroidei: development and relationships, pp. 542–546. In: H. G. Moser et al. (eds.), Ontogeny and systematics of fishes. Am. Soc. Ichthyol. Herp., Spec. Publ. No. 1.

Richards, W. J. and K. C. Lindeman. 1986. Recruitment dynamics of reef fishes: planktonic processes, settlement and demersal ecologies, and fishery analysis. Bull. Mar. Sci. 41: 392–410.

Richardson, S. L. and C. Bond. 1978. Two unusual cottoids fishes from the Northeast Pacific. Paper presented at Annu. Meet., Am. Soc. Ichthyol. Herpetol., Tempe, Arizona.

Richardson, S. L. and W. A. Laroche. 1979. Development and occurrence of larvae and juveniles of the rockfishes, *Sebastes crameri*, *Sebastes pinniger* and *Sebastes helvomaculatus* (Family Scorpaenidae) off Oregon. Fish. Bull. 77: 1–41.

Richardson, S. L. and B. B. Washington. 1980. Guide to identification of some sculpin (Cottidae) larvae from marine and brackish waters off Oregon and adjacent areas in the northeast Pacific. NOAA Tech. Rep. NMFS Circ. 430.

Ricker, W. E. 1954. Stock and recruitment. J. Fish. Res. Bd. Can. 11: 559–623.

Ricker, W. E. 1958. Handbook of computations for biological statistics of fish populations. Bull. Fish. Res. Bd. Can. 119.

Riley, J. D. 1966. Marine fish culture in Britain VII. Plaice (*Pleuronectes platessa* L.) post-larval feeding on *Artemia salina* L. nauplii and the effects of varying feeding levels. J. Cons. Perm. Int. Explor. Mer. 30: 204–221.

Roberts, C. M. 1995. Rapid build up of fish biomass in a Caribbean marine reserve. Conserv. Biol. 9: 816–826.

Roberts, C. M. 1998. Sources, sinks and the design of marine reserve networks. Fisheries 23 (7): 16–19.

Roberts, C. M., J. A. Bohnsack, F. Gell, J. P. Hawkins, and R. Goodridge. 2001. Effects of marine reserves on adjacent fisheries. Science 294: 1920–1923.

Roberts, C. M., B. Halpern, S. R. Palumbi, and R. R. Warner. 2001. Designing marine reserve networks: why small, isolated protected areas are not enough. Conserv. Biol. In Pract. 2: 11–17.

Roberts, C. M. and J. P. Hawkins. 2000. Fully protected marine areas: a guide. WWF: http://www.panda.org/resources/publications/water/mpreserves/ma_dwnld.htm.

Robinson, S. M. C. and D. M. Ware. 1988. Ontogenetic development of growth rates in larval Pacific herring, *Clupea harengus pallasi*, measured with RNA-DNA ratios in the Strait of Georgia, British Columbia. Can. J. Fish. Aquat. Sci. 45: 1422–1429.

Rocha-Olivares, A. 1998. Multiplex haplotype-specific PCR: a new approach for species identification of the early life stages of rockfishes of the species-rich genus *Sebastes* Cuvier. J. Exp. Mar. Biol. Ecol. 231: 279–290.

Roemmich, D. and J. McGowan. 1995. Climatic warming and the decline of zooplankton in the California Current. Science 267: 1324–1326.

Rooker, J. R. and G. J. Holt. 1996. Application of RNA:DNA ratios to evaluate the condition and growth of larval and juvenile red drum (*Sciaenops ocellatus*). Mar. Freshwater Res. 47: 283–290.

Rothschild, B. J. 1986. Dynamic of marine fish populations. Harvard University Press, Cambridge, MA.

Rothschild, B. J. and M. J. Fogarty. 1989. Spawning-stock biomass: A source of error in recruitment/stock relationships and management advice. ICES J. Mar. Sci. 45: 131–135.

Rothschild, B. J. and T. R. Osborn. 1988. Small-scale turbulence and plankton contact rates. J. Plankton Res. 10: 465–474.

Ruple, D. 1984. Gobioidei: development, pp. 582–587. In: H. G. Moser et al. (eds.), Ontogeny and systematics of fishes. Am. Soc. Ichthyol. Herp., Spec. Publ. No. 1.

Russ, G. R. and A. C. Alcala. 1996. Do marine reserves export adult fish biomass? Evidence from Apo Island, Central Philippines. Mar. Ecol. Prog. Ser. 132: 1–9.

Russell, F. S. 1973. A summary of the observations on the occurrence of planktonic stages of fish off Plymouth, 1924–1972. J. Mar. Biol. Assoc. U.K. 53: 347–355.

Russell, F. S. 1976. The eggs and planktonic stages of British marine fishes. Academic Press, London.

Ryland, J. S. and J. H. Nichols. 1975. Effect of temperature on the embryonic development of the plaice, *Pleuronectes platessa* L. (Teleostei). J. Exp. Mar. Biol. Ecol. 18: 121–137.

Sakurai, Y. 1983. Reproductive behavior of walleye pollock in captivity. Aquabiology 5: 1–7.

Sameoto, D. D. and L. O. Jaroszynski. 1969. Otter surface sampler: a new neuston net. J. Fish. Res. Bd. Can. 25: 2240–2244.

Sameoto, D. D., L. O. Jaroszynski, and W. B. Fraser. 1977. A multiple opening and closing plankton sampler based on the MOCNESS and N.I.O. Nets. J. Fish. Res. Bd. Can. 34: 1230–1235.

Sameoto, D. D., L. O. Jaroszynski, and W. B. Fraser. 1980. BIONESS, a new design in multiple net zooplankton samplers. Can. J. Fish. Aquat. Sci. 37: 722–724.

Sanzo, L. 1956. Divisione: Zeomorphi Regan, pp. 461–470. In: Uova, larve estadi giovanili de Teleostei. Fauna Flora Golfo Napoli Monogr. 38.

Saville, A. 1963. Estimation of the abundance of a fish stock from egg and larval surveys. Rapp. P.-v. Réun. Cons. Int. Explor. Mer. 155: 164–170.

Schanck, D. 1974. On the reliability of methods for quantitative surveys of fish larvae, pp. 201–212. In: J. H. S. Blaxter (ed.), The early life history of fish. Springer-Verlag, New York.

Scheiber, H. N. 1990. California marine research and the founding of modern fisheries oceanography: CalCOFI's early years, 1947–1964. CalCOFI Rep. 31: 63–83.

Schmidt, J. 1932. Danish eel investigations during 25 years (1905–1930). Copenhagen: Carlsberg Found.

Schmitt, R. J. and S. J. Holbrook. 1996. Local-scale patterns of larval settlement in a planktivorous damselfish—do they predict recruitment? Mar. Freshwater Res. 47: 449–463.

Schultz, E. T. and R. K. Cowen. 1994. Recruitment of coral-reef fishes to Bermuda: local retention or long-distance transport? Mar. Ecol. Prog. Ser. 109: 15–28.

Schumacher, J. D., N. A. Bond, R. D. Brodeur, P. A. Livingston, J. M. Napp, and P. J. Stabeno. 2003. Climate change in the southeastern Bering Sea and some consequences for biota, pp. 17–40. In: G. Hempel and K. Sherman (eds.), Large marine ecosystems of the world—trends in exploitation, protection and research. Elsevier Science Publishers, Amsterdam, The Netherlands.

Scofield, W. L. 1957. Marine fisheries dates. Unpubl. rep. on file at Calif. State Fish. Lab., Long Beach, California.

Secor, D. H., J. M. Dean, and E. H. Laban. 1991. Manual for otolith removal and preparation for microstructural examination. EPRI and Belle W. Baruch Inst. Mar. Biol. and Coast. Res.

Secor, D. H. and E. D. Houde. 1998. Use of larval stocking in restoration of Chesapeake Bay striped bass. ICES J. Mar. Sci. 55: 228–239.

Seeb, L. W. and A. W. Kendall, Jr. 1991. Allozyme polymorphisms permit the identification of larval and juvenile rockfishes of the genus *Sebastes*. Environ. Biol. Fish. 30: 191–201.

Seikai, T. 2002. Flounder culture and its challenges in Asia. Rev. Fish. Sci. 10: 421–432.

Serebryakov, V. P. 1990. Population fecundity and reproductive capacity of some food fishes in relation to year-class-strength fluctuations. ICES J. Mar. Sci. 47: 267–272.

Sette, O. E. 1943a. Biology of the Atlantic mackerel (*Scomber scombrus*) of North America. Pt. 1: Early life history, including growth, drift, and mortality of the egg and larval populations. U.S. Fish Wildl. Serv., Fish. Bull. 50: 149–237.

Sette, O. E. 1943b. Studies on the Pacific pilchard or sardine (*Sardinops caerulea*). 1. Structure of a research program to determine how fishing affects the resource. U.S. Fish Wildl. Serv., Spec. Sci. Rep. 19.

Sette, O. E. and E. H. Ahlstrom. 1948. Estimations of abundance of the eggs of the Pacific pilchard (*Sardinops caerulea*) off southern California during 1940 and 1941. J. Mar. Res. 7: 511–542.

Shelbourne, J. E. 1964. The artificial propagation of marine fish. Adv. Mar. Biol. 2: 1–83.

Shelbourne, J. E. 1965. Rearing marine fish for commercial purposes. CalCOFI Rep. 10: 53–63.

Shepherd, J. G. 1982. A versatile new stock-recruitment relationship for fisheries, and the construction of sustainable yield curves. J. Cons. Int. Explor. Mer. 39: 160–167.

Sherman, K., W. Smith, W. Morse, M. Berman, J. Green, and L. Ejsymont. 1984. Spawning strategies of fishes in relation to circulation, phytoplankton production, and pulses in zooplankton off the northeastern United States. Mar. Ecol. Prog. Ser. 18: 1–19.

Shi, Y., D. R. Gunderson, and P. J. Sullivan. 1996. Growth and survival of 0+ English sole, *Pleuronectes vetulus*, in estuaries and adjacent waters off Washington. Fish. Bull. 95: 161–173.

Sibunka, J. D. and M. J. Silverman. 1984. MARMAP surveys of the continental shelf from Cape Hatteras, North Carolina, to Cape Sable, Nova Scotia (1977–1983). Atlas No. 1. Summary of operations. U.S. Dep. Commer., NOAA Tech. Mem. NMFS-F/NEC-33.

Sidell, B. D., R. G. Otto, and D. A. Powers. 1978. A biochemical method for distinction of striped bass and white perch larvae. Copeia 1978: 340–343.

Sinclair, M. 1988. Marine populations: an essay on population regulation and speciation. Washington Sea Grant, University of Washington Press, Seattle.

Sinclair, M. and M. J. Trembley. 1984. Timing of spawning of Atlantic herring (*Clupea harengus harengus*) populations and the match-mismatch theory. Can. J. Fish. Aquat. Sci. 41: 1055–1065.

Skud, B. E. 1982. Dominance in fishes: the relation between environment and abundance. Science 216: 144–149.

Smith, C. L. 1971. A revision of the American groupers: *Epinephelus* and allied genera. Bull. Am. Mus. Nat. Hist. 146: 67–241.

Smith, P. E. and R. W. Eppley. 1992. Primary production and the anchovy population in the Southern California Bight: comparison of time series. Limnol. Oceanogr. 27: 1–17.

Smith, P. E. and H. G. Moser. 1988. CalCOFI time series: an overview of fishes. CalCOFI Rep. 29: 66–78.

Smith, P. E. and S. L. Richardson. 1977. Standard techniques for pelagic fish egg and larval survey. FAO Fisheries Tech. Pap. 175.

Smith, P. J., P. G. Benson, and A. A. Frentzos. 1980. Electrophoretic identification of larval and 0-group flounders (*Rhombosolea* spp.) from Wellington Harbour. N. Z. J. Mar. Freshwater Res. 14: 401–404.

Smith, W. G. and W. W. Morse. 1993. Larval distribution patterns: early signals for the collapse/recovery of Atlantic herring *Clupea harengus* in the Georges Bank Area. Fish. Bull. 91: 338–347.

Sogard, S. M. 1997. Size–selectivity mortality in the juvenile stage of teleost fishes: a review. Bull. Mar. Sci. 60: 1129–1157.

Solemdal, P. 1981. Overview. Enclosure studies. Rapp. Proc. Verb. 178: 117–120.

Solemdal, P., E. Dahl, D. S. Danielssen, and E. Moksness. 1984. The cod hatchery in Flødevigen—background and realities. Flødevigen rapportser. 1: 17–45.

Solemdal, P. and B. Ellertsen. 1984. Sampling fish larvae with large pumps: quantitative and qualitative comparisons with traditional gear, pp. 335–364. In: E. Dahl, D. S. Danielssen, E. Moksness, and P. Solemdal (eds.), The propagation of cod *Gadus morhua* L. Inst. Mar. Res. Flødevigen Biol. Sta.

Song, J. and L. R. Parenti, 1995. Clearing and staining whole fish specimens for simultaneous demonstration of bone, cartilage, and nerves. Copeia: 114–118.

Soutar, A. and J. D. Isaacs. 1974. Abundance of pelagic fish during the 19th and 20th centuries as recorded in anaerobic sediments off the Californias. Fish. Bull., U.S. 72: 257–273.

Sparta, A. 1933. Ordine: Allotriognathi, pp. 266–279. In: Uova, large e stadi giovanili di Teleostei. Fauna Flora Golfo Napoli Monogr. 38.

Springer, V. G. and G. D. Johnson. 2000. Use and advantage of ethanol solution of alizarin red S dye for staining bone in fishes. Copeia 2000: 300–301.

Stearns, D. E., G. J. Holt, R. B. Forward, Jr., and P. L. Pickering. 1994. Ontogeny of phototactic behavior in red drum larvae (Sciaenidae: *Sciaenops ocellatus*). Mar. Ecol. Prog. Ser. 104: 1–11.

Steedman, H. F. (ed.). 1976. Zooplankton fixation and preservation. UNESCO Manager. Oceanogr. Method. 4.

Steele, J. H. 1996. Regime shifts in fisheries management. Fish. Res. 25: 19–23.

Stephens, J. S., Jr., P. A. Morris, D. J. Pondella, T. A. Koonce, and G. A. Jordan. 1994. Overview of the dynamics of an urban artificial reef fish assemblage at King Harbor, California, USA, 1974–1991: a recruitment driven system. Bull. Mar. Sci. 55: 1224–1239.

Stevens, E. G., A. C. Matarese, and W. Watson. 1984b. Ammodytoidei: development and relationships, pp. 574–575. In: H. G. Moser et al. (eds.), Ontogeny and systematics of fishes. Am. Soc. Ichthyol. Herp., Spec. Publ. No. 1.

Stevens, E. G., W. Watson, and A. C. Matarese. 1984a. Notothenioidea: development and relationships, pp. 561–564. In: H. G. Moser et al. (eds.), Ontogeny and systematics of fishes. Am. Soc. Ichthyol. Herp., Spec. Publ. No. 1.

Stockhausen, W. T., R. N. Lipcius, and B. M. Hickey. 2000. Joint effects of larval dispersal, population regulation, marine reserve design, and exploitation on production and recruitment in the Caribbean spiny lobster. Bull. Mar. Sci. 66: 957–990.

Stoecker, D. K. and J. J. Govoni. 1984. Food selection by young larval gulf menhaden (*Brevoortia patronus*). Mar. Biol. 80: 299–306.

Stoner, A. W. and M. Ray. 1996. Queen conch, *Strombus gigas*, in fished and unfished locations of the Bahamas: effects of a marine reserve on adults, juveniles, and larval production. Fish. Bull. 94: 551–565.

Strauss, R. E. 1993. Relationships among the cottid genera *Artedius*, *Clinocottus*, and *Oligocottus* (Teleostei: Scorpaeniformes). Copeia 1993: 518–522.

Svåsand, T. 1998. Cod enhancement studies in Norway—background and results with emphasis on releases in the period 1983–1990. Bull. Mar. Sci. 62: 313–324.

Swearer, S. E., J. E. Caselle, D. W. Lea, and R. R. Warner. 1999. Larval retention and recruitment in an island population of a coral reef fish. Nature. 402: 799–802.

Taning, A. V. 1952. Experimental study of meristic characters in fishes. Biol. Rev. Camb. Philos. Soc. 27: 169–193.

Taylor, W. R. and G. C. Van Dyke. 1985. Revised procedures for staining and clearing small fishes and other vertebrates for bone and cartilage study. Cybium. 9: 107–119.

Theilacker, G. H. 1987. Feeding ecology and growth energetics of larval northern anchovy, *Engraulis mordax*. Fish. Bull. 85: 213–228.

Theilacker, G. H., A. S. Kimbrell, and J. S. Trimmer. 1986. Use of an ELISPOT immunoassay to detect euphausiid predation on anchovy larvae. Mar. Ecol. Prog. Ser. 30: 127–131.

Theilacker, G. H. and S. M. Porter. 1995. Condition of larval walleye pollock, *Theragra chalcogramma*, in the western Gulf of Alaska assessed with histological and shrinkage indices. Fish. Bull. 93: 333–344.

Theilacker, G. H. and W. Shen. 1993. Calibrating starvation-induced stress in larval fish using flow cytometry. Am. Fish. Soc. Sym. 14: 85–94.

Theilacker, G. H. and W. Shen. 2001. Evaluating growth of larval walleye pollock, *Theragra chalcogramma*, using cell cycle analysis. Mar. Biol. 136: 897–907.

Theilacker, G. H. and Y. Watanabe. 1989. Midgut cell height defines nutritional status of laboratory raised larval northern anchovy, *Engraulis mordax*. Fish. Bull. 87: 457–469.

Thompson, W. F. and R. Van Cleve. 1936. Life history of the Pacific halibut. 2. Distribution and early life history. Rep. Int. Fish. Comm.

Thorisson, K. 1994. Is metamorphosis a critical interval in the early life of marine fishes? Aquacult. Fish. Manage. F. 40: 23–36.

Trantor, D. J. and J. H. Fraser (eds.). 1968. Zooplankton sampling. UNESCO Manager. Oceanogr. Method. 2.

Travis, J., F. Coleman, C. Grimes, D. Conover, T. Bert, and M. Tringali. 1998. Critically assessing stock enhancement: an introduction to the Mote Symposium. Bull. Mar. Sci. 62: 305–311.

Tucker, J. W., Jr. 1998. Marine fish culture. Kluwer, Boston.

Tucker, J. W., Jr. and J. L. Laroche. 1984. Radiographic techniques in studies of young fishes. In Moser, H. G. al. (eds.), Ontogeny and systematics of fishes, p. 37–39. Spec. Publ. 1, Am. Soc. Ichthyol. Herpetol. Allen Press, Lawrence, KS.

Tupper, M. and R. G. Boutilier. 1997. Effects of habitat on settlement, growth, predation risk and survival of a temperate reef fish. Mar. Ecol. Prog. Ser. 151: 225–236.

Tupper, M. and W. Hunte. 1994. Recruitment dynamics of coral reef fishes in Barbados. Mar. Ecol. Prog. Ser. 108: 225–235.

Tupper, M. and F. Juanes. 1999. Effects of a marine reserve on recruitment of grunts (Pisces: Haemulidae) at Barbados, West Indies. Environ. Biol. Fish. 55: 53–63.

Turrell, W. R. 1992. New hypotheses concerning the circulation of the northern North Sea and its relation to North Sea fish stock recruitment. ICES J. Mar. Sci. 49: 107–123.

Tyler, J. C. 1980. Osteology, phylogeny, and higher classification of the fishes of the order Plectognathi (Tetraodontiformes). NOAA Tech. Rep. NMFS Circ. 434.

Urho, L. 1989. Fin damage in larval and adult fishes in a polluted inlet in the Baltic. pp. 493–494. In: J. H. S. Blaxter et al. (eds.), The early life history of fish. Rapp. P.- v. Réun. Cons. Int. Explor. Mer., 191.

Valles, H., S. Spongaules, and H. A. Oxenford. 2001. Larval supply to a marine reserve and adjacent fished area in the Soufriere Maine Management Area, St Lucia, West Indies. J. Fish. Biol. 59: 152–177.

van der Meeren, T. and K. E. Naas. 1997. Development of rearing techniques using large enclosed ecosystems in the mass production of marine fish fry. Rev. Fish. Sci. 5: 367–390.

van der Veer, H. W., R. Berghaln, J. M. Miller, and A. D. Rijnsdorp. 2000. Recruitment in flatfish, with special emphasis on North Atlantic species: progress made by the flatfish symposia. ICES J. Mar. Sci. 57: 202–215.

van der Veer, H. W., L. Pihl, and M. J. N. Bergman. 1990. Recruitment mechanisms in North Sea plaice *Pleuronectes platessa*. Mar. Ecol. Prog. Ser. 64: 1–12.

Walford, L. A. 1938. Effect of currents on distribution and survival of the eggs and larvae of the haddock (*Melanogrammus aeglefinus*) on Georges Bank. Bull. U.S. Bur. Fish. 49: 1–73.

Walker, B. 1952. A guide to the grunion. Cal. Fish and Game 38: 409–420.

Wallace, D. 1978. Two anomalies of fish larval transport and their importance in environmental assessment. NY Fish Game J. 25: 59–71.

Walline, P. D. 1985. Growth of larval walleye pollock related to domains within the southeast Bering Sea. Mar. Ecol. Prog. Ser. 21: 197–203.

Walters, C. 2000. Impacts of dispersal, ecological interactions, and fishing effort dynamics on efficacy of marine protected areas: How large should protected areas be? Bull. Mar. Sci. 66: 745–757.

Wang, J. C. 1986. Fishes of the Sacramento-San Joaquin estuary and adjacent waters, California: a guide to early life histories. Interagency Ecological Studies Program of the Sacramento-San Joaquin Estuary. Tech. Rep. 9.

Warner, R. R. 1975. The adaptive significance of sequential hermaphroditism in animals. Amer. Nat. 109: 61–82.

Warner, R. R., S. E. Swearer, and J. E. Caselle. 2000. Larval accumulation and retention: implications for the design of marine reserves and essential fish habitat. Bull. Mar. Sci. 66: 821–830.

Washington, B. B. 1986. Systematic relationships and ontogeny of the sculpins *Artedius*, *Clinocottus*, and *Oligocottus* (Cottidae: Scorpaeniformes). Proc. Calif. Acad. Sci. 44: 157–224.

Watson, W. 1996. Serranidae: sea basses, pp. 876–899. In: H. G. Moser (ed.), The early stages of fishes of the California Current region. CalCOFI Atlas 33.

Watson, W., R. L. Charter, H. G. Moser, R. D. Better, D. A. Ambrose, S. R. Charter, L. O. Robertson, E. M. Sandknop, E. H. Lynn, and J. Standard. 1999. Fine-scale distributions of planktonic fish eggs in the vicinities of Big Sycamore Canyon and Vandenberg ecological reserves, and Anacapa and San Miguel Islands, California. CalCOFI Rep. 40: 128–153.

Watson, W. and R. Davis. 1989. Larval fish diets in shallow coastal waters off San Onofre, California. Fish. Bull. 87: 569–591.

Weibe, P. H., K. H. Burt, S. H. Boyd, and A. W. Morton. 1976. A multiple opening/closing net and environmental sensing system for sampling zooplankton. J. Mar. Res. 34: 313–326.

Western, D. 1995. The role of marine protected areas in the conservation of marine fisheries. Sustainable Devel. Fish. Africa: 90–91.

Wiles, G. C., R. D. D'Arrigo, and G. C. Jacoby. 1998. Gulf of Alaska atmosphere-ocean variability over recent centuries inferred from coastal tree-ring records. Climate Change 38: 289–306.

Williams, M. A. and C. C. Coutant. 2003. Modification of schooling behavior in larval atherinid fish, *Atherina mochon*, by heat exposure of eggs and larvae. Trans. Am. Fish. Soc. 132: 638–645.

Wourms, J. P. 1981. Viviparity: the maternal–fetal relationship in fishes. Am. Zool. 21: 473–515.

Wyatt, T. 1972. Some effects of food density on the growth and behaviour of plaice larvae. Mar. Biol. 14: 210–216.

Yoklavich, M. M. (ed.). 1998. Marine harvest refugia for west coast rockfish: a workshop. NOAA TM NMFS SWFSC 255.

Yoshimura, K., A. Hagiwara, T. Yoshimatsu, and C. Kitajima. 1996. Culture technology of marine rotifers and the implications for intensive culture of marine fish in Japan. Mar. Freshw. Res. 47: 217–222.

TAXONOMICAL INDEX

A

Acanthistius, 105
Acanthuridae, 85
Acanthurus sp., 85, 98
Achirus lineatus, 201
Acipenser spp., 31, 232
Actinopterygians, 10
Agonidae (poachers), 84
Alaska plaice *(Pleuronectes quadrituberculatus)*, 64
Albuliforms, 80
Alosa sapidissima, 35, 232, 246
Amblyopsids (cavefishes), 186
American and European freshwater eels (*Anguilla* spp.), 36
American lobster *(Homarus americanus)*, 246
American plaice *(Hippoglossoides platessoides)*, 200
American shad *(Alosa sapidissima)*, 35, 232
Ammodytes, 13, 75, 84, 193
Ammodytes hexapterus, 84, 97
Ammodytidae (sand lances), 13, 75, 84, 193
Anchoa mitchelli, 201
Anchovies (Engraulidae), 13, 51, 81, 186
Anglerfishes (Lophiiformes), 82
Anguilla, 14
Anguilla anguilla, 100
Anguillidae, 2, 36
Anguillids, 20, 188, 230
Anguilliformes, 81
Anguilliforms, 65
Anoplarchus purpurescens, 84, 96

Anoplopoma fimbria, 83, 93, 186
Anoplopomatidae, 83
Anthias gordensis, 111
Anthias nicholsi, 110
Anthias tenuis, 112
Anthias woodsi, 111
Anthiinae, 108
Apricot bass *(Plectranthias garrupellus)*, 108, 113
Ara (Niphon spinosus), 113, 117
Archosargus rhomboidalis, 201
Arctic flounder *(Liopsetta glacialis)*, 17, 18
Ariids (marine catfishes), 11, 64
Arrowhead soapfish *(Belonoperca chabanaudi)*, 115, 120
Artedius harringtoni, 83, 94
Atherinidae, 83
Atheriniforms, 65
Atherinopsis californiensis, 83, 91
Atka mackerel *(Pleurogrammus monopterygius)*, 267
Atlantic and Pacific cod (*Gadus* spp.), 26
Atlantic cod *(Gadus morhua)*, 3, 33, 190, 192–193, 235, 245
Atlantic herring *(Clupea harengus)*, 26, 34–35, 130, 134, 183, 190, 193, 194, 207, 231
Atlantic mackerel *(Scomber scombrus)*, 4, 212
Atlantic menhaden *(Brevoortia tyrannus)*, 211
Atlantic sole *(Solea solea)*, 30
Atlantic spiny lumpsucker *(Eumicrotremus spinosus)*, 18

B

Barracudinas (Paralepididae), 82
Barred soapfish *(Diploprion bifasciatum)*, 115, 120
Basslet *(Liopropoma* sp.), 121
Bathylagidae, 81, 100
Bathylagus ochotensis, 81, 86
Bathylagus wesethi, 64
Bathymasteridae, 84
Bay anchovy *(Anchoa mitchilli)*, 201
Belonoperca chabanaudi, 115, 120
Beryciforms, 52, 80
Bigeye bass *(Pronotogrammus eos)*, 109
Black prickleback *(Xiphister atropurpureus)*, 35
Black sea bass *(Centropristis striata)*, 106
Black-belly dragonfish *(Stomias atriventer)*, 57
Blackfin soapfish *(Rypticus nigripinis)*, 122
Blenniids, 19
Bloater *(Coregonus hoyi)*, 194
Blue hamlet *(Hypoplectrus gemma)*, 107
Blue lanternfish *(Tarletonbeania crenularis)*, 82, 88
Bluefish *(Pomatomus saltatrix)*, 52, 209, 211, 232
Boarfish *(Capros aper)*, 83, 92
Boarfishes (Caproidae), 83
Bonefishes (albuliforms), 80
Bony, ray-finned vertebrates (actinopterygians), 10
Boxfishes (Ostraciidae), 57
Bregmaceros macclellandi, 82, 90
Brevoortia tyrannus, 211, 232
Bristlemouths (*Cyclothone* spp.), 187
Brosmophycis marginata (?), 82, 89
Brown trout *(Salmo trutta)*, 134
Buffalo sculpin *(Enophrys bison)*, 35
Butterfishes (Stromateidae), 85
Bythitidae, 82

C

Cabezon *(Scorpaenichthys marmoratus)*, 52
California clingfish *(Gobiesox rhessodon)*, 84, 97
California grunion *(Leuresthes tenuis)*, 32–33, 200
Callionymidae, 85
Callionymus (Repomucenus) beniteguri, 85, 97
Capelin *(Mallotus villosus)*, 81, 86, 130
Caproidae, 83
Capros aper, 83, 92
Carapidae, 52, 82
Careproctus reinhardti, 84, 95
Catfishes (siluriforms), 17, 18
Cavefishes (amblyosids), 186
Centrarchids, 15, 19, 189
Centropristis-Paralabrax, 105
Centropristis striata, 106
Cephalopholis, 115
Cephalopholis sp. 119
Ceratioids, 37
Chain pearlfish *(Echiodon drummondi)*, 82
Channichthyidae *(Pagetopsis macropterus)*, 84, 97
Chauliodontids (viperfishes), 27
Chauliodus macouni, 57, 64
Chlopsidae, 81
Chonrichthyans, 10
Cichlidae, 17
Cichlids (Cichlidae), 17
Citharichthys spp., 64
Clingfishes (Gobiesocidae), 84
Clingfishes (Gobiesociforms), 75
Clinocottus, 104
Clupea harengus, 26, 34–35, 130, 134, 183, 190, 193–207
Clupea pallasi, 18, 32, 34–35, 64, 81, 86, 130, 131, 188, 231, 233
Clupea spp., 17
Clupeidae, 27, 30, 81
Clupeids, 19, 51, 186
Clupeiformes, 81
Clupeiforms, 65, 75
Cod (*Gadus* spp.), 13, 17, 187
Cod icefishes (nototheniids), 18
Codfishes (Gadidae), 82
Codfishes and relatives (gadiforms), 47
Codlets (Bregmacerotidae), 82
Coelacanth (*Latimeria* sp.), 187
Coho salmon *(Oncorhynchus kisutch)*, 52, 53
Cololabis saira, 57, 64, 83, 91
Combtooth blennies (blenniids), 19
Common carp *(Cyprinus carpio)*, 4, 246
Common whitefish *(Coregonus lavaretus)*, 211
Coregonus clupeaformis, 212
Coregonus hoyi, 194
Coregonus lavaretus, 211
Cottidae, 17, 19, 75, 83, 103–104
Cratinus, 105
Creolefish *(Paranthias furcifer)*, 119
Cryptacanthodes aleutensis, 84, 97
Cryptacanthodidae, 84
Cunner *(Tautogolabrus adspersus)*, 13
Curlfin sole *(Pleuronichthys decurrens)*, 64
Cyclopteridae, 75, 84

Cyclothone spp., 187
Cyprinodon spp., 186
Cyprinus carpio, 4, 246

D

Damselfishes (Pomacentrids), 15
Darkblotched rockfish *(Sebastes crameri)*, 83, 93
Darkfin sculpin *(Malacocottus zonurus)*, 84, 95
Deepsea smelts (Bathylagidae), 81, 100
Desert pupfishes (*Cyprinodon* spp.), 186
Dicentrarchus, 104
Diplectrum sp., 105, 107
Diploprion bifasciatum, 115, 120
Diploprionini, 108, 115–116
Dipnoians, 10
Dorosoma cepedianus, 236
Dover sole *(Microstomus pacificus)*, 64, 85, 99
Dragonets (Callionymidae), 85
Dragonet *(Callionymus (Repomucenus) beniteguri)*, 97
Dules, 105
Dwarf wrymouth *(Cryptacanthodes aleutensis)*, 84, 97

E

Eared blacksmelt *(Bathylagus ochotensis)*, 81, 86
Echiodon drummondi, 82, 89
Eel pouts (zoarcids), 75, 230
Eels (Anguilliformes), 81
Eels (Anguilliforms), 65
Eels and relatives (elopomorphs), 74
Elasmobranchs, 22
Eleginus gracilis, 82, 90
Elopidae, 81
Elopiforms, 80
Elopomorphs (eels and relatives), 74
Elops sp., 81, 86
Embiotoca lateralis, 13
Embiotocids, 11–13, 15, 30, 188
English sole *(Parophrys vetulus)*, 130, 143, 232
Engraulidae, 13, 51, 81, 186
Engraulis, 14
Engraulis mordax, 33, 86, 129, 137–138, 159, 188, 189, 201–202, 221
Enophrys bison, 35
Eopsetta jordani, 35
Epinephelinae, 108, 115
Epinipheline serranids (groupers), 51
Epinephelini, 108, 110–124
Epinephelus sp., 119
Epinephelus spp. (groupers), 104, 115

Etrumeus teres, 64
Eumicrotremus orbis, 57, 84, 95
Eumicrotremus spinosus, 18
European freshwater eel *(Anguilla anguilla)*, 36, 100
Euthynnus alletteratus, 85, 98
Exocoetins, 65

F

False morays (Chlopsidae), 81
Filefishes (Tetraodontiformes), 85
Flagelloserranus, 117
Flatfishes (pleuronectiforms), 52, 80, 85, 188, 257
Flathead sole *(Hippoglossoides elassodon)*, 64–65
Flatheads (platycephalids), 12
Flounders (pleuronectids), 13
Flying fishes (exocoetins), 65
Forcipiger longirostris, 52, 53
Fourhorn poacher *(Hypsagonus quadricornis)*, 84, 94
Freshwater eels (anguillids, *Anguilla* spp.), 2, 20, 36, 188, 230

G

Gadidae, 82
Gadiformes, 47, 82
Gadus macrocephalus, 14, 62, 226
Gadus morhua, 3, 13, 33, 190, 235, 245
Gadus ogac, 17
Gadus spp., 26, 187
Gag *(Mycteroperca microlepis)*, 119
Gargaropteron pterodactylops, 52, 53
Gizzard shad *(Dorosoma cepedianus)*, 236
Glyptocephalus zachirus, 64
Goatfishes (mullids), 52
Gobiesocidae, 84
Gobiesociforms, 75
Gobiesox rhessodon, 84, 97
Gobiid (goby), 18, 64
Goby (gobiid), 18, 64
Golden redfish *(Sebastes norvegicus)*, 16
Gonioplecturus, 115
Gonioplectrus hispanicus, 118
Gonostomatidae, 12, 81, 103, 230
Goosefish *(Lophius americanus)*, 83, 91
Goosefishes (*Lophius* spp.), 19
Grammistes sexilineatus, 122
Grammistini, 110, 117, 122–124
Green goby *(Microgobius thalassinus)*, 85, 98
Greenland cod *(Gadus ogac)*, 17, 18

Greenland turbot *(Reinhardtius hippoglossoides)*, 264
Greenlings (hexagrammids, Hexagrammidae), 17, 19, 52, 83
Greenlings *(Hexagrammos* spp.), 52
Grenadier (Macrouronae *(Nezumia* ?)), 90
Grenadiers (Macrouridae), 82
Groupers (epinipheline serranids), 51
Grunt sculpin *(Rhamphocottus richardsoni)*, 83, 94
Gunnels *(Pholis* spp.), 18
Guppy *(Poecilia reticulatus)*, 15
Gymnocephalus cernuus, 233

H
Haddock *(Melanogrammus aeglefinus)*, 192–193, 194, 201
Hagfishes (myxinids), 12
Hakes (Merlucciidae), 82
Hakes *(Urophycis* spp.), 52
Halfbeaks (hemiramphins), 65
Harlequin bass *(Serranus tigrinus)*, 107
Hatchetfishes (sternoptychids), 103
Helicolenus percoides, 14
Hemanthias leptus, 112
Hemanthias peruanus, 109
Hemanthias signifer, 112
Hemanthias vivanus, 109
Hemilepidotus spp., 104
Hemiptripteridae, 84
Hemiramphins, 15, 65
Herring *(Clupea* spp.), 17
Herring-like fishes (Clupeiformes, clupeiforms), 65, 75, 81
Herrings (Clupeidae), 19, 30, 51, 81, 186
Hexagrammidae, 17, 83
Hexagrammos lagocephalus, 52, 53, 83, 93
Hexagrammos spp., 52
High cockscomb *(Anoplarchus purpurescens)*, 84, 96
Highsnout melamphid *(Melamphaes lugubris)*, 83, 91
Hippoglossoides elassodon, 64–65
Hippoglossoides platessoides, 200
Hippoglossus stenolepis, 31, 128, 192
Holanthias, 108
Holocentridae, 52, 83
Homarus americanus, 246
Hypomesus pretiosus, 32, 236
Hypoplectrus, 105
Hypoplectrus gemma, 107
Hypsagonus quadricornis, 84, 94

I
Icosteus aenigmaticus, 64
Idiacanthus antrostomus, 81, 87
Irish lords *(Hemilepidotus* spp.), 104

J
Jack mackerel *(Trachurus symmetricus)*, 46, 64–65, 213
Jacksmelt *(Atherinopsis californiensis)*, 83, 91
Japanese flounder *(Paralichthys olivaceus)*, 248
Jeboehlkia, 122, 124
Jeboehlkia gladifer, 123

K
Kali macrodon, 52, 53
Kelp bass *(Paralabrax clathratus)*, 106
King-of-the-salmon *(Trachipterus altivelus)*, 57, 82, 88

L
Labridae, 84
Labrids, 11–12
Ladyfishes *(Elops* sp.), 86
Lake whitefish *(Coregonus clupeaformis)*, 212
Lampanyctus regalis, 82, 88
lampreys (petromyzonids), 232
Lanternfishes (Myctophidae), 51, 82, 100, 103, 230
lanternfishes and their relatives (myctophiforms), 52, 75
Lateolabrax, 104
Latimeria sp., 187
Leiostomus xanthurus, 223, 232
Leopard coralgrouper *(Plectropomus leopardus)*, 118
Lepidopsetta, 14
Lestidiops ringens, 82, 88
Leuresthes tenuis, 32, 33, 200
Leuroglossus schmidti, 64
Lightfishes (Gonostomatidae), 12, 81, 103, 230
Limanda aspera, 266
Lined sole *(Achirus lineatus)*, 201
Liopropomini, 110, 117, 121
Lingcod *(Ophiodon elongatus)*, 18, 239
Liopropoma sp., 121
Liopsetta glacialis, 17
Liparidae, 84
Little tunny *(Euthynnus alletteratus)*, 85, 98
Live-bearers (poecilliids), 11, 189
Lizardfishes (synodontids), 65
Longfin escolar *(Scombrolabrax heterolepis)*, 48

Taxonomical Index

Longnose butterflyfish *(Forcipiger longirostris)*, 52
Longspine thornyhead *(Sebastolobus altivelis)*, 52, 53
Longtail bass *(Hemanthias leptus)*, 112
Lophius americanus, 83, 91
Lophius spp. (goosefishes), 19
Lophiidae, 83
Lophiiformes, 82
Lower percoid, 11
Lungfishes (dipnoians), 10
Lutjanidae, 84

M

Mackerel *(Scomber* spp.), 186
Mackerels (Scombridae), 85
Macrouridae, 82
Macrourids, 65
Mail-cheeked fishes (Scorpaeniformes), 52, 75, 83
Malacocottus zonurus, 84, 95
Mallotus villosus, 81, 86, 130
Marine catfishes, 18, 64
Maurolicus spp., 65
Melamphaes lugubris, 83, 91
Melanogrammus aeglefinus, 192, 201, 246
Menhaden *(Brevoortia tyrannus)*, 232
Merlucciidae, 82
Merluccius productus, 57, 82, 90, 211
Microgadus proximus, 48
Microgobius thalassinus, 85, 98
Microstomus pacificus, 64, 85, 99
Mola mola, 17
Molidae, 85
Molids, 74–75
Morone, 104, 105
Morone americanus, 232
Morone americanus/M. saxatilis, 58
Morone saxatilis, 36, 232
Mugil cephalus, 57
Mugilids, 52
Mullets (mugilids), 52
Mullids, 52
Mycteroperca, 115
Mycteroperca microlepis, 119
Myctophidae (lanternfishes), 51, 55, 82, 100, 103, 230
Myctophiforms (lanternfishes and relatives), 52, 75
Myctophinae, 103
Myripristis sp., 83, 92
Myxinids (hagfishes), 12

N

Nautichthys oculofasciatus, 84, 94
New Zealand flounders *(Rhombosolea* spp.), 58
New Zealand snapper *(Pagurus auratus)*, 210
Nezumia, 82, 90
Niphon spinosus, 113, 117
Niphonini, 108, 113–115
Northen anchovy *(Engraulis mordax)*, 33, 86, 129, 137–138, 159, 188, 189, 201, 202, 221
Northern ronquil *(Ronquilis jordani)*, 84, 96
Northern smoothtongue *(Leuroglossus schmidti)*, 64
Norway pout *(Trispterus esmarkii)*, 194
Nototheniids (cod icefishes), 18

O

Oarfish *(Regalecus glesne)*, 186
Oarfishes (Regalidae), 82
Ocean sunfish *(Mola mola)*, 17
Ocean sunfishes (molids), 74, 75
Oligocottus, 104
Onchorhynchus, 14
Onchorhynchus spp., 2, 19–20, 35, 188, 230, 247
Oncorhynchus gorbuscha, 234
Oncorhynchus kisutch, 52, 53
Ophiodon elongatus, 18, 239
Ophiodon, 15
Opsanus spp., 2
Osmeridae, 17, 19, 81
Ostraciidae, 57

P

Pacific blackdragon *(Idiacanthus antrostomus)*, 81, 87
Pacific cod *(Gadus macrocephalus)*, 62, 226
Pacific hake *(Merluccius productus)*, 57, 82, 90, 211
Pacific halibut *(Hippoglossus stenolepis)*, 31, 128, 192
Pacific herring *(Clupea pallasi)*, 18, 32, 34–35, 64, 81, 86, 130, 131, 188, 231, 233
Pacific pompano *(Peprilus simillimus)*, 85, 99
Pacific reef bass *(Pseudogramma thaumasium)*, 123
Pacific rockfishes *(Sebastes* spp.), 231
Pacific salmon *(Onchorhynchus* spp.), 2, 19, 20, 35, 188
Pacific sand lance *(Ammodytes hexapterus)*, 84, 97

Pacific sardine *(Sardinops sagax)*, 4, 32, 57, 159, 192, 220
Pacific saury *(Cololabis saira)*, 57, 64, 83, 91
Pacific spiny lumpsucker *(Eumicrotremus orbis)*, 57, 84, 95
Pacific tomcod *(Microgadus proximus)*, 49
Pacific viperfish *(Chauliodus macouni)*, 57, 64
Pagetopsis macropterus, 84, 97
Pagurus auratus, 210
Panama lightfish *(Vinciguerria lucetia)*, 81, 87
Paralabrax clathratus, 106
Paralabrax spp. 58
Paralepididae, 82
Paralichthys olivaceus, 248
Paranthias, 115
Paranthias furcifer, 119
Parophrys vetulus, 130, 232
Parrotfishes (scarids), 11
Pearlfish *(Echiodon drummondi)*, 89
Pearlfishes (Carapidae), 52, 82
Pearlsides *(Maurolicus)*, 65
Peprilus simillimus, 85, 99
Perch-like fishes (Perciformes), 52, 75, 80, 84
Perciformes, 52, 75, 80, 84
Petrale sole *(Eopsetta jordani)*, 35
Petromyzonids (lampreys), 232
Pholis spp., 18
Pink salmon *(Oncorhynchus gorbuscha)*, 234
Pinpoint lampfish *(Lampanyctus regalis)*, 82, 88
Plaice *(Pleuronectes platessa)*, 4, 33, 129, 194, 201, 247
Plainfin midshipman *(Porichthys notatus)*, 35
Platichthys/Parophrys, 64
Platycephalids, 12
Plectranthias garrupellus, 108, 113
Plectropomus, 115
Plectropomus leopardus, 118
Pleurogrammus monopterygius, 267
Pleuronectes platessa, 4, 33, 129, 194, 201, 247
Pleuronectes quadrituberculatus, 64
Pleuronectidae, pleuronectids (right-eyed flounders), 13, 18, 65, 85
Pleuronectiformes, 85
Pleuronectiforms, 52, 80, 188, 257
Pleuronichthys decurrens, 64
Pleuronichthys spp., 65, 85
Poachers (Agonidae), 84
Poecilia reticulatus, 15
Poecilliids, 11, 15, 189
Pomacentrids, 15
Pomatomus saltatrix, 52, 209, 211, 232
Pomoxis annularis, 136
Porgies (sparids), 12
Porichthys notatus, 35
Pricklebacks (Stichaeidae), 84
Pronotogrammus, 108
Pronotogrammus aureorubens, 109, 112
Pronotogrammus eos, 109, 112
Pronotogrammus martinicensis, 108, 110
Pronotogrammus multifasciatus, 108, 111
Prowfish *(Zaprora silenus)*, 84, 97
Prowfishes (Zaproridae), 84
Psychrolutidae, 84
Pseudogramma, 124
Pseudogramma thaumasium, 123
Pseudopleuronectes americanus, 62, 246
Psychrolutidae, 84
Puffers (tetraodontiformes), 85
Pygmy sea bass *(Serraniculus pumilio)*, 106

Q
Queen conch *(Strombus gigas)*, 241

R
Ragfish *(Icosteus aenigmaticus)*, 64
Ranzania laevis, 85, 99
Rattails (Macrouridae), 65
Razorfish *(Myripristis* sp.), 83
Razorfish *(Xyrichthys* sp.), 84, 96
Red barbier *(Hemanthias vivanus)*, 109
Red brotula *(Brosmophycis marginata* (?)), 82, 89
Red drum *(Sciaenops ocellatus)*, 248
Redfish *(Sebastes)*, 13, 58
Red gurnard perch *(Helicolenus percoides)*, 14
Regalecus glesne, 186
Regalidae, 82
Reinhardtius hippoglossoides, 264
Rhamphocottidae, 83
Rhamphocottus richardsoni, 83, 94
Rhomboplites aurorubens, 84, 96
Rhombosolea spp., 58
Ribbonfishes (Trachipteridae), 82
Right-eyed flounders (pleuronectids), 18
Right-eyed flounders *(Pleuronichthys* spp.), 65, 85
Rock greenling *(Hexagrammos lagocephalus)*, 52, 83, 93
Rockfish *(Sebastes* spp.), 16, 58, 75, 185
Ronquils (Bathymasteridae), 84
Ronquilis jordani, 53, 84, 96
Roughtongue bass *(Pronotogrammus martinicensis)*, 108, 110
Round herring *(Etrumeus teres)*, 64

Taxonomical Index

Ruffe *(Gymnocephalus cernuus)*, 233
Rypticus nigripinis, 122
Rypticus spp., 104

S

Sablefish *(Anoplopoma fimbria)*, 83, 93, 186
Sablefishes (Anoplopomatidae), 83
Saffron cod *(Eleginus gracilis)*, 82, 90
Sailfin sculpin *(Nautichthys oculofasciatus)*, 84, 94
Salmo trutta, 134
Salmon *(Oncorhynchus* spp.), 230, 247
Salmon and their relatives (salmoniforms), 36, 65
Salmon and trout (salmonids), 197
Salmonids (salmon and trout), 197
Salmoniforms, 65
Sand lances *(Ammodytes* spp.), 13, 75, 84, 193
Sand perches *(Diplectrum* spp.), 107
Sandbasses *(Paralabrax* spp.), 58
Sanddab *(Citharichthys* spp.), 64
Sardines, 219
Sardinops sagax, 4, 32, 57, 159, 192, 220
Sauries (Scomberesocidae), 83
Scalloped ribbonfish *(Zu cristatus)*, 82, 89
Scalyhead sculpin *(Artedius harringtoni)*, 83, 94
Scarids, 11
Schultzea, 105
Schlinderia, 186
Sciaenops ocellatus, 248
Scomber scombrus, 4, 212
Scomberesocidae, 83
Scomber spp., 186
Scombridae, 85
Scombrolabrax heterolepis, 48
Scorpaenichthys marmoratus, 52
Scorpaena, 14
Scorpaenichthys, 15
Scorpaenidae, 83
Scorpaenids, 80
Scorpaeniformes, 52, 75, 83
Scorpionfishes (Scorpaenidae), 80, 83
Sculpins (Cottidae), 17, 18, 19, 75, 83, 103–104
Sculpins and scorpionfishes (scorpaenids), 80
Sea bream *(Archosargus rhomboidalis)*, 201
Sea horses and pipefishes (syngnathids), 16, 17, 18
Sea tadpole *(Careproctus reinhardti)*, 84, 95
Seabasses (Serranidae), 11, 104
Sebastes, 13, 15, 75

Sebastes crameri, 83, 93
Sebastes flavidus, 232
Sebastes norvegicus, 16
Sebastes spp., 16, 58, 185, 231
Sebastolobus alascanus, 263
Sebastolobus altivelis, 52, 53
Sebastolobus spp., 19, 64
Serraniculus, 105
Serraniculus pumilio, 106
Serranidae, 11, 104
Serraninae, 104–105
Serranus, 105
Serranus tigrinus, 107
Shad *(Alosa sapidissima)*, 246
Sharks and rays (chondrichthyans), 10
Sharks and rays (elasmobranchs), 22
Shortspine thornyhead *(Sebastolobus alascanus)*, 263
Siluriforms, 17–18
Silversides (Atherinidae), 83
Sixline soapfish *(Grammistes sexilineatus)*, 122
Slender barracudina *(Lestidiops ringens)*, 82, 88
Slender sunfish *(Ranzania laevis)*, 85, 99
Smelts (Osmeridae), 17, 19, 81
Snailfishes (Cyclopteridae), 75, 84
Snappers (Lutjanidae), 84
Snubnose blacksmelt *(Bathylagus wesethi)*, 64
Soapfishes *(Rypticus* spp.), 104
Soldierfish *(Myripristis* sp.), 92
Solea solea, 30
Spanish flag *(Gonioplectrus hispanicus)*, 118
Sparids, 12
Splittail bass *(Hemanthias signifer)*, 112
Spot *(Leiostomus xanthurus)*, 223, 232
Spotted codlet *(Bregmaceros macclellandi)*, 82, 90
Squirrelfishes (Holocentridae), 52, 83
Squirrelfishes and their relatives (beryciforms), 80
Sternoptychids, 103
Stichaeidae, 84
Stizostedion vitreum, 254
Stomias atriventer, 57
Streamer bass *(Pronotogrammus aureorubens)*, 109
Striped bass *(Morone saxatilis)*, 36, 58, 232
Striped mullet *(Mugil cephalus)*, 57
Striped seaperch *(Embiotoca lateralis)*, 13
Stromateidae, 85
Strombus gigas, 241
Sturgeon *(Acipenser* spp.), 31, 232
Sunfishes (centrarchids), 19, 189

Surf smelt *(Hypomesus pretiosus)*, 32, 236
Surfperch (embiotocids), 11, 12, 13, 30, 188
Surgeonfish *(Acanthurus* sp.), 98
Surgeonfishes (Acanthuridae), 85
Swallowtail bass *(Anthias woodsi)*, 111
Syngnathids, 16
Synodontids, 65

T
Tarletonbeania crenularis, 82, 88
Tarpons (Elopidae), 81
Tarpons (elopiforms), 80
Tautogolabrus adsperus, 13
Teleosts, 16, 19, 20, 65, 101
Tetraodontiformes, 85
Thalassenchelys coheni, 86
Theragra chalcogramma, 23–24, 42–43, 64, 127, 175, 184, 186–187, 211, 212, 223, 255, 273
Thornyhead (*Sebastolobus* spp.), 19, 64
Threadfin bass *(Pronotogrammus multifasciatus)*, 108, 111
Threadnose bass *(Anthias tenuis)*, 112
Thunnini (tuna), 230
Thunnus spp., 13, 186
Toadfishes (*Opsanus* spp.), 2
Trachipteridae, 82
Trachipterus altivelus, 57, 82, 88
Trachurus symmetricus, 46, 64, 65, 213
Trispterus esmarkii, 194
Tuna (*Thunnus* spp.), 13, 186
Tunas (Thunnini), 230

U
Urophycis spp. (hakes), 52

V
Vermillion snapper *(Rhomboplites aurorubens)*, 84, 96
Vinciguerria lucetia, 81, 87
Viperfishes (chauliodontids), 27
Viviparous blenny *(Zoarces viviparous)*, 15
Viviparous brotulus (Bythitidae), 82

W
Walleye *(Stizostedion vitreum)*, 254
Walleye pollock *(Theragra chalcogramma)*, 23, 24, 42, 43, 64, 127, 175, 184, 186, 187, 211, 212, 223, 255, 273
White crappie *(Pomoxis annularis)*, 136
White perch *(Morone americanus)*, 232
White perch/striped bass *(Morone americanus/M. saxatilis)*, 58
Winter flounder *(Pseudopleuronectes americanus)*, 62, 246
Witch flounder *(Glyptocephulus zachirus)*, 64
Wrasses (Labridae), 84
Wrasses (labrids), 11–12
Wrymouths (Cryptacanthodidae), 84

X
Xiphister atropurpureus, 35
Xyrichthys sp., 84, 96

Y
Yellowfin bass *(Anthias nicholsi)*, 110
Yellowtail rockfish *(Sebastes flavidus)*, 232
Yellowfin sole (*Limanda aspera*), 266

Z
Zaprora silenus, 84, 97
Zaproridae, 84
Zoarces viviparous, 15
Zoarcids, 75, 230
Zu cristatus, 82, 89

GENERAL INDEX

A
Able, K.W., 84, 95, 155, 215, 216, 232
Aboussouan, A., 60
Absolute fecundity, 25, 26
Activation, 40
Adult stage, 2
Aeration, behavior for, 19
Agardy, T., 237, 238
Ahlstrom, E.H., 4, 6, 46, 56, 65, 80, 81, 82, 83, 86, 87, 88, 90, 93, 100, 103, 150, 172
"Ahlstrom volume," 102, 103
Albinism, 11
Alcala, A.C., 238
Alderdice, D.F., 45, 129, 130, 210, 253
Allen, D.M., 97, 216
Allen, L.G., 233
Allison, G.W., 237
Allometric growth, 293
Allometry, 293
Almatar, S.M., 171
Anadromous and catadromus, 232
Anderson, J.T., 210
Anderson, W.W., J.W. Gehringer, and E. Cohen, 221
Angell, C.L., 131
Annual egg production method, 173
Anthropogenic physical changes in habitat, 232–233
Apstein, C., 129
Aquaculture, 4–5
Artificial fertilization, 250
Artificially spawning, fertilizing, and rearing planktonic fish eggs, 271–272
Assemblage analysis, 178–179
Asynchronous, consecutive, successive, or sequential hermaphrodites, 11
Auer, N.A., 60
Auster, P.J., 238

B
Babcock, R.C., S. Kelly, N.T. Shears, J.W. Walker, and T.J. Willis, 238
Bagenal, T.B., 155, 194
Bailey, K.M., 50, 144, 198, 211, 212, 253, 254, 257
Bailey, K.M., H. Nakata, and H. Van der Veer, 214, 223
Baird, S.F., 246
Baldwin, C.C., 104, 105, 108, 109, 110, 111, 112, 113, 114, 115, 116, 117, 120, 122, 123, 124
Ball, O.P., 46
Balon, E.K., 41
Barham, E.G., 156, 157, 158
Barker, D.L., 216
Barlow, G.W., 189
Barnes, S.H., 65
Barnett, M., 27, 28
Barrowman, N.J., 194
Battaglene, S.C., 248
Batty, R.S., 253

General Index

Baumgartner, T.R., A. Soutar, and V. Ferreira-Bartrina, 193, 225
Baxter, J.L., 219
Beamish, R., 20
Beamish, R.J., D.J. Noakes, G.A. McFarlane, L. Klyashtorin, V.V. Ivanov, and V. Kurashov, 191, 226
Behavior, reproductive, 36–37
Behavior studies, 253–254
Berrien, P.L., 150, 218
Beverton, R.J.H., 197
Beverton-Holt curve, 196, 197
Bigelow, H., 217
Bigger is better hypothesis, 208
Biochemical genetics, 58–59
Biological factors, 213–214
Biological zero, 129
Bisexuality, 12
Blastocoel, 42
Blastopore, 43
Blaxter, J.H. S., 7, 40, 50, 51, 130, 132, 253, 254
Blom, G., J.T. Nordeide, T. Svåsand, and A. Borge, 248
Blood, D.M., A.C. Matarese, and M.M. Yoklavich, 40, 42, 43, 66
Boehlert, G.W., 16, 217, 230, 231
Bond, C., 84, 95
Bongo nets, 155–156
Bony fishes, 10
Bortone, S.A., 104
Botsford, L.W., A. Hastings, and S.D. Gaines, 238, 241
Brachionus plicatilis, 252
Branching diagrams (cladograms), 101, 102
Brander, K., 170
Breder, C., 10, 11, 18
Breed, 30
Breitburg, D.L., K.A. Rose, and J.H. Cowan, 235
Broad dispersal, 242
Brodeur, R.D., 143, 158, 226
Brodeur, R.D., C.E. Mills, J.E. Overland, G.E. Walters, and J.D. Schumacher, 227
Brogan, M.W., 162, 163
Brood, 30
Brothers, E.B., C.P. Mathews, and R. Lasker, 7, 142
Browman, H.I., 8, 40, 136
Brown, D.L., 156, 158
Brown, D.M., 155
Brownell, C.L., 143, 254

Buchanan-Wollaston, H.J., 4
Buckley, L.J., 141, 255
Buckley, L.J., A.S. Smigielski, T.A. Halavik, and G.C. Laurence, 256
Burst swimming speeds of larvae, 51
Busby, M.S., 84, 94
Butler, J.L., H.G. Moser, G.S. Hageman, and L.L. Nordgren, 105, 106

C

Cada, G.F., 154
Cadigan, N.G., 214
California Cooperative Oceanic Fisheries Investigations (CalCOFI), 218, 219
CalVET (*Cal* COFI Vertical *Egg Tow*), 159–160
Cambray, J.A., 186
Camera-net system to sample plankton, 164
Campana, S.E., 143, 209, 257
Canino, M.F., K.M. Bailey, and L.S. Incze, 211
Canino, M.F., 253, 256
Canino, M.F. and K.M. Bailey, 253
Cannibalism, 189, 212
Capture success as a function of predator to prey length ratio, 208
Caputi, N., W.J. Fletcher, A. Pearce, and C.F. Chubb, 211
Carls, M.G., 235
Carson-Ewart, B.M., 60, 80
Catadromus, 232
Caudal region development, 48, 49
Causes of population fluctuations, 193–194
Chambers, M.C., 8
Chapman, M.R., 239
Characters, 62–66
Chase, J., 218
Checkley, D.M., 139, 161, 211
Chen, D., S.E. Zebiak, A.J. Busalacchi, and M.A. Cane, 226
Cheng, L., 156, 158
Choate, J.H., P.J. Doherty, B.A. Kerrigan, and J.M. Leis, 154
Choosing sampling methods, 165
Chorion texture, 65
Chu, Y.T., 115
Cladistics, 101
Cladogram, 101, 102
   of American anthiines based on adult and larval characters, 114
   of epinephelines based on adult and larval characters, 116
   of genera of Epinephelini, 117

Clark, F.N., 32
Clark, J., W.G. Smith, A.W. Kendall, Jr., and M.P. Fahay, 221
Clearing and staining larval fish, 61–62, 291–292
Clemmesen, C., 200, 255, 256, 257
Climatic variation, 225–227
Coastal landscape changes, 233
Coastal ocean pelagic and bottom, 231
Cohen, D.M., 100
Collette, B.B., 85, 98
Colton, J.B., Jr., 169, 218, 221
Commission cod, 247
Conceptual model of the effects of marine reserves, 239
Concordant and disconcordant fluctuation, 192–193
Condition index validation, 254–257
Conducting ichthyoplankton tows, 276–277
Conover, D.O., 215
Conover, D.O., J. Travis, and F.C. Coleman, 238, 239
Conservation biology, 236
Conservation studies, 5–6
Continuous plankton recorders, 160
Continuous Underway Fish Egg Sampler (CUFES), 161, 224
Convergence, 101
Conway, D.V.P., I.R.B. McFadzen, and P.R.G. Tranter, 140
Coombs, S.H., C.A. Fosh, and M.A. Keen, 130, 172
Cordell, J., 5, 234
Cote, I.M., I.Mosqueira, and J.D. Reynolds, 240
Counts, R.C., 82, 90
Coutant, C.C., 234
Cowan, J.H., Jr., 209, 254
Cowen, R.K., 209, 211, 216
Craik, J.C.A., 128
Creswell, R.L., 242
Critical period theory, 198–205
Crowder, L.B., 200, 223
Crowder, L.B., S.J. Lyman, W.F. Figueira, and J. Priddy, 208, 242
Crowder, L.B., and R.G. Werner, 223
Cruising speeds of larvae of various species of fishes, 50
Culturing marine fish larvae, 273–275
Cunha, M.E., I. Figueiredo, A. Farinha, and M. Santos, 150, 174
Currents and flow features, 211, 216

Cury, P., 211
Cushing, D.H., 3, 151, 192, 195, 196, 197, 202, 203, 210, 211, 212, 225
Cushing curves, 196, 197
Cyr, H., 152
Cytometry, 257

D

Daan, N.A., 143
Dahlberg, M.D., 155
Dahlgren, C.P., J.A. Sobel, and D.E. Harper, 241
Daily egg production method, 176
Daily fecundity reduction method, 177
Daily growth ring analysis, 199
Data analysis, 167–169
Data records, 166
Davies, I.E., 156, 157, 158
Davis, G.E., 238, 239, 241
Davis, R., 5
De Lafontaine, Y., 247
Deblois, E., 199, 203, 210, 212
Degree-days, 129
DeMartini, E., 10, 11
Demaster, D.P., 227
Demersal egg spawners, 17
Demersal eggs, 17–19, 232
Dempsey, C.H., 150
Dennis, G.G., 162
Density dependence, 194–195
Depth determination, 164–165
Derived (apomorphic), 101
Diffusive dispersal, 242
Dight, I.J., 238
Direct approach, 56
Direct development, 39
Distribution and abundance, 171–172
Diversity/Identification of planktonic fish eggs and larvae, 281–282
Doan, T., 257
Doherty, P.J., 162, 216
Dovel, W.L., 162
Dower, J.F., 211
Doyle, M.J., 178
Drewry, G.E., 60, 83, 91
Dugan, J.E., 238, 239, 241
Duguay, L.E., 254
Duker, G.J., 217, 245
Dunn, J.R., 50, 82, 90
Duplication phase, 21
D'Vincent, S., H.G. Moser, and E.H. Ahlstrom, 85, 99

E

Early history traits, 189–190
Early life history, 2
Early life history studies, 2–3
Eckmann, R., U. Gaedke, and H.J. Wetzlar, 211
Ecology of fish eggs and larvae
  biological zero, 129
  degree-days, 129
  ecology defined, 125–126
  effects of disease on fish eggs, 131
  egg ecology, 126, 128
  environmental influences on feeding, 140–141
  factors affecting egg survival, 129–131
  factors influencing rates of predation, 144
  feeding and condition, 135–142
  feeding mechanism, 135–136
  food organisms, 140
  functional development, 133–134
  functional morphology of larvae, 131–134
  growth, 142–143
  length frequency diagrams, 142
  measuring condition, 141–142
  measuring predation, 144–145
  modes of predation on larval fishes, 144
  otoliths, 142–143
  oxygen consumption of fish eggs, 130
  position of fish eggs and larvae in the ecosystem, 126
  predation, 143–145
  prey size in relation to larval fish size, 139–140
  relationship between body size and development and the characteristics of water, 131–133
  relationship between prey width and larval length, 140
  spring bloom, 134–135
  tagesgrade, 129
  temperature, 129
  vertical distribution, 128–129
Economou, A.N., 140
Ecotourism, 238
Effluents and organochemicals, 235–236
Egg, 1
Egg ecology, 126, 128
Eggs and larvae, development of, 42, 43
Ehrenbaum, E., 6, 59
Electrophoresis, 58

Ellertsen, B., 160, 211
Ellertsen, B., P. Fossum, P. Solemdal, and S. Sundby, 203
Embryo characters, 65–66
Embryonic development
  activation, 40
  blastocoel, 42
  blastopore, 43
  development, 44–45
  eggs and larvae, development of, 42, 43
  embryonic shield, 42
  events immediately after spawning, 40
  fertilization, 40
  germ ring, 42
  myomeres, 43
  neural keel, 42
  notochord, 43
  optic vesicle, 43
  periblast, 42
  perivitelline space, 42–44
  rate of development and aging of eggs, 45
  terminology, 41
Embryonic development and endogenous nutrition, studies of, 253
Embryonic shield, 42
Embryonic stage, 1
Embryonic stage of development, determining, 66
Empirical relationships, 195–197
Empirical stock/recruitment curves, 196
Environmental data, 165
Environmental influences, 195
Environmental influences on feeding, 140–141
Epinepheline tribes, 108–124
Eppley, R.W., 202
Estuaries, 231–232
Estuarine dependent, 232
Eutrophication and oxygen depletion, 235
Evidence for starvation of larvae at sea, 199–200
Evidence for the sensitivity of larval fishes to lack of food, 200, 202–205
Exotic introductions, 233–234

F

Factors affecting survival of eggs and larvae, 209–213
Factors affecting survival of juveniles, 215–217
Fahay, M.P., 59, 60, 80, 81, 82, 86, 90, 118, 155, 216, 232

General Index

Fanning, 19
Fay, L.A., M.L. Gessner, and P.C. Stromberg, 236
Fecundity, 25–29, 184–185
Fecundity determination, 269–270
Feeding mechanism, 135–136
Fertilization, 40
Fin rays, 74–75
Fish eggs and larvae: identification and systematics
  "Ahlstrom volume," 102, 103
  biochemical genetics, 58–59
  branching diagrams (cladograms), 101, 102
  characters, 62–66
  chorion texture, 65
  cladistics, 101
  cladogram of American anthiines based on adult and larval characters, 114
  cladogram of epinephelines based on adult and larval characters, 116
  cladogram of genera of Epinephelini, 117
  clearing and staining, 61–62
  convergence, 101
  derived (apomorphic), 101
  direct approach, 56
  electrophoresis, 58
  embryo characters, 65–66
  embryonic stage of development, determining, 66
  epinepheline tribes, 108–124
  fin rays, 74–75
  groups of American anthiines based on larval characters, 114
  higher-level characters, 80, 81–85
  historical perspective, 80, 100
  homologous, 101
  identification and staging of fish eggs, 62–66
  indirect approach, 56
  laboratory, 60–61
  larval characters, 75–80
  meristic characters, 74–75
  microsopes, 61
  monophyletic, 102
  morphological identification, 56–58
  morphology, 67–73
  myomeres, 74
  oil globules, 65
  ontogenetic criterion, 102
  ontogeny recapitulates phylogeny, 101
  Ordinal/Subordinal eggs and larval characters of Northeast Pacific fishes, 68–71
  Ordinal/Subordinal fin and meristic characters of Northeast Pacific fishes, 72–73
  outgroup comparison, 101
  perivitelline space, 65
  pigment, 74
  polarity of larval characters, 101, 102
  primitive (pleisiomorphic), 101
  radiography, 62
  shape, 64
  shared derived characters (synapomorphies), 101
  size, 64
  systematic methods, 101–102
  theory of systematics, 100–101
  tools, 61
  transformation series, 101
  use of larval fish characters in systematic studies, 102–124
  vertebral counts, 76–79
  video and image analysis, 61
  yolk characters, 65
Fisheries mismanagement, 237
Fisheries-Oceanography Coordinated Investigations (FOCI), 222–223
Fitch, J.E., 105
Flowmeters, 163–164
Fogarty, M.J., 7, 185
Fogarty, M.J., M.P. Sissenwine, and E.B. Cohen, 182
Fontaine, C.T., 252
Food availability, 215
Food density thresholds for marine fish larvae, 201
Food for larvae, 251–252
Food organisms, 139–140
Food requirements and feeding studies, 253
Formulas for estimating density and abundance of eggs or larvae in a plankton tow, 168
Forrester, C.R., 45, 129, 130, 210, 253
Fortier, L., 143, 170, 203
Fortuño, J.M., 60
Foucher, R., 20
Foundation of the critical period concept, 191
Frame trawl, 159
Francis, M.P., 210

Francis, R.C., S.R. Hare, A.B. Hollowed, and W.S. Wooster, 226
Frandsen, M., 235
Frank, K.T., 143
Franzin, W.G., 161
Fraser, J.H., 148
Freeberg, M.H., W.W. Taylor, and R.W. Brown, 212
Fritzsche, R.A., 60
Fuiman, L.A., 8, 150, 247, 254
Fukuhara, O., 47, 253
Furuyu, K., 248

G
Gagne, J.A., 203
Gallagher, R.P., 154
Gamble, J.C., 247
Gametes, obtaining, 249
Garcia Rubies, 243
Gehringer, J.W., 161
Germ ring, 42
Global Ocean Ecosystem Dynamics (GLOBEC), 224–225
Goksoyr, A., T.S. Solberg, and B. Serigstad, 257
Gonochorism, 12
Goode, G.B., 245
Gordon, D.J., D.F. Markle, and J.E. Olney, 82, 89
Gosline, W.A., 104
Gould, S.J., 100
Govoni, J.J., 8, 140, 211
Govoni, J.J., G.W. Boehlert, and Y. Watanabe, 50, 136, 253
Graham, J.J., S.R. Chenoweth, and C.W. Davis, 142
Granmo, A., R. Ekelund, J.A. Sneli, M. Berggren, and J. Svavarsson, 236
Grant, A., 8
Grave, H., 212
Graves, J.E., M.J. Curtis, P.A. Oeth, and R.S. Waples, 58, 59
Gregory, R.S., 154
Grimes, C.B., 211
Gross, M.R., 18
Groups of American anthiines based on larval characters, 114
Growth-mortality theory, 208–209
Growth rate, 186
Guarding, 19
Guenette, S., T. Lauk, and C. Clark, 238, 241
Gulec, I., and D.A. Holdway, 235
Gunderson, D.R., 148, 173, 241, 242, 264
Gunderson, D.R., D.A. Armstrong, Y.B. Shi, and R.A. McConnaughey, 232
Gynogenesis, 12

H
Habitats
  anadromous and catadromus, 232
  anthropogenic physical changes in habitat, 232–233
  coastal landscape changes, 233
  coastal ocean pelagic and bottom, 231
  demersal eggs and pelagic larvae, 232
  effluents and organochemicals, 235–236
  estuaries, 231–232
  estuarine dependent, 232
  eutrophication and oxygen depletion, 235
  exotic introductions, 233–234
  metals, 236
  modified runoff, 232–233
  oil and oil dispersants, 235
  open-ocean pelagic, 230
  pollutants, 234–236
  shoreline armoring, 233
  variety of habitats occupied by fish eggs and larvae, 230–232
Haegele, C., 34
Hales, L.S., 215
Halpern, B.S., 240
Harbicht, S.M., 161
Hardy, J.D., Jr., 60
Hare, J.A., 209, 211, 216, 230
Harris, R.P., 160, 231, 232
Harvey, S.M., 128
Haryu, T., 84, 97
Hatching, 45, 47
Hattori, S., 41
Haury, L.R., J.A. McGowan, and P.H. Weibe, 149, 158
Hauser, J.P., 169
Hay, D.E., 166
Heath, M.W., 148, 150, 152, 155, 184
Hempel, G., 17, 18, 27, 30, 34, 133
Hermaphroditism, 11–12
Hermes, R., N.N. Navaluna, and A.C. del Norte, 148
Herring hypothesis, 205
Hewitt, R.P., 192, 219, 220

Hewitt, R.P., G.H. Theilacker, and N.C. Lo, 213
High-speed plankton samplers, 161
Higher-level characters, 80, 81–85
Hill, R.L., 227, 242
Hinckley, S., 24
Hinckley, S., A.J. Hermann, K.L. Meir, and B.A. Megrey, 188
Hislop, J.R.G., 215
Hissmann, K., H. Fricke, and J. Schauer, 187
Histological condition index, 289–290
Histology to determine if multiple spawnings are occurring, use of, 263–268
Hjort, J., 3, 135, 183, 190, 191, 198, 202, 205, 209, 217, 218, 245, 246
Holbrook, S.J., 214
Holdway, D.A., 235
Hollowed, A.B., K.M. Bailey, and W.S. Wooster, 192
Holm, E.R., 214
Holt, G.J., 141
Holt, S.H., 197
Homologous, 101
Horn, M.H., 233
Houde, E.D., 50, 82, 85, 90, 97, 144, 149, 150, 162, 164, 171, 189, 190, 201, 208, 212, 248, 253, 254
Hourston, A., 34
Howell, B.R., 249
Howell, B.R., E. Moksness, and T. Svåsand, 248
Howell, W.H., 45, 47, 253
Hubbs, C.L., 41, 52, 115
Hunt von Herbing, I., 210
Hunte, W., 214
Hunter, J.R., 4, 137, 138, 173, 201, 264
Hunter J.R., N.C.H. Lo, and L.A. Fuiman, 8
Hurst, T.P., 215
Hybridogenesis, 12

I

Ichthyoplankton, 148, 150
Identification and staging of fish eggs, 62–66
Identification of fish eggs and larvae, 59–62
Identification of fish larvae, 66–80
Iles, T.C., 196
Indirect acoustic and optical, 162
Indirect approach, 56
Indirect development, 39
Individual size, 186
Influence of age-related changes in encounter rates on the gross vulnerability of larval fish of different sizes to predation, 209

Isaacs, J.D., 155, 157, 221
Isaacs-Kidd depressor, 159
Iwata, N., 248

J

Jahn, A.E., 169
Jameson, G.S., 239
Jaroszynski, L.O., 156
Jennings, S., 240
Johns, D.M., 47
Johnson, G.D., 27, 28, 60, 104, 105, 108, 110, 113, 115, 116, 117, 123, 124
Johnson, M.W., J.R. Rooker, D.M. Gatlin, III, and G.J. Holt, 256
Jones, C.M., 143
Jones, P.J.S., 238
Jones, P.W., F.D. Martin, and J.D. Hardy, Jr., 60
Jones, S., 83, 92
Jørstad, K.E., V. Øiestad, O.I. Paulsen, K. Naas, and O. Skaala, 248
Juanes, F., 215, 243
Jude, D.J., 162
Juvenile, 40
Juvenile development, 52–54
Juvenile habitat requirements, 216–217
Juvenile stage, importance of, 213–217
Juveniles, 2

K

Kalmer, E., 140
Kaufman, L., J. Ebersole, J. Beets, and C.C. McIvor, 214
Kawaguchi, K., 81, 87
Keene, M.J., 83, 91
Keener, P., 115, 161
Kellermann, A., 60
Kelso, J.R.M., 236
Kelso, W.A., 148, 152
Kendall, A.W., Jr., 40, 41, 52, 58, 65, 66, 83, 93, 105, 106, 107, 108, 109, 110, 111, 112, 113, 117, 118, 119, 121, 127, 172, 217, 245
Kendall, A.W., Jr., L.S. Incze, P.B. Ortner, S.R. Cummings, and P.K. Brown, 222
Kidd, L.W., 155, 157
Kikuchi, K., 248
Kingsford, M.J., I.M. Suthers, and C.A. Gray, 236
Kiørboe, T., 253, 254
Klyashtorin, L., 225
Knutsen, G.M., 253

Koening, C.C., F.C. Coleman, C.B. Grimes, G.R. Fitzhugh, K.M. Scanlon, C.T. Gledhill, and M. Grace, 238
Koslow, J.A., 150, 183, 192, 228
Koslow, J.A., S. Brault, J. Dugas, and F. Page, 215, 218
Kotthaus, A., 117
Kramer, D., 81, 86
Kramer, D.L., 239
Kumaran, M., 83, 92
Kuntz, Y., 10, 40

L

La Bolle, L.D., Jr., H.W. Li, and B.C. Mundy, 162
Laboratory exercises
  artificially spawning, fertilizing, and rearing planktonic fish eggs, 271–272
  clearing and staining larval fish, 291–292
  conducting ichthyoplankton tows, 276–277
  culturing marine fish larvae, 273–275
  diversity/identification of planktonic fish eggs and larvae, 281–282
  fecundity determination, 269–270
  histological condition index, 289–290
  histology to determine if multiple spawnings are occurring, use of, 263–268
  larval fish gut contents, 286–288
  midgut cell height, 289–290
  ova measurement to determine number of spawnings and introduction to writing a scientific paper, 261–262
  preparing and examining larval fish otoliths for daily growth, 283–285
  sorting and identifying fish eggs and larvae in ichthyoplankton samples, 278–280
Laboratory setup, 60–61
Landry, F., T. J. Miller, and W.C. Leggett, 211
Laprise, R., 231
Large scale culturing and rearing, 248–249
Laroche, J., 33, 83, 84, 96
Laroche, W.A., 93
Larva, 40
Larval characters, 75–80
Larval development
  burst swimming speeds of larvae, 51
  caudal region development, 47, 48
  cruising speeds of larvae of various species of fishes, 50
  hatching, 45, 47
  notochord flexion and stages of larval development, 47, 50
  photophore formation, 51
  transformation stage, 51–52
Larval fish gut contents, 286–288
Larval stage, 1
Lasker, R., 3, 7, 150, 173, 200, 203, 204, 205, 211, 219
Last, J.M., 139, 140
Lauck, T., C.W. Clark, M. Mangel, and G.R. Munro, 237
Laurence, G.C., 45, 201, 253
Leggett, W.C., 143, 170, 199, 203, 209, 210, 212, 247
Lehtinen, K.J., K. Mattsson, J. Tana, C. Engstrom, O. Lerche, and J. Hemming, 236
Leis, J.M., 8, 59, 60, 80, 84, 85, 96, 98, 99, 100, 115, 117, 118, 119
Leithiser, R.L., K.F. Ehrlich, and A.B. Thum, 160
Length frequency diagrams, 142
Leslie, J.K., 236
Levin, P., 216
Levings, C.O., 239
Lewis, R.M., W.F. Hettler, Jr., E.H. Wilkins, and G.N. Johnson, 161
Li, S., 254
Life history stages of fish, 1–2
Light for rearing, 251
Light traps, 162, 163
Lindeman, K.C., 216
Lindeman, K.C., R. Pugliese, G.T. Waugh, and J.S. Ault, 239
Lindholm, J.B., P.J. Auster, M. Ruth, and L. Kaufman, 242
Lindquist, A., 230, 231
Litvak, M.K., 209
Liu, H.W., R.R. Stickney, W.W. Dickhoff, and D.A. McCaughran, 253
Lluch-Belda, D., R.A. Schwartzlose, R. Serra, R. Parrish, T. Kawasaki, D. Hedgecock, and R.J.M. Crawford, 193, 225
Lo, N.C.H., 4, 173, 175, 221
Loar, J.M., 154
Longevity, 185
Longhurst, A., 158, 185
Lough, R.G., 141
Lough, R.G., and D.G. Mountain, 141
Love, M.S., M. Yoklavich, and L. Thorsteinson, 185

M

Mabee, P.M., 101, 102
MacDonald, M., 246
Macewicz, B.J., 264
MacKenzie, B.R., 253, 254
MacKenzie, B.R., T.J. Miller, S. Cyr, and W.C. Leggett, 211
Macklin, S.A., 211
Madenjian, C.P., 162
Man, A., R. Law, and N.V.C. Polunin, 238
Mangor-Jensen, A., 235
Mantua, N.J., S.R. Hare, Y. Zhang, J.M. Wallace, and R.C. Francis, 226
Marak, R.R., 155
Marcy, B.C., Jr., 155
Margulies, D., 253
Marine protected areas, 5, 236, 237–243
Marine reserves, 237–239
Marine reserves and recruitment, 240–243
Markle, D.F., 56, 82, 90
Marliave, J.B., 35, 216
MARMAP (Marine Resources Monitoring, Assessment, and Prediction), 222
Marr, J.C., 219
Marsden, J.E., C.C. Krueger, and H.M. Hawkins, 154
Martell, S.J.D., C.J. Walters, and S.S. Wallace, 239
Martin, F.D., 60, 83, 91
Mason, J.C., 155, 162
Mass releases of yolk sac larvae, 245–247
Masuda, R., 248
Matarese, A.C., 56, 57, 60, 63, 64, 67, 68–73, 77, 80, 81, 82, 83, 84, 85, 86, 88, 91, 95, 96, 97, 99
Matarese, A.C., A.W. Kendall, Jr., D.M. Blood, and B.M. Vinter, 60, 63, 64
Matarese, A.C., S.L. Richardson, and J.R. Dunn, 49
Mathias, J.A., 254
May, R.C., 198, 199, 200
McDermott, S.F. and S.A. Lowe, 265
McEachron, L.W., R.L. Colura, B.W. Bumguardner, and R. Ward, 248
McEvoy, J., 253
McEvoy, L.A., 253
McGowan, J., 227, 228
McGowan, J.A., 155
McGurk, M.S., 199
Mearns, A.J., 65
Megrey, B.A., 173, 175
Megrey, B.A., A.B. Hollowed, S.R. Hare, S.A. Macklin, and P.J. Stabeno, 223
Member-vagrant hypothesis, 207
Meristic characters, 74–75
Merrett, N.R., 65
Mertz, G., 203, 211, 212
Mesocosm studies, 247–248
Mesoscate dispersal, 242
Messieh, S.N., 126
Metals, 236
Methot, R.D., 155, 157, 158, 159, 204, 221
Methot trawl, 157–158, 159
Microscopes, 61
Midgut cell height, 289–290
Milicich, M.J., 216
Miller, D., 234
Miller, J.M., W. Watson, and J.M. Leis, 60
Miller, T.J., L.B. Crowder, J.A. Rice, and E.A. Marschall, 3, 226
Miller high-speed sampler, 161
Milt, 20
Misitano, D.A., E. Casillas, and C.R. Haley, 236
Modes of predation on larval fishes, 144
Modified runoff, 232–233
Moksness, E., 247, 257
Monophyletic, 102
Monteleone, D.M., 254
Morgan, R.P., Jr., 58, 236
Mork, J., P. Solemdal, and G. Sundnes, 58
Morphological identification, 56–58
Morphology, 67–73
Morse, W.W., 150, 169
Mortality estimation, 178
Moser, H.G., 8, 16, 56, 58, 60, 80, 81, 82, 87, 88, 100, 102, 103, 150, 178, 221
Moser, H.G., R.L. Charter, P.E. Smith, D.A. Ambrose, S.R. Charter, C.A. Myers, E.M. Sandknop, and W. Watson, 6
Mountain, D.G., 141
Müller, U.K., 131
Müller, U.K., and J.J. Videler, 131
Mulligan, T.J., F.D. Martin, R.A. Smucker, and D.A. Wright, 7

Multiple trophic levels, 228
Multispecies focus, 228
Mundy, B.C., 217, 230
Munk, P., 59, 136, 158, 253
Murawski, S.A., R. Brown, H.L. Lai, P.J. Rago, and L. Hindrickson, 237
Murphy, G.I., 219
Myers, R.A., 187, 194, 203, 211, 212, 214
Myers, R.A., G. Mertz, and N.J. Barrowman, 196
Myomeres, 43, 74

N
Naas, K.T., 248
Naevdal, G., 58
Nagahama, Y., M. Yoshikuni, M. Yamashita, T. Tokumoto, and Y. Katsu, 22
Nedreaas, K., 58
Neilson, J.D., 143, 172
Neira, F.J., A.G. Miskiewicz, and T. Trnski, 60
Nelson, J.S., 104
Nest building, 19
Nest cleaning, 19
Neural keel, 42
Neuston nets, 156–157
Nichols, J.H., 253
Nielsen, J.G., 59
Night lights, 162
Nishiyama, T., 84, 97
Nissling, A., 128
Noble, R.L., 161
Nondispersing, 242
Norcross, B.L., 235
Nordeide, J.T., J.H. Fossa, A.G.V. Salvanes, and O.M. Smedstad, 215, 248
Notochord, 43
Notochord flexion and stages of larval development, 47, 50
Nursery ground, 216–217

O
O'Brien, W.J., 136
O'Connell, C.P., 201
Offshore transport hypothesis, 190, 198
Offshore transport-retention theory, 205–209
Øiestad, V., 247
Oil and oil dispersants, 235
Oil globules, 65
Okiyama, M., 59, 60, 82, 88
Olivar, M.P., 60

Olla, B.L., M. Davies, and C.H. Ryer, 249
Olney, J.E., 231
Ontogenetic criterion, 102
Ontogeny, 39
Ontogeny recapitulates phylogeny, 101
Oocyte growth stages in teleosts, 20
Oogenesis, 20, 31
Oozeki, Y., 50, 253
Open-ocean pelagic, 230
Optic vesicle, 43
Ordinal/Subordinal eggs and larval characters of Northeast Pacific fishes, 68–71
Ordinal/Subordinal fin and meristic characters of Northeast Pacific fishes, 72–73
Orton, G.L., 101
Osborn, T.R., 211
Otolith increment validation, 257
Otoliths, 142–143
Outgroup comparison, 101
Ova measurement to determine number of spawnings and introduction to writing a scientific paper, 261–262
Ovoviviparous, 16
Oxygen consumption of fish eggs, 129–130
Ozawa, T., 60

P
Pacific Decadal Oscillation (PDO), 226
Palsson, W., 131, 238, 241
Parental investment in individual offspring, 188–189
Parenti, L.R., 62
Paul, A.J., 136
Pearson, K.E., 264
Pentilla, D., 32
Pepin, P., 187, 189, 212, 213, 231
Periblast, 42
Perivitelline space, 42–44, 65
Perry, R.I., 172
Peterson, C.H., 5
Phillips, A.C., 155, 162
Phillips, A.C., and J.C. Mason, 162
Photophore formation, 51
Physical factors, 210–211
Physoclistous swim bladder, 11
Picquelle, S.J., 173, 175
Pigment, 74
Pinder, A.C., 60
Pipe, R.P., S.H. Coombs, and K.F. Clarke, 158

Planes, S.,R. Galzin, A. Garcia Rubies,
R. Goñi, J.G. Harmelin, L. Le Diréach,
P. Lenfant, and A. Quetglas, 241, 243
Plankton pumps, 160
Platforms (ships), 153–154
Pleisiomorphic, 101
Polacheck, T., D. Mountain, D. McMillan,
W. Smith, and P. Berrien, 211
Polarity of larval characters, 101, 102
Pollutants, 234–236
Pollution effects, 257
Polovina, J.J., G.T. Mitchum, N.E. Graham,
M.P. Craig, E.E. Demartini, and
E.N. Flint, 228
Polychlorinated biphenyls (PCBs), 236
Polynuclear aromatic hydrocarbons
(PAHs), 236
Population
Beverton-Holt curve, 196, 197
bigger is better hypothesis, 208
capture success as a function of predator
to prey length ratio, 208
causes of population fluctuations,
193–194
characteristics, 184–190
concordant and disconcordant
fluctuation, 192–193
critical period theory, 198–205
Cushing curves, 196, 197
daily growth ring analysis, 199
density dependence, 194–195
early-history traits, 189–190
empirical relationships, 195–197
empirical stock/recruitment curves, 196
environmental influences, 195
evidence for starvation of larvae at
sea, 199–200
evidence for the sensitivity of larval fishes
to lack of food, 200, 202–205
fecundity, 184–185
fluctuations, 190–194
food density thresholds for marine fish
larvae, 201
growth-mortality theory, 208–209
growth rate, 186
herring hypothesis, 205
individual size, 186
influence of age-related changes
in encounter rates on the gross
vulnerability of larval fish of different
sizes to predation, 209

longevity, 185
match or mismatch of larval production
to that of their larval food, 203
member-vagrant hypothesis, 207
offshore transport-retention theory,
205–209
parental investment in individual
offspring, 188–189
population fecundity, 196
population size, 186–187
recruitment process in fishes, 190
recruitment theories, 198–205
Ricker curve, 195–196, 197
spawner-recruit relationship, 194–209
spawning frequency, 188
specific fecundity, 185
specificity of habitate requirements, 188
stage duration hypothesis, 208
stock-recruitment relationships, 197
survival curves for natural populations
of fishes, 198–199
variability in world population
abundance of fishes, 187
Population assessment, 3–4
Population dynamics, 182–184
Population dynamics and recruitment,
183–184
Population fecundity, 25, 196
Population size, 186–187
Porter, S.M., 141, 255
Posgay, J.A., 155
Position of fish eggs and larvae in the
ecosystem, 126
Potthoff, T., 61, 74, 292
Potthoff, T., W.J. Richards, and
S. Ueyanagi, 48
Powell, P., 219
Powles, H., 56, 154
Predation, 143–145, 254
Predator removal, 19
Predators, 212–213, 215–216
Preparing and examining larval fish otoliths
for daily growth, 283–285
Prey size in relation to larval fish size, 139
Primary growth phase, 21
Primitive, 101
Prince, R.D., 236
Propagation of fish population, 245–246
Propagule dispersal models and population
structure, 242

Protandric hermaphrodites, 12
Protandrous hermaphrodites, 20
Protogynous hermaphrodites, 20
Protogyny, 12
Pryor-Connaughton, V., C.E. Epifanio, and R. Thomas, 254
Purcell, J.E., 143
Purcell, J.E., T.D. Siferd, and J.B. Marliave, 254

Q
Quantz, G., 253

R
Radiography, 62
Radovich, J., 219
Rate of development and aging of eggs, 45
Ray, M., 241
Ray-finned vertebrates, 10
Raymond, L.P., 201
Rearing and culturing of marine fishes
   artificial fertilization, 250
   behavior studies, 253–254
   Brachionus plicatilis, 252
   Commission cod, 247
   condition index validation, 254–257
   cytometry, 257
   embryonic development and endogenous nutrition, studies of, 253
   food for larvae, 251–252
   food requirements and feeding studies, 253
   gametes, obtaining, 249
   histology, 255
   large scale culturing and rearing, 248–249
   light for rearing, 251
   mass releases of yolk sac larvae, 245–247
   mesocosm studies, 247–248
   otolith increment validation, 257
   pollution effects, 257
   predation, 254
   propagation of fish population, 245–247
   rearing containers, 250
   RNA/DNA, 255–256
   rotifiers, culturing, 252
   temperature control, 251
   water for rearing, 250–251
   wild food, obtaining, 252
Rearing containers, 250
Recruitment, setting, 214
Recruitment fluctuations, 3
Recruitment of targeted populations and marine reserves, 240–243
Recruitment process in fishes, 190
Recruitment studies
   Fisheries-Oceanography Coordinated Investigations (FOCI), 222–223
   future of, 225–228
   Global Ocean Ecosystem Dynamics (GLOBEC), 224–225
   history of, 217–222
   multiple trophic levels, 228
   multispecies focus, 228
   new paradigm for, 227–228
   South Atlantic Bight Recruitment Experiment (SABRE), 223–224
   timescales, 227–228
   timescales of variability and biological responses, 225–227
   U.S. GLOBEC program, 225
Recruitment theories, 198–205
Reed, D.C., 241
Regime shifts, 150
REGROUP, 178–179
Relationship between prey width and larval length, 140
Relationship of marine reserves to recruitment of targeted populations, 240–243
Releases of yolk sac larvae, 245–247
Rennis, D.S., 59, 60, 119
Reporting results, 171–179
Reproductive process
   absolute fecundity, 25, 26
   behavior, reproductive, 36–37
   bony fishes, 10–11
   breed, 30
   brood, 30
   demersal egg spawners, 17
   demersal eggs, 17–19
   demersal spawners, 17
   developing, 23
   differences between demersal and pelagic eggs, 17
   dimorphic mating pairs with demersal eggs, 18
   duplication phase, 21
   egg cases, 10
   egg laying, 16–19
   Egg Mass Ratio (EMR) and Gonadal-Somatic Index (GSI), 24–25
   envelope formation, 21–22

General Index

estimating fecundity, 29
factors triggering maturation and
    spawning, 30–35
fecundity, 25–29
fecundity estimation techniques, 29
fertilization, 30
follicle development, 21
gestation, 30
gonadal development, 20
gross anatomical examination, 23
gross maturation stages, 22–25
habitats, 36
hatching, 30
heterochronal spawn, 14
histological examination, 29
hormonal regulation of vitellogenesis
    in fishes, 22
immature, 23
incubation time, 30
indeterminate spawn, 14
internal fertilization, 10
ischronal spawn, 14
iteroparous, 20, 35
iteroparous timing, 14
larval survival and natural selection, 28
live-bearing, 12–16
mating, 30
mating associations, 18
maturation, 22
maturation phase, 21
mature, 23
mean ova diameter determination, 29
migrations, 36
moon cycles, 32
mouth brooding, 18
oogenesis, 20–22
oviparous fishes, 16
ovulation, 22
pelagic eggs, 19
pelagic spawners, 17
photoperiod and periodicity, 31–32
population fecundity, 25
primary growth phase, 21
recovery, 23
redds, 19
relative fecundity, 25
reproductive patterns, 12–16
secondary sexual characters, 37
semelparous, 35
semelparous timing, 14
sexual dimorphism, 37
single spawn, 14
size distribution of oocytes of
    discontinuous and serial spawners, 34
spawning, 23, 29–37
spawning sites, 35–36
spawning variation with latitude, 32–34
spent, 23
spermatogenesis, 20–22
teleosts with diverse reproductive
    patterns, 14–15
vitellogenesis, 21, 22
viviparity, 16
viviparous fishes, 12
yolk vesicle formation, 21
Revera, D.B., 252
Rice, J.A., L.B. Crowder, and
    F.P. Binkowski, 194
Richards, W.J., 59, 60, 81, 84, 85, 86, 96, 98,
    105, 106, 107, 216
Richardson, S.L., 33, 83, 84, 93, 94, 95, 148,
    155, 165, 166, 168, 172
Ricker, W.E., 197
Ricker curve, 195–196, 197
Riley, J.D., 201
RNA/DNA, 255–256
Roberts, C.M., 238, 242
Roberts, C.M., J.A. Bohnsack, F. Gell,
    J.P. Hawkins, and R. Goodridge, 238, 242
Robinson, S.M.C., 255
Rocha-Olivares, A., 58, 59
Roemmich, D., 227, 228
Role of stimuli and behavior in fish movement
    to estuarine nursery areas, 217
Rooker, J.R., 141
Rosen, D., 10, 11, 18
Rosenthal, H., 126
Rothschild, B.J., 151, 185, 192, 211
Rotifers, culturing, 252
Roy, C., 211
Ruple, D., 85, 98
Russ, G.R., 238
Russell, F.S., 59, 60, 150
Rutherford, D.A., 148, 152
Ryland, J.S., 253

S
Sakurai, Y., 34
Salinity, 210–211
Sameoto, D.D., 156, 159
Sample collecting, 166
Sample preservation, 166

General Index

Sampling fish eggs and larvae
   annual egg production method, 173
   assemblage analysis, 178–179
   bongo nets, 155–156
   CalVET (*Cal* COFI *Vertical Egg Tow*), 159–160
   camera-net system to sample plankton, 164
   choosing sampling methods, 165
   continuous plankton recorders, 160
   continuous underway fish egg sampler, 161
   daily egg production method, 176
   daily fecundity reduction method, 177
   data analysis, 167–169
   data records, 166
   depth determination, 164–165
   distribution and abundance, 171–172
   environmental data, 165
   flowmeters, 163–164
   formulas for estimating density and abundance of eggs or larvae in a plankton tow, 168
   frame trawl, 159
   high-speed plankton samplers, 161
   ichthyoplankton, 148
   indirect acoustic and optical, 162
   Isaacs-Kidd depressor, 159
   light traps, 162, 163
   Methot trawl, 157–158, 159
   Miller high-speed sampler, 161
   mortality estimation, 178
   neuston nets, 156–157
   night lights, 162
   plankton pumps, 160
   platforms (ships), 153–154
   REGROUP, 178–179
   reporting results, 171–179
   sample collecting, 166
   sample preservation, 166
   sampling gear, 154–162
   serial net samplers, 158–159
   shoreside sample processing, 166–167
   sleds, 162
   spawning biomass distribution, 173–175
   survey design, 152–153
   tidal and channel nets, 161
   time and space scales, 148–149
   Tucker plankton trawls, 156, 157
   types of studies, 150
   use of studies, 150–152
   variability in ichthyoplankton data, 169–171
   vertical distribution, 172

Sampling gear, 154–162
Sandknop, E.M., 56, 57
Sanzo, L., 83, 92
Sars, G.O., 4, 245
Saville, A., 173
Schanck, D., 155, 161
Schekter, R.C., 50, 253
Schmitt, R.J., 214
Schreiber, H.N., 219
Schultz, E.T., 216
Schumacher, J.D., N.A. Bond, R.D. Brodeur, P.A. Livingston, J.M. Napp, and P.J. Stabeno, 227
Scofield, W.L., 219
Secor, D.H., 248
Seikai, T., 248
Self-fertilization, 11
Semelparity, 188
Semelparous, 2, 20
Senescence, 2
Serebryakov, V.P., 183
Serial net samplers, 158–159
Sette, O.E., 4, 41, 172, 217, 218, 219, 220
Sexuality in fishes, 11
   asynchronous, consecutive, successive, or sequential hermaphrodites, 11
   bisexuality, 12
   gonochorism, 12
   gynogenesis, 12
   hermaphroditism, 11–12
   hybridogenesis, 12
   protandric hermaphrodites, 12
   protogyny, 12
   range of, 11
   self-fertilization, 11
   synchronous or simultaneous hermaphroditism, 11
   unisexuality or parthenogenesis, 12
Shackell, N.L., 238
Shape, 64
Shared derived characters (synapomorphies), 101
Shaw, R.F., 209
Shears, T.H., 213
Shelbourne, J.E., 4, 247
Shen, W., 141, 257
Shepherd, J.G., 196, 197
Shine, R., 18
Shoreline armoring, 233
Shoreside sample processing, 166–167
Sibunka, J.D., 150, 222
Silverman, M.J., 222

Sinclair, M., 3, 205, 206, 207
Sissenwine, M.P., 169
Size, 64
Skiftesvik, A.B., 8, 40
Skud, B.E., 225
Sleds, 162
Smirnov, B., 225
Smith, P.E., 148, 155, 165, 166, 168, 172, 178, 202, 221
Smith, P.J., P.G. Benson, and A.A. Frentzos, 58, 115
Smith, W.G., 150
Sogard, S.M., 54, 215
Solemdal, P., 160, 247
Song, J., 62
Sorting and identifying fish eggs and larvae in ichthyoplankton samples, 278–280
Soutar, A., 221
South Atlantic Bight Recruitment Experiment (SABRE), 223–224
Sparta, A., 82, 89
Spawner-recruit relationship, 194–209
Spawning biomass distribution, 173–175
Spawning frequency, 188
Specific fecundity, 185
Specificity of habitat requirements, 188
Spermatogenesis, 20, 31
Spring bloom, 134–135
Springer, V.G., 61
Squamation, 2, 40
Stable ocean hypothesis, 203–204
Stage duration hypothesis, 208
Stearns, D.E., G.J. Holt, R.B. Forward, Jr., and P.L. Pickering, 254
Steedman, H.F., 148
Steele, J.H., 191, 227
Stehr, C.L., 20, 257
Stephens, J.S., Jr., P.A. Morris, D.J. Pondella, T.A. Koonce, and G.A. Jordan, 233
Steroidogenesis, 31
Stevens, E.G., 83, 93
Stevens, E.G., A.C. Matarese, and W. Watson, 84
Stevens, E.G., W. Watson, and A.C. Matarese, 97
Stock-recruitment relationships, 197
Stockhausen, W.T., R.N. Lipcius, and B.M. Hickey, 238
Stoecker, D.K., 140
Stoner, A.W., 241
Storminess and turbulence, role of, 211
Strauss, R.E., 104

Survey design, 152–153
Survival curves for natural populations of fishes, 198–199
Survival of eggs and larvae
  biological factors, 213–214
  currents and flow features, 211, 216
  factors affecting survival of eggs and larvae, 209–213
  factors affecting survival of juveniles, 215–217
  feeding, and, 212
  food availability, 215
  juvenile habitat requirements, 216–217
  juvenile stage, importance of, 213–217
  nursery ground, 216–217
  physical factors, 210–211
  predators, 212–213, 215–216
  recruitment, setting, 214
  role of stimuli and behavior in fish movement to estuarine nursery areas, 217
  salinity and, 210–211
  sources of mortality of larvae by stage, 213
  storminess and turbulence, role of, 211
  temperature, role of, 210
Svåsand, T., 248
Swearer, S.E., J.E. Caselle, D.W. Lea, and R.R. Warner, 241
Synapomorphies, 101
Synchronous hermaphrodites, 20
Synchronous or simultaneous hermaphroditism, 11

T

Tagesgrade, 129
Taning, A.V., 134
Taylor, W.R., 61
Temperature, 129, 210
Temperature control, 251
Theilacker, G.H., 141, 200, 255, 257
Theilacker, G.H., A.S. Kimbrell, and J.S. Trimmer, 145
Thompson, A.B., 170
Thompson, W.F., 128
Thorisson, K., 215
Tidal and channel nets, 161
Tighe, K.A., 83, 91
Tilseth, S., 253
Time and space scales, 148–149
Timescales, 227–228

Timescales of variability and biological responses, 225–227
Transformation series, 101
Transformation stage, 51–52
Trantor, D.J., 148
Travis, J., F. Coleman, C. Grimes, D. Conover, T. Bert, and M. Tringali, 248
Tremblay, M.J., 205, 207
Trippel, E.A., 8
Trnski, T., 60
Tsukamoto, K., 248
Tucker, J.W., Jr., 4, 62
Tucker plankton trawls, 156, 157
Tupper, M., 214, 215, 216, 243
Turrell, W.R., 193

U
Unisexuality or parthenogenesis, 12
Urho, L., 235
U.S. GLOBEC program, 225
Use of larval fish characters in systematic studies, 102–124

V
Valles, H., S. Spongaules, and H.A. Oxenford, 242
Vallin, L., 128
Van Cleve, R., 128
Van der Meeren, T., 248
Van der Veer, H.W., R. Berghaln, J.M. Miller, and A.D. Rijnsdorp, 214, 215, 216
Van Dyke, G.C., 61
Variability in ichthyoplankton data, 169–171
Variability in world population abundance of fishes, 187
Variety of habitats occupied by fish eggs and larvae, 230–232
Vertebral counts, 76–79
Vertical distribution, 128–129, 172
Vetter, R.D., 241, 242

Videler, J.J., 131
Video and image analysis, 61
Vinter, B.M., 82, 83, 90, 93
Vitellogenesis, 21, 22

W
Walford, L.A., 217, 218
Walker, B., 32, 33
Wallace, D., 5
Walline, P.D., 212
Walters, C., 237
Wang, J.C., 60
Ware, D.M., 226, 255
Warner, R.R., 12, 240
Warner, R.R., S.E. Swearer, and J.E. Caselle, 241
Washington, B.B., 83, 84, 94
Water for rearing, 250–251
Watson, W., 5, 105, 122, 123, 150, 161, 178
Weibe, P.H., K.H. Burt, S.H. Boyd, and A.W. Morton, 158
Werner, F.E., 223
Werner, R.G., 8
Western, D., 238
Wild food, obtaining, 252
Wiles, G.C., R.D. D'Arrigo, and G.C. Jacoby, 227
Williams, M.A., 234
Wilson, M.T., 158
Wourms, J.P., 16
Wyatt, T., 201

Y
Year-class phenomenon in fishes, 182–183
Yoklavich, M., 16, 239
Yolk characters, 65
Yolk-sac larvae, 40
Yolk sac larvae, releases of, 245–247
Yoshimura, K., A. Hagiwara, T. Yoshimatsu, and C. Kitajima, 252